Electronics for Vinyl

Electronics for Vinyl is the most comprehensive book ever produced on the electronic circuitry needed to extract the best possible signal from grooves in vinyl. What is called the "vinyl revival" is in full swing, and a clear and comprehensive account of the electronics you need is very timely. Vinyl reproduction presents some unique technical challenges; the signal levels from moving-magnet cartridges are low, and those from moving-coil cartridges lower still, so a good deal of high-quality low-noise amplification is required.

Some of the features of *Electronics for Vinyl* include:

- integrating phono amplifiers into a complete preamplifier;
- differing phono amplifier technologies; covering active, passive, and semi-passive RIAA equalisation and transconductance RIAA stages;
- the tricky business of getting really accurate RIAA equalisation without spending a fortune on expensive components, such as switched-gain MM/MC RIAA amplifiers that retain great accuracy at all gains, the effects of finite open-loop gain, cartridge-preamplifier interaction, and so on;
- noise and distortion in phono amplifiers, covering BJTs, FETs, and opamps as input devices, hybrid phono amplifiers, noise in balanced MM inputs, noise weighting, and cartridge load synthesis for ultimately low noise;
- archival and nonstandard equalisation for 78s etc.;
- building phono amplifiers with discrete transistors;
- subsonic filtering, covering all-pole filters, elliptical filters, and suppression of subsonics by low-frequency crossfeed, including the unique Devinyliser concept;
- ultrasonic and scratch filtering, including a variety of variable-slope scratch filters;
- line output technology, including on zero-impedance outputs, on level indication for optimal setup, and on specialised power supplies; and
- description of six practical projects which range from the simple to the highly sophisticated, but all give exceptional performance.

Electronics for Vinyl brings the welcome news that there is simply no need to spend huge sums of money to get performance that is within a hair's breadth of the best theoretically obtainable. But you do need some specialised knowledge, and here it is.

Douglas Self studied engineering at Cambridge University, then psychoacoustics at Sussex University. He has spent many years working at the top level of design in both the professional audio and hi-fi industries, and has taken out a number of patents in the field of audio technology. He currently acts as a consultant engineer in the field of audio design.

Electronics for Vinyl

Douglas Self

Routledge
Taylor & Francis Group

LONDON AND NEW YORK

First published 2018
by Routledge
2 Park Square, Milton Park, Abingdon, Oxon OX14 4RN
52 Vanderbilt Avenue, New York, NY 10017

Routledge is an imprint of the Taylor & Francis Group, an informa business

Library of Congress Cataloging-in-Publication Data
Names: Self, Douglas, author.
Title: Electronics for vinyl / Douglas Self.
Description: New York ; London : Routledge, 2017.
Identifiers: LCCN 2017006870 | ISBN 9781138705449 (hardback) |
    ISBN 9781138705456 (pbk.)
Subjects: LCSH: Audio amplifiers. | Phonograph—High-fidelity systems.
Classification: LCC TK7871.58.A9 S45224 2017 | DDC 621.389/33—dc23
LC record available at https://lccn.loc.gov/2017006870

ISBN 13: 978-1-138-70544-9 (hbk)
ISBN 13: 978-1-138-70545-6 (pbk)

Typeset in Times New Roman
by Apex CoVantage, LLC

To Julie, with all my love

Contents

Acknowledgements

My heartfelt thanks to:

Gareth Connor of The Signal Transfer Company for unfailing encouragement, providing the facilities with which some of the experiments in this book were done, and with much appreciation of our long collaboration in the field of audio.

Other Titles

Preface

"The End . . . is the Knowledge of Causes and secret motions of things, and the enlarging of the bounds of human empire, to the effecting of all things possible."

Francis Bacon, New Atlantis

"Another damned thick book! Always scribble, scribble, scribble! Eh, Mr. Gibbon?"

Attributed to Prince William Henry, Duke of Gloucester, in 1781 upon receiving the second volume of The History of the Decline and Fall of the Roman Empire *from its author.*

This book aims to give a comprehensive account of the electronics needed to extract the best possible signal from grooves in vinyl. This is timely, as what is called the "vinyl revival" is still in full swing. Vinyl reproduction presents some fascinating technical challenges; the signal levels from moving-magnet cartridges are low, and those from moving-coil cartridges lower still, so a good deal of low-noise amplification is required. In the moving-magnet case this is much complicated by the high inductance of the cartridge. RIAA equalisation is required for a flat response, and achieving this both accurately and economically is a major study in itself. Vinyl produces large amounts of subsonic noise, and this has to be filtered out both effectively and unobtrusively to prevent bad things happening in amplifiers and loudspeakers.

Some of the basic material was published in my *Small Signal Audio Design* (2nd edition). When the time came to consider updating the book, there was the problem that its length was already pushing the limits for practical publishing. I have heard it said you should never write a book you cannot pick up with one hand. This issue was solved by splitting off everything related to vinyl reproduction and adding a great deal of new material to make a new and separate book focused on vinyl electronics. Here it is, containing more than three times as much text and more than three times as many illustrations; most of the content is new material. There are new chapters on phono amplifier architectures and ultrasonic and scratch filtering. The new chapter on subsonic filtering is, I think I can say without fear of successful contradiction, the most comprehensive account of it ever published. I have tried to make it easy to dip into this book—not everyone will want to read it straight through. A few important concepts are therefore explained more than once, when they are particularly relevant. I hope you will not accuse me of hesitation, repetition, or deviation.

The focus is on the analogue domain, where the processing is done with opamps or discrete transistors, usually working at a nominal level of a volt or less. There are good reasons for this. While you can do almost anything in the digital domain, you first have to get the signal *into* the digital domain. Since the nominal signal level from a moving magnet is only 5 mV rms and the output from a moving-coil cartridge generally much lower than that, clearly you are going to have to quite a lot of analogue amplification before you can apply it to the input of an A-to-D converter. Most of them require between 3 and 6 V peak to peak for maximum digital output, usually called FSD. Also, the subsonic disturbances that come from disc warps and irregularities can occur at frighteningly high levels, barely 20 dB below maximum, and it is an excellent idea to remove them before they erode headroom in the ADC. Only 18 months ago I was involved in a telemetry project where some quite sophisticated analogue processing, including allpass filters for time-compensation, was required before the signal could be accurately digitised and transmitted.

In the pursuit of high quality at low cost, there are certain principles that pervade this book. Low-impedance design reduces the effects of Johnson noise and current noise without making voltage noise worse; the only downside is that a low impedance requires an opamp capable of driving it effectively, and sometimes more than one. The most ambitious application of this approach so far has been in the ultra-low noise Elektor 2012 Preamplifier.

Another principle is that of using multiple components to reduce the effects of random noise. This may be electrical noise; the outputs of several amplifiers are averaged (very simply with a few resistors), and the noise from them partially cancels. Multiple amplifiers are also very useful for driving the low impedances just mentioned. Alternatively it may be numerical noise, such as tolerances in a component value; making up the required value with multiple parts in series or parallel also makes errors partially cancel. This technique has its limits because of the square-root way it works; four amplifiers or components are required to half the errors, sixteen to reduce them to a quarter, and so on.

RIAA equalisation is a prime example of a requirement for non-standard component values (crossover design and filter design in general being the others in the audio business). As a general rule, in a series-feedback RIAA network only one component can be a preferred value from the widely used E24 resistor series. While resistors are freely available in the E96 series, if you are faced with a required value that is effectively random, you will do much better by using two E24 values in series or, preferably, in parallel. The nominal value will on average be three times more accurate than a single E96. With three E24 values combined the accuracy of the nominal value is further improved by ten times compared with a single E96. This may appear profligate with components, but resistors are cheap. There is another advantage to the multiple component approach that is less obvious; the effective tolerance is reduced. By how much depends on how much the values vary; the maximum improvement with 2xE24 is $\sqrt{2}$ times and with 3xE24 is $\sqrt{3}$ times. The problem is more difficult with capacitors as they not usually available in as many values as the E24 series. Finding the best 3xE24 values is a non-trivial problem, and I have been much helped in this by the work of Gert Willmann, which he has generously shared with me. You will find much on his contribution in the body of this book.

There is also the Principle of Optimisation, which may sound imposing but just means that each circuit block is closely scrutinised to see if it is possible to improve it by doing a bit more thinking rather than a bit more spending. One example is the optimisation of RIAA equalisation networks. There are four ways to connect resistors and capacitors to make an RIAA network, and I have shown that one of them requires significantly smaller values of expensive precision capacitors than the others. This new finding is presented in detail in this book, along with related techniques of optimising resistor values to get convenient capacitor values.

And now what there is not. I have no time for faith-based audio, so there will be no discussion of esoteric components with insulting price-tags that actually achieve nothing. There will be no truck with the anti-science of cables that are alleged to know which way the information is supposed to be flowing. Passive RIAA equalisation is not recommended; it is thoroughly deprecated with good reason. I have spent more time than I care to contemplate in double-blind listening tests—properly conducted ones, with rigorous statistical analysis—and every time the answer was that if you couldn't measure it you couldn't hear it. Very often if you could measure it you still couldn't hear it

You may be surprised to find that there is not a great deal in this book about balanced phono preamplifiers. This is because a magnetic cartridge is a floating winding—it does not have a centre-tap or a ground reference until you connect it to an amplifier. There also should be no unwanted currents flowing in the ground conductor unless something is miswired. A balanced input therefore gains you nothing—a view that was substantiated in long discussions on the DIYaudio forum—and loses you some signal-to-noise ratio because of the extra electronics required for balancing.

So much has been added to make this book that it is difficult to summarise it, but the new material includes:

- More on resistor and capacitor selection for awkward values
- Preamplifier architecture and interfacing with phono amplifiers
- Optimal use of capacitors in RIAA equalisation stages
- Switched-gain MM amplifiers that retain accurate RIAA
- Hybrid phono amplifiers
- Noise in balanced MM inputs
- Noise weighting: A-weighting and ITU-R weighting, practical filter designs
- More on discrete-transistor circuitry for phono amplifiers

- Butterworth subsonic filters from 1st to 6th order
- Elliptical subsonic filters from 3rd to 6th order
- Elliptical subsonic filter optimisation
- Subsonic filtering by VLF crossfeed: The Devinyliser
- Butterworth ultrasonic filters from 2nd to 6th order
- Combining subsonic and ultrasonic filters in one stage
- A wholly new chapter giving six practical phono amplifier projects

If you have made the decision to use vinyl as your music-delivery medium, I believe this book will help you get the best possible audio off your discs. Whatever you think of vinyl, there is no doubt that it presents some fascinating technical challenges. The circuitry presented here is not merely good enough for home construction but sound enough for manufacture; I have done a lot of that. All measurements were performed with an Audio Precision SYS-2702.

A good deal of thought and experiment has gone into this book, and I dare to hope that I have moved analogue audio design a bit further forward. I hope you find it both useful and enjoyable.

To the best of my knowledge no supernatural assistance was received in the making of this book.

All suggestions for its improvement that do not involve its combustion will be gratefully received. You can find my email address on the front page of my website at douglas-self.com.

Further information and PCBs, kits and built circuit boards of some of the designs described here can be found at: www.signaltransfer.freeuk.com

Douglas Self

London, December 2016

Chapter 1

The Basics

Signals

An audio signal can be transmitted either as a voltage or a current. The construction of the universe is such that almost always the voltage mode is more convenient; consider for a moment an output driving more than one input. Connecting a series of high-impedance inputs to a low-impedance output is simply a matter of connecting them in parallel, and if the ratio of the output and input impedances is high there will be negligible variations in level. To drive multiple inputs with a current output it is necessary to have a series of floating current-sensor circuits that can be connected in series. This can be done,[1] as pretty much anything in electronics can be done, but it requires a lot of hardware and probably introduces performance compromises. The voltage-mode connection is just a matter of wiring.

Obviously, if there's a current, there's a voltage, and vice versa. You can't have one without the other. The distinction is in the output impedance of the transmitting end (low for voltage mode, high for current mode) and in what the receiving end responds to. Typically, but not necessarily, a voltage input has a high impedance; if its input impedance was only 600 Ω, as used to be the case in very old audio distribution systems, it is still responding to voltage, with the current it draws doing so a side issue, so it is still a voltage amplifier. In the same way, a current input typically, but not necessarily, has a very low input impedance. Current outputs can also present problems when they are not connected to anything. With no terminating impedance, the voltage at the output will be very high, and probably clipping heavily; the distortion is likely to crosstalk into adjacent circuitry. An open-circuit voltage output has no analogous problem.

Current-mode connections are not common. One example is the Krell Current Audio Signal Transmission, (CAST) technology, which uses current-mode to interconnect units in the Krell product range. While it is not exactly audio, the 4–20 mA current loop format is widely used in instrumentation. The current-mode operation means that voltage drops over long cable runs are ignored, and the zero offset of the current (i.e. 4 mA = zero) makes cable failure easy to detect: if the current suddenly drops to zero, you have a broken cable.

The old DIN interconnection standard was a form of current-mode connection in that it had voltage output via a high output impedance, of 100 kΩ or more. The idea was presumably that you could scale the output to a convenient voltage by selecting a suitable input impedance. The drawback was that the high output impedance made the amount of power transferred very small, leading to a poor signal-to-noise ratio. The concept is now wholly obsolete.

Amplifiers

At the most basic level, there are four kinds of amplifier, because there are two kinds of signal (voltage and current) and two types of port (input and output). The handy word "port" glosses over whether the input or output is differential or single ended. Amplifiers with differential input are very common—such as all opamps and most power amps—but differential outputs are rare and normally confined to specialised telecoms chips.

Table 1.1 summarises the four kinds of amplifier:

Table 1.1 The four types of amplifier

Amplifier type	Input	Output	Application
Voltage amplifier	Voltage	Voltage	General amplification
Transconductance amplifier	Voltage	Current	Voltage control of gain
Current amplifier	Current	Current	???
Transimpedance amplifier	Current	Voltage	Summing amplifiers, DAC interfacing

Voltage Amplifiers

These are the vast majority of amplifiers. They take a voltage input at a high impedance and yield a voltage output at a low impedance. All conventional opamps are voltage amplifiers in themselves, but they can be made to perform as any of the four kinds of amplifier by suitable feedback connections. Figure 1.1a shows a high-gain voltage amplifier (e.g. opamp) with series voltage feedback. The closed-loop gain is $(R1 + R2)/R2$.

Transconductance Amplifiers

The name simply means that a voltage input (usually differential) is converted to a current output. It has a transfer ratio $A = I_{OUT}/V_{IN}$, which has dimensions of I/V or conductance, so it is referred to as a transconductance amplifier. It is possible to make a very simple, though not very linear, voltage-controlled amplifier with transconductance technology; differential-input operational transconductance amplifier (OTA) ICs have an extra pin that gives voltage control of the transconductance, which when used with no negative feedback gives gain control. Performance falls well short of that required for quality hi-fi or professional audio. Figure 1.1b shows an OTA used without feedback; note the current-source symbol at the output.

Current Amplifiers

These accept a current in and give a current out. Since, as we have already noted, current-mode operation is rare, there is not often a use for a true current amplifier in the audio business. They should not be confused with current-feedback amplifiers (CFAs) which have a voltage output, the "current" bit referring to the way the feedback is applied in current-mode.[2] The bipolar transistor is sometimes described as a current amplifier, but it is nothing of the kind. Current may flow in the base circuit, but this is just an unwanted side effect. It is the *voltage* on the base that actually controls the transistor. I have seen it stated that the Hall-effect multiplier is a current amplifier; this is wholly untrue, as the output is a voltage. A true current amplifier can be built by following a transimpedance amplifier with a transconductance amplifier, but this uses two separate stages, with a voltage as an intermediate quantity.

Transimpedance Amplifiers

A transimpedance amplifier accepts a current in (usually single ended) and gives a voltage out. It is sometimes called an I-V converter. It has a transfer ratio $A = V_{OUT}/I_{IN}$, which has dimensions of V/I or resistance. That is why it is referred to as a transimpedance or transresistance amplifier. Transimpedance amplifiers are usually made by applying shunt voltage feedback to a high-gain voltage amplifier. The voltage amplifier stage (VAS) in most power amplifiers is a transimpedance amplifier. They are used for I-V conversion when interfacing to DACs with current outputs. Transimpedance amplifiers are sometimes incorrectly described as "current amplifiers".

Figure 1.1c shows a high-gain voltage amplifier (e.g. opamp) transformed into a transimpedance amplifier by adding the shunt voltage feedback resistor *R1*. The transimpedance gain is simply the value of *R1*, though it is normally expressed in V/mA rather than ohms.

Negative Feedback

Negative feedback is one of the most useful and omnipresent concepts in electronics. It can be used to control gain, to reduce distortion, and improve frequency

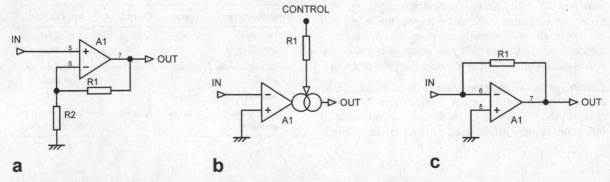

a   **b**   **c**

Figure 1.1 a) A Voltage amplifier, b) a transconductance amplifier, c) a transimpedance amplifier

response, and to set input and output impedances, and one feedback connection can do all these things at the same time. Negative feedback comes in four basic modes, as in the four basic kinds of amplifier. It can be taken from the output in two different ways (voltage or current feedback) and applied to the amplifier input in two different ways (series or shunt). Hence there are four combinations.

However, unless you're making something exotic like an audio constant-current source, the feedback is always taken as a voltage from the output, leaving us with just two feedback types, series and shunt, both of which are extensively used in audio. When series feedback is applied, as in Figure 1.1a, the following statements are true:

> Negative feedback reduces voltage gain.
> Negative feedback increases gain stability.
> Negative feedback increases bandwidth.
> Negative feedback increases amplifier input impedance.
> Negative feedback reduces amplifier output impedance.
> Negative feedback reduces distortion.
> Negative feedback does not directly alter the signal-to-noise ratio.

If shunt feedback is applied to a voltage amplifier to make a transimpedance amplifier, as in Figure 1.1c, all the above statements are still true, except since we have applied shunt rather than series negative feedback, the input impedance is reduced.

The basic feedback relationship is Equation 1.1 is dealt with at length in any number of textbooks, but it is of such fundamental importance that I feel obliged to include it here. The open-loop gain of the amplifier is A, and β is the feedback fraction, such that if in Figure 1.1a $R1$ is 2 kΩ and $R2$ is 1 kΩ, β is 1/3. If A is very high, you don't even need to know it; The 1 on the bottom becomes negligible, and the A's on top and bottom cancel out, leaving us with a gain of almost exactly three.

$$\frac{Vout}{Vin} = \frac{A}{1 + A\beta} \qquad 1.1$$

Negative feedback can however do much more than stabilising gain. Anything unwanted occurring in the amplifier, be it distortion or DC drift, or almost any of the other ills that electronics is heir to, is also reduced by the negative-feedback factor. (NFB factor for short) This is equal to:

$$NFBfactor = \frac{1}{1 + A\beta} \qquad 1.2$$

What negative feedback cannot do is improve the noise performance. When we apply feedback the gain drops, and the noise drops by the same factor (or less), leaving the signal-to-noise ratio the same (or worse). Negative feedback and the way it reduces distortion is explained in much more detail in one of my other books.[3]

Nominal Signal Levels and Dynamic Range

The absolute level of noise in a circuit is not of great significance in itself—what counts is how much greater the signal is than the noise—in other words the signal to noise ratio. An important step in any design is the determination of the optimal signal level at each point in the circuit. Obviously a real signal, as opposed to a test sine wave, continuously varies in amplitude, and the signal level chosen is purely a nominal level. One must steer a course between two evils:

> If the signal level is too low, it will be contaminated unduly by noise.

> If the signal level is too high there is a risk it will clip and introduce severe distortion.

The wider the gap between them the greater the dynamic range. You will note that the first evil is a certainty, while the second is more of a statistical risk. The consequences of both must be considered when choosing a level. If the best possible signal-to-noise is required in a studio recording, then the internal level must be high, and if there is an unexpected overload you can always do another take. In live situations it will often be preferable use a lower nominal level and sacrifice some noise performance to give less risk of clipping.

If you seek to increase the dynamic range, you can either increase the maximum signal level or lower the noise floor. The maximum signal levels in opamp-based equipment are set by the voltage capabilities of the opamps used, and this usually means a maximum signal level of about 10 Vrms or +22 dBu. Discrete-transistor technology removes the absolute limit on supply voltage and allows the voltage swing to be at least doubled before the supply rail voltages get inconveniently high. For example, ±40V rails are quite practical for small-signal transistors and permit a theoretical voltage swing of 28 Vrms or +31 dBu. However, in view of the complications of designing your own discrete circuitry and the greater space and power it requires, those nine extra dB of headroom are dearly bought. You must also consider the maximum signal capabilities of stages downstream—they might get damaged.

The dynamic range of human hearing is normally taken as 100 dB, ranging from the threshold of hearing at 0 dB SPL to the usual "Jack hammer at 1 m" at +100 dB SPL; however, hearing damage is generally reckoned to begin with long exposures to levels above +80 dB SPL. There is, in a sense, a physical maximum to the loudest possible sound. Since sound is composed of cycles of compression and rarefaction, this limit is reached when the rarefaction creates a vacuum, because you can't have a lower pressure than that. This corresponds to about +194 dB SPL. I thought this would probably be instantly fatal to a human being, but a little research showed that stun grenades generate +170 to +180 dB SPL, so maybe not. It is certainly possible to get asymmetrical pressure spikes higher than +194 dB SPL, but it is not clear that this can be defined as sound.

Compare this with the dynamic range of a simple piece of cable. Let's say it has a resistance of 0.5 Ω; the Johnson noise from that will be −155 dBu. If we comply with the European Low Voltage Directive the maximum voltage will be 50 Vpeak = 35 Vrms = +33 dBu, so the dynamic range is 155 + 33 = 188 dB, which purely by numerical coincidence is close to the maximum sound level of 194 dB SPL.

Gain Structures

There are some very basic rules for putting together an effective gain structure in a piece of equipment. Like many rules, they are subject to modification or compromise when you get into a tight corner. Breaking them reduces the dynamic range of the circuitry, either by worsening the noise or restricting the headroom; whether this is significant depends on the overall structure of the system and what level of performance you are aiming at. Three simple rules are:

1) Don't amplify then attenuate.
2) Don't attenuate then amplify.
3) The signal should be raised to the nominal internal level as soon as possible to minimise contamination with circuit noise.

There are rare exceptions. For an example, see Chapter 11 on moving-coil disc inputs, where attenuation after amplification does not compromise headroom because of a more severe headroom limit downstream.

Amplification Then Attenuation

Put baldly it sounds too silly to contemplate, but it is easy to thoughtlessly add a bit of gain to make up for a loss somewhere else, and immediately a few dB of precious and irretrievable headroom are gone for good. This assumes that each stage has the same power rails and hence the same clipping point, which is usually the case in opamp circuitry.

Figure 1.2a shows a system with a gain control designed to keep 10 dB of gain in hand. In other words, the expectation is that the control will spend most of its working life set somewhere around its "0 dB" position, where it introduces 10 dB of attenuation, as is typically the case for a fader on a mixer. To maintain the nominal signal level at 0 dBu we need 10 dB of gain, and a +10 dB amplifier (Stage 2) has been inserted just before the gain control. This is not a good decision. This amplifier will clip 10 dB before any other stage in the system and introduces what one might call a headroom bottleneck.

There are exceptions. The moving-coil phono head amp described in Chapter 11 appears to flagrantly break this rule, as it always works at maximum gain even when this is not required. But when considered in conjunction with the following RIAA stage, which also has considerable gain, it makes perfect sense, for the stage gains are configured so that the second stage always clips first, and there is actually no loss of headroom.

Attenuation Then Amplification

In Figure 1.2b the amplifier is now after the gain control, and noise performance rather than headroom suffers. If the signal is attenuated, any active device will inescapably add noise in restoring the level. Every conventional gain-control block has to address this issue. If we once

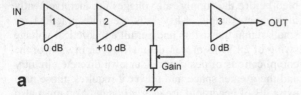
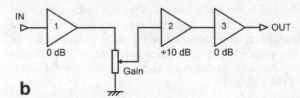

Figure 1.2 a) Amplification then attenuation. Stage 2 will always clip first, reducing headroom. b) Attenuation then amplification. The noise from Stage 2 degrades the S/N ratio. The lower the gain setting, the worse the degradation.

more require a gain variable from +10 dB to off, i.e. minus infinity dB, as would be typical for a fader or volume control, then usually the potentiometer is placed before the gain stage, as in Figure 1.2b, because as a rule some loss in noise performance is more acceptable than a permanent 10 dB reduction in system headroom. If there are options for the amplifier stages in terms of a noise/cost trade-off (such as using the 5532 versus a TL072) and you can only afford one low-noise stage, then it should be Stage 2.

If all stages have the same noise performance this configuration is 10 dB noisier than the previous version when gain is set to 0 dB.

Raising the Input Signal to the Nominal Level

Getting the incoming signal up to the nominal internal level right away in one jump is almost always preferable as it gives the best noise performance. Sometimes when large amounts of gain are required it is better done in two amplifier stages; typical examples are microphone preamps with wide gain ranges and phono preamps that insist on performing the RIAA equalisation in several goes. (The latter are explored in Chapter 5.) In these cases the noise contribution of the second stage may be significant.

Consider a signal path which has, say, an input of −10 dBu and a nominal internal level of 0 dBu, and so needs an overall gain of +10 dB:

1) The first version in Figure 1.3a has an input amplifier, Amp 1, with +10 dB of gain followed by two

unity-gain circuit blocks, A and B. These might for example be lowpass and highpass Sallen and Key filters for bandwidth definition; see Chapter 13. All circuit blocks are assumed to have equivalent input noise at −100 dBu, so the first stage in Figure 1.3a has a noise output of −100 + 10 = −90 dBu. At the output of A the noise is the rms-sum of the −90 dBu input and the −100 dBu from A, giving −89.6 dBu. At the output of B another −100 dBu has been rms-added, so the final noise output is −89.2 dBu. It is clear that A and B have contributed little to the final noise, due to the raised level of the signal when it passes through them.

2) Now take a second version of the signal path that has an input amplifier Amp 1 with +5 dB of gain, followed by block A, another amplifier with 5 dB of gain, then block B. See Figure 1.3b. The noise output is now −87.6 dB, 1.6 dB worse, because the noise from A has now been amplified by +5 dB in Amp 2. There are also more parts, and the second version appears to be clearly an inferior design. Usually it would be, but there can be good reasons for splitting up gain into two stages; if the distortion performance is critical, then using two stages with +5 dB of closed-loop gain rather than +10 dB means that each stage has 5 dB more negative feedback and lower distortion. This approach can be refined by not splitting the gain in half but putting more in the first stages where the signal levels and hence the distortion will be lower. This technique was used very successfully in my multitude-of-opamps power amplifier, published in Elektor;[4] the first stage had a gain of

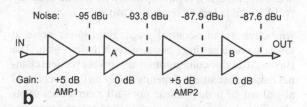

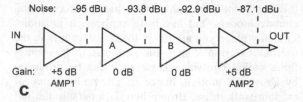

Figure 1.3 Four ways of arranging two unity-gain blocks, A and B, and +10 dB of amplification, showing that noise build-up depends on stage order

+11 dB and the second +6 dB. If amplification is done in two stages, then for the lowest noise they should come first in the signal path and not have block A put between them.

3) Third version in Figure 1.3c puts both A and B unity-gain stages between Amp 1 and Amp 2. The noise from both A and B is now amplified by +5 dB by Amp 2, and so the noise output is increased to −87.1 dBu. This is only 0.5 dB worse than Case 2, as a consequence of how rms-addition works; putting A before Amp 2 has already done most of the damage.

4) It is now fairly clear that putting A and B first, and then following them with a single +10 dB amplifier stage, as in Figure 1.3d, will give the highest output noise yet, and so it does, as the noise from both A and B is amplified by +10 dB. The noise out is now −84.0 dBu, 5.2 dB worse than Case 1. And all we have done is alter the order of the stages.

I think this demonstrates effectively that signals should always be amplified to their nominal internal level as soon as possible.

Active Gain Controls

The previous section should not be taken to imply that noise/headroom performance must always be sacrificed when a gain control is included in the signal path. This is not so. If we move beyond the idea of a fixed-gain block and recognise that the amount of gain present can be varied, then less gain when the maximum is not required will reduce the noise generated. For volume-control purposes it is essential that the gain can be reduced to near zero, though it is not necessary for it to be as firmly "off" as the faders or sends of a mixer.

An active volume-control stage gives lower noise at lower volume settings because there is less gain. The Baxandall active configuration also gives excellent channel balance because it depends solely on the mechanical alignment of a dual linear pot—all mismatches of its electrical characteristics are cancelled out, and there are no quasi-log dual slopes to induce anxiety.

Active gain controls are looked at in depth in Chapter 4.

Noise and Its Colours

Noise here refers only to the random noise generated by resistances and active devices. The term is sometimes used to include mains hum, spurious signals from demodulated RF, and other nonrandom sources, but this threatens confusion, and I prefer to call the other unwanted signals "interference". In one case we strive to minimise the random variations arising in the circuit itself, in the other we are trying to keep extraneous signals out, and the techniques are wholly different.

When noise is referred to in electronics it means white noise unless it is specifically labelled as something else, because that is the form of noise that most electronic processes generate. There are two elemental noise mechanisms which make themselves felt in all circuits and active devices. These are Johnson noise and Shot noise, which are both forms of white noise. Both have Gaussian probability density functions. These two basic mechanisms generate the noise in both BJTs and FETs, though in rather different ways.

There are other forms of noise that originate from less fundamental mechanisms such as device processing imperfections, which do not have a white spectrum; examples are $1/f$ (flicker) noise and popcorn noise. These noise mechanisms are described later in this chapter.

Nonwhite noise is given a colour which corresponds to the visible spectrum; thus red noise has a larger low-frequency content than white noise, while pink is midway between the two.

White noise has equal power in equal absolute bandwidth, i.e. with the bandwidth measured in Hz. Thus there is the same power between 100 and 200 Hz as there is between 1100 and 1200 Hz. It is the type produced by most electronic noise mechanisms.[5]

Pink noise has equal power in equal ratios of bandwidth, so there is the same power between 100 and 200 Hz as there is between 200 and 400 Hz. The energy per Hz falls at 3 dB per octave as frequency increases; i.e. the power density is proportional to $1/f$. Pink noise is widely used for acoustic applications like room equalisation and loudspeaker measurement as it gives a flat response when viewed on a third-octave or other constant-percentage-bandwidth spectrum analyser.[6]

Red noise has energy per Hz falling at 6 dB per octave rather than 3, the power density proportional to $1/f^2$. It is important in the study of stochastic processes and climate models, but has little application in audio. The only place you are likely to encounter it is in the oscillator section of analogue synthesisers. It is sometimes called Brownian noise as it can be produced by Brownian motion, hence its alternative name of random-walk noise. Brown here is a person and not a colour.[7]

Blue noise has energy per Hz rising at 3 dB per octave. The power density is proportional to f. Blue noise is used for dithering in image anti-aliasing, but has, as far as I am aware, no application to audio. The spectral density of blue noise is proportional to the frequency. It appears that the light-sensitive cells in the retina of the mammalian eye are arranged in a pattern that resembles blue noise.[8] Great stuff, this evolution.

Violet noise has energy per Hz rising at 6 dB per octave. (I imagine you saw that one coming.) The power density is proportional to f^2. It is also known as "differentiated white noise" as a differentiator circuit has a frequency response rising at 6 dB per octave. Sometimes called purple noise. A real-life source of violet noise is the acoustic thermal noise of water; at high frequencies it dominates hydrophone reception.

Grey noise is pink noise modified by a psychoacoustic equal loudness curve, such as the inverse of the A-weighting curve, to give the perception of equal loudness at all frequencies.

Green noise really does exist, though not in the audio domain. It is used for stochastic half-toning of images and consists of binary dither patterns composed of homogeneously distributed minority pixel clusters. Another definition is pink noise with increased levels around 500 Hz; for background noise generators and the like, this is supposed to more closely resemble the noise of the natural environment (i.e. without man-made noise like road traffic, aeroplanes, etc.).

Black noise also has some kind of existence. One definition of black noise is the absence of noise, i.e. silence; I do not think this is very useful. Another definition is noise with the spectrum of black-body radiation; it has nothing to do with audio.

Johnson Noise

Johnson noise is produced by all resistances, including those real resistances hiding inside transistors (such as r_{bb}, the base spreading resistance). It is not generated by the so-called intrinsic resistances, such as r_e, which are an expression of the Vbe/Ic slope and not a physical resistance at all. Given that Johnson noise is present in every circuit and often puts a limit on noise performance, it is perhaps a bit surprising that it was not discovered until 1928 by John B. Johnson at Bell Labs.[9] The likely reason is that the valves of the day were very much noisier than the resistors.

The rms amplitude of Johnson noise is easily calculated with the classic formula:

$$v_n = \sqrt{4kTRB} \qquad\qquad 1.3$$

Where:
  v_n is the rms noise voltage
  T is absolute temperature in °K
  B is the bandwidth in Hz
  k is Boltzmann's constant
  R is the resistance in Ohms

The only thing to be careful with here (apart from the usual problem of keeping the powers of ten straight) is to make sure you use Boltzmann's constant (1.380662×10^{-23}), and NOT the Stefan-Boltzmann constant ($5.67 \, 10^{-08}$), which relates to black-body radiation and will give some spectacularly wrong answers. Often the voltage noise is left in its squared form for ease of summing with other noise sources. Table 1.2 gives a feel for how resistance affects the magnitude of Johnson noise. The temperature is 25 °C and the bandwidth is 22 kHz.

Johnson noise theoretically goes all the way to daylight, and presumably even further up to gamma-ray frequencies, but in the real world is ultimately band-limited by the shunt capacitance of the resistor. Johnson noise is not produced by circuit reactances—i.e. pure capacitance and inductance. In the real world, however, reactive components are not pure, and the winding resistances of transformers can produce significant Johnson noise; this is an important factor in the design of moving-coil cartridge step-up transformers. Capacitors with their very high leakage resistances approach perfection much more closely, and the capacitance has a filtering effect. They usually have no detectable effect on noise performance, and in some circuitry it is possible to reduce noise by using a capacitive potential divider instead of a resistive one.[10]

The noise voltage is of course inseparable from the resistance, so the equivalent circuit is of a voltage source in series with the resistance present. While Johnson noise is usually represented as a voltage, it can also be treated as a Johnson noise current, by means of the Thevenin-Norton transformation, which gives the alternative equivalent circuit of a current source in shunt with the resistance. The equation for the noise current is simply the Johnson voltage divided by the value of the resistor it comes from $i_n = v_n/R$.

Table 1.2 Resistances and their Johnson noise

Resistance Ohms Ω	Noise voltage uV	Noise voltage dBu	Application
1	0.018	−152.2 dBu	Moving-coil cartridge impedance (low output)
3.3	0.035	−147.0 dBu	Moving-coil cartridge impedance (medium output)
10	0.060	−142.2 dBu	Moving-coil cartridge impedance (high output)
47	0.13	−135.5 dBu	Line output isolation resistor
68	0.16	−133.9 dBu	Line output isolation resistor
100	0.19	−132.2 dBu	Output isolation or feedback network
150	0.23	−130.4 dBu	Dynamic microphone source impedance
200	0.27	−129.2 dBu	Dynamic microphone source impedance (older)
250	0.30	−128.2 dBu	Worst-case output impedance of 1 kΩ pot
300	0.33	−127.4 dBu	Typical value in low-impedance design
400	0.38	−126.2 dBu	Typical value in low-impedance design
500	0.43	−125.2 dBu	Worst-case output impedance of 2 kΩ pot
600	0.47	−124.4 dBu	The ancient matched-line impedance
700	0.50	−123.7 dBu	Typical MM cartridge resistance
800	0.54	−123.2 dBu	Typical value in low-impedance design
1000	0.60	−122.2 dBu	A nice round number
1200	0.66	−121.4 dBu	Typical value in low-impedance design
1250	0.67	−121.2 dBu	Worst-case output impedance of 5 kΩ pot
1500	0.74	−120.4 dBu	Typical value in low-impedance design. E12
2000	0.85	−119.2 dBu	Typical value in low-impedance design. E24
2500	0.95	−118.2 dBu	Worst-case output impedance of 10 kΩ pot
5000	1.35	−115.2 dBu	Worst-case output impedance of 20 kΩ pot
12500	2.13	−111.2 dBu	Worst-case output impedance of 50 kΩ pot
25000	3.01	−108.2 dBu	Worst-case output impedance of 100 kΩ pot
1 mega (10^6)	19.0	−92.2 dBu	Another nice round number
1 giga (10^9)	601	−62.2 dBu	As used in capacitor microphone amplifiers
1 tera (10^{12})	1900	−32.2 dBu	Insulation testers read in tera-ohms
1 peta (10^{15})	601,281	−2.2 dBu	OK, it's getting silly now

When it is first encountered, this ability of resistors to generate electricity from out of nowhere seems deeply mysterious. You wouldn't be the first person to think of connecting a small electric motor across the resistance and getting some useful work out—and you wouldn't be the first person to discover it doesn't work. If it did, then by the First Law of Thermodynamics (the law of conservation of energy) the resistor would have to get colder, and such a process is flatly forbidden by . . . the Second Law of Thermodynamics. The Second Law is no more negotiable than the First Law, and it says that energy cannot be extracted by simply cooling down one body. If you could, it would be what thermodynamicists call a Perpetual Motion Machine of the Second Kind, and they are no more buildable than the more familiar Perpetual Motion Machine of the First Kind, which if it existed would make energy out of nowhere.

It is interesting to speculate what happens as the resistor is made larger. Does the Johnson voltage keep increasing, until there is a hazardous voltage across the resistor terminals? Obviously not, or picking up any piece of plastic would be a lethal experience. Johnson noise comes from a source impedance equal to the resistor generating it, and this alone would prevent any problems. Table 1.2 ends with a couple of silly values to see just how this works; the square root in the equation means that you need a petaohm resistor (1×10^{15} Ω) to reach even 600 mV rms of Johnson noise. Resistors are made up to at least 100 GΩ, but petaohm resistors (PΩ?) would really be a minority interest.

Shot Noise

It is easy to forget that an electric current is not some sort of magic fluid but is actually composed of a finite though usually very large number of electrons, so current is in effect quantised. Shot noise is so called because it allegedly sounds like a shower of lead shot being poured onto a drum, and the name emphasises the discrete nature of the charge carriers. Despite the picturesque description the spectrum is still that of white noise, and the noise current amplitude for a given steady current is described by a surprisingly simple equation (as Einstein said, the most incomprehensible thing about the universe is that it is comprehensible) that runs thus:

$$\text{Noise current } i_n = \sqrt{2qI_{dc}B} \qquad 1.4$$

Where:
  q is the charge on an electron (1.602×10^{-19} coulomb)
  I_{dc} is the mean value of the current
  B is the bandwidth examined

As with Johnson noise, often the shot noise is left in its squared form for ease of summing with other noise sources. Table 1.3 helps to give a feel for the reality of shot noise. As the current increases, the shot noise increases too, but more slowly as it depends on the square root of the DC current; therefore the *percentage* fluctuation in the current becomes less. It is the small currents which are the noisiest.

The actual level of shot noise voltage generated if the current noise is assumed to flow through a 100 ohm resistor is rather low, as the last column shows. Certainly there are many systems which will be embarrassed by an

extra noise source of −99 dBu, but to generate this level of shot noise requires 1 amp to flow through 100 ohms, which naturally means a voltage drop of 100 V and 100 watts of power dissipated. These are not often the sort of circuit conditions that exist in preamplifier circuitry. This does not mean that shot noise can be ignored completely, but it can usually be ignored unless it is happening in an active device where the shot noise is amplified.

1/f Noise (Flicker Noise)

This is so called because it rises in amplitude proportionally as the frequency examined falls. Unlike Johnson noise and shot noise, it is not a fundamental consequence of the way the universe is put together, but the result of imperfections in device construction. Flicker noise appears in all kinds of active semiconductors, and also in some types of resistor. When 1/f noise exists, as frequency falls the total noise density stays level down to the 1/f corner frequency, after which it rises at 6 dB/octave. This can frequently be seen in opamp spec-sheets. For a discussion of flicker noise in resistors see Chapter 2.

Popcorn Noise

This form of noise is named after the sound of popcorn being cooked, not eaten. It is also called burst noise or bistable noise and is a type of low-frequency noise that is found primarily in integrated circuits, appearing as low-level step changes in the output voltage, occurring at random intervals. Viewed on an oscilloscope this type of noise shows bursts of changes between two or more discrete levels. The amplitude stays level up to a corner frequency, at which point it falls at a rate of $1/f^2$. Different burst-noise mechanisms within the same device can exhibit different corner frequencies. The exact mechanism is poorly understood, but is known to be related to the presence of heavy-metal ion contamination, such

Table 1.3 How shot noise varies with current

Current	Current noise	Fluctuation	R	Voltage	Voltage
DC	nA rms	%	Ohms	noise uV	noise dBu
1 pA	0.000084 nA	8.4%	100	8.4×10^{-6}	−219.3 dBu
1 nA	0.0026 nA	0.27%	100	0.000265	−189.3 dBu
1 uA	0.084 nA	0.0084%	100	0.0084	−159.3 dBu
1 mA	2.65 nA	0.00027%	100	0.265	−129.3 dBu
1 A	84 nA	0.000008%	100	8.39	−99.3 dBu

as gold. As for $1/f$ noise, the only measure that can be taken against it is to choose an appropriate device. Like $1/f$ noise, popcorn noise does not have a Gaussian amplitude distribution.

Summing Noise Sources

When random noise from different sources is summed, the components do not add in a $2 + 2 = 4$ manner. Since the noise components come from different sources, with different versions of the same physical processes going on, they are uncorrelated and will partially reinforce and partially cancel, so root-mean-square (rms) addition holds, as shown in Equation 1.5. If there are two noise sources with the same level, the increase is 3 dB rather than 6 dB. When we are dealing with two sources in one device, such as a bipolar transistor, the assumption of no correlation is slightly dubious, because some correlation is known to exist, but it does not seem to be enough to cause significant calculation errors.

$$Vntot = \sqrt{(Vn1^2 + Vn2^2 + ...)} \qquad 1.5$$

Any number of noise sources may be summed in the same way by simply adding more squared terms inside the square root, as shown by the dotted lines. When dealing with noise in the design process, it is important to keep in mind the way that noise sources add together when they are not of equal amplitude. Table 1.4 shows how this works in decibels. Two equal voltage noise sources give a sum of +3 dB, as expected. What is notable is that when the two sources are of rather unequal amplitude, the smaller one makes very little contribution to the result.

Table 1.4 The summation of two uncorrelated noise sources

Source 1 dB	Source 2 dB	dB sum
0	0	+3.01
0	−1	+2.54
0	−2	+2.12
0	−3	+1.76
0	−4	+1.46
0	−5	+1.19
0	−6	+0.97
0	−10	+0.41
0	−15	+0.14
0	−20	+0.04

If we have a circuit in which one noise source is twice the rms amplitude of the other, (a 6 dB difference) then the quieter source only increases the rms-sum by 0.97 dB, a change barely detectable on critical listening. If one source is 10 dB below the other, the increase is only 0.4 dB, which in most cases could be ignored. At 20 dB down, the increase is lost in measurement error. This mathematical property of uncorrelated noise sources is exceedingly convenient, because it means that in practical calculations we can neglect all except the most important noise sources with minimal error. Since all semiconductors have some variability in their noise performance, it is rarely worthwhile to make the calculations to great accuracy.

Noise in Amplifiers

There are basic principles of noise design that apply to all amplifiers, be they discrete or integrated, single ended or differential. Practical circuits, even those consisting of an opamp and two resistors, have multiple sources of noise. Typically one source of noise will dominate, but this cannot be taken for granted and it is essential to evaluate all the sources and the ways that they add together if a noise calculation is going to be reliable. Here I add the complications one stage at a time.

Figure 1.4 shows that most useful of circuit elements, the perfect noiseless amplifier (these seem to be unaccountably hard to find in catalogues). It is assumed to have a definite gain A, without bothering about whether it is achieved by feedback or not, and an infinite input impedance. To emulate a real amplifier noise sources are concentrated at the input, combined into one voltage noise source and one current noise source. These can represent any number of actual noise sources inside the real amplifier. Figure 1.4 shows two ways of drawing the same situation.

It does not matter on which side of the voltage source the current source is placed; the "perfect" amplifier has an

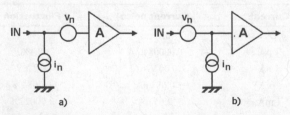

Figure 1.4 The noise sources of a perfect amplifier. The two circuits are exactly equivalent.

infinite input impedance, and the voltage source a zero impedance, so either way all of the current noise flows through whatever is attached to the input.

Figure 1.5 shows the first step to a realistic situation, with a signal source now connected to the amplifier input. The signal source is modelled as a perfect zero-impedance voltage source, with added series resistance R_s. Many signal sources are modelled accurately enough for noise calculations in this way. Examples are low-impedance dynamic microphones, moving-coil phono cartridges, and most electronic outputs. In others cases, such as moving-magnet phono cartridges and capacitor microphone capsules, there is a big reactive component which has a major effect on the noise behaviour and cannot be ignored or treated as a resistor. The magnitude of the reactances tends to vary from one make to another, but fortunately the variations are not usually large enough for the circuit approach for optimal noise to vary greatly. It is pretty clear that a capacitor microphone will have a very high source impedance at audio frequencies and will need a special high-impedance preamplifier to avoid low-frequency roll-off. It is perhaps less obvious that the series inductance of a moving-magnet phono cartridge becomes the dominating factor at the higher end of the audio band, and designing for the lowest noise with the 600 Ω or so series resistance alone will give far from optimal results. This is dealt with in Chapter 11.

There are two sources of voltage noise in the circuit of Figure 1.5.

1) The amplifier voltage noise source v_n at the input.
2) The Johnson noise from the source resistance Rs.

These two voltage sources are in series and sum by rms-addition because they are uncorrelated.

There is only one current noise component; the amplifier noise current source i_n across the input. This generates

a noise voltage when its noise current flows through Rs. (It cannot flow into the amplifier input because we are assuming an infinite input impedance.) This third source of voltage noise is also added in by rms-addition, and the total is amplified by the voltage gain A and appears at the output. The noise voltage at the input is the equivalent input noise (EIN). This is impossible to measure, so the noise at the amplifier output is divided by A to get the EIN. Having got this, we can compare it with the Johnson noise from the source resistance Rs; with a noiseless amplifier there would be no difference, but in real life the EIN will be higher by a number of dB, which is called the noise figure (NF). This gives a concise way of assessing how noisy our amplifier is and if it is worth trying to improve it. Noise figures very rarely appear in hi-fi literature, probably because most of them wouldn't look very good; some would look the utter rubbish that they are. For the fearless application of noise figures to phono cartridge amplifiers see Chapters 9 and 11.

Noise in Phono Amplifiers

There are two basic noise situations in phono circuitry. A moving-magnet (MM) cartridge has high inductance, and as a result at HF much of the noise is generated by input device current noise and Johnson noise from the 47 kΩ loading resistor Rin rather than the resistance of the cartridge. The frequency-dependent impedance of the inductance makes things complicated. A reasonable rule is to design for minimum noise using the cartridge impedance at 3852 Hz, which will give near-optimal RIAA-equalised noise.[11] At this frequency a typical MM cartridge will have an impedance of around 10 kΩ. Chapter 9 gives much more detail on MM phono amplifier noise.

In contrast, moving-coil (MC) cartridges generally act as low-value resistances with minimal series inductance, and so the design approach for low noise is quite different. Voltage noise is all-important, and the effect of current noise and Johnson noise from any loading resistor is usually negligible. Chapter 11 gives much more detail on MC phono amplifier noise.

All the other parts of a phono amplifier system, such as flat amplification or subsonic filtering, usually operate under favourable impedance conditions comparable with MC inputs.

Noise in Bipolar Transistors

An analysis of the noise behaviour of discrete bipolar transistors can be found in many textbooks, so this is

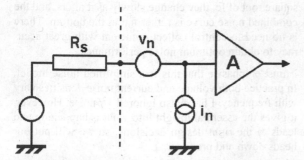

Figure 1.5 The perfect amplifier and noise sources, with a signal source now connected

something of a quick summary of the vital points. Two important transistor parameters for understanding noise are r_{bb}, the base spreading resistance, and r_e, the intrinsic emitter resistance. The first, r_{bb}, is a real physical resistance—what is called an *extrinsic* resistance. The second parameter, r_e, is an expression of the Vbe/Ic slope and not a physical resistance at all, so it is called an *intrinsic* resistance.

Noise in bipolar transistors, as in amplifiers in general, is best dealt with by assuming a noiseless transistor with a theoretical noise voltage source in series with the base and a theoretical noise current source connected from base to emitter. These sources are usually described simply as the "voltage noise" and the "current noise" of the transistor.

Bipolar Transistor Voltage Noise

The voltage noise v_n is made up of two components:

1) The Johnson noise generated in the base spreading resistance r_{bb}.
2) The collector current (Ic) shot noise creating a noise voltage across r_e, the intrinsic emitter resistance.

These two components can be calculated from the equations given earlier and rms-summed thus:

Voltage noise density $v_n = \sqrt{4kTr_{bb} + 2(kT)^2 / (qI_c)}$
in V/rtHz (usually nV/rtHz)   1.6

Where:
   k is Boltzmann's constant (1.380662×10^{-23})
   q is the charge on an electron (1.602×10^{-19} coulomb)
   T is absolute temperature in °K
   I_c is the collector current
   r_{bb} is the base resistance in ohms

The first part of this equation is the usual expression for Johnson noise and is fixed for a given transistor type by the physical value of r_{bb}, so the lower this is the better. The only way you can reduce this is by changing to another transistor type with a lower r_{bb} or using paralleled transistors. The absolute temperature is a factor; running your transistor at 25 °C rather than 125 °C reduces the Johnson noise from r_{bb} by 1.2 dB. Input devices usually run cool, but this may not be the case with moving-coil preamplifiers, where a large I_c is required, so it is not impossible that adding a heatsink would give a measurable improvement in noise.

The second (shot noise) part of the equation decreases as collector current Ic increases; this is because as Ic increases, r_e decreases proportionally, following $r_e = 25/$

Ic where Ic is in mA. The shot noise however is only increasing as the square root of Ic, and the overall result is that the total v_n falls—though relatively slowly—as collector current increases, approaching asymptotically the level of noise set by first part of the equation. There is an extra voltage noise source resulting from flicker noise produced by the base current flowing through r_{bb}; this is only significant at high collector currents and low frequencies due to its $1/f$ nature and is not usually included in design calculations unless low-frequency quietness is a special requirement.

Bipolar Transistor Current Noise

The current noise, i_n, is mainly produced by the shot noise of the steady current I_b flowing through the transistor base. This means it increases as the square root of Ib increases. Naturally Ib increases with Ic. Current noise is given by

Current noise density $i_n = \sqrt{2qI_b}$ in A/rtHz
(usual values are in pA/rtHz)   1.7

Where:
   q is the charge on an electron
   I_b is the base current

So, for a fixed collector current, you get less current noise with high-beta transistors because there is less base current.

The existence of current noise as well as voltage noise means it is not possible to minimise transistor noise just by increasing the collector current to the maximum value the device can take. Increasing Ic reduces voltage noise, but it increases current noise, as in Figure 1.6. There is an optimum collector current for each value of source resistance, where the contributions are equal. Because both voltage and current noise are proportional to the square root of Ic, they change slowly as it alters, and the combined noise curve is rather flat at the bottom. There is no need to control collector current with great accuracy to obtain optimum noise performance.

I must emphasise that this is a simplified noise model. In practice both voltage and current noise densities vary with frequency. I have also ignored $1/f$ noise. However, it gives the essential insight into what is happening and leads to the right design decisions, so we will put our heads down and press on.

A quick example shows how this works. In a voltage amplifier we want the source impedances seen by the

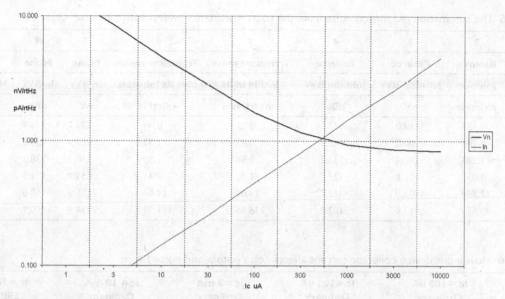

Figure 1.6 How voltage noise density v_n and current noise density i_n vary with collector current Ic in a generic transistor. As Ic increases, voltage noise asymptotes to a limit while current noise continuously increases.

input transistors to be as low as possible, to minimise Johnson noise from them and to minimise the effects of input device current noise flowing through them. In a typical bit of circuitry using low-impedance design it may be 100 Ω. How do we minimise the noise from a single transistor faced with a 100 Ω source resistance?

We assume the temperature is 25 °C, the bandwidth is 22 kHz, the r_{bb} of our transistor is 40 Ω and its h_{fe} (beta) is a healthy 150. Set Ic to 1 mA, which is plausible for an amplifier input stage, step the source resistance from 1 to 100,000 Ω in decades, and we get Table 1.5.

Column 1 shows the source resistance, and Column 2 the Johnson noise density it generates by itself. Factor in the bandwidth, and you get Columns 3 and 4, which show the voltage in nV and dBu respectively.

Column 5 is the noise density from the transistor, the rms-sum of the voltage noise and the voltage generated by the current noise flowing in the source resistance. Column 6 gives total noise density when we sum the source resistance noise density with the transistor noise density. Factor in the bandwidth again, and the resultant noise voltage is given in Columns 7 and 8. The final column (9) gives the noise figure (NF), which is the amount by which the combination of transistor and source resistance is noisier than the source resistance alone. In other words, it tells how close we have got to perfection, which would be a noise figure of 0 dB. The results for

the 100 Ω source show that the transistor noise is less than the source resistance Johnson noise; there is little scope for improving things by changing transistor type or operating conditions.

The results for the other source resistances are worth looking at. The lowest source resistance considered is 1 Ω, representing a low-output MC cartridge. This gives the lowest noise output, (−134.9 dBu) as you would expect, but the NF is very poor at 17.3 dB, because the r_{bb} at 40 Ω is generating a lot more noise than the 1 Ω source. This gives you some idea why it is hard to design quiet moving-coil head amplifiers. The best noise figure and the closest approach to theoretical perfection is with a 1000 Ω source, attained with a *greater* noise output than 100 Ω; it is essential to remember that the lowest NF does *not* mean the lowest noise output. As source resistance increases further, NF worsens again; a transistor with Ic = 1 mA has relatively high current noise and performs poorly with high source resistances.

Since Ic is about the only thing we have any control over here, let's try altering it. If we increase Ic to 3 mA we find that for a 100 Ω source resistance, our amplifier is only a marginal 0.2 dB quieter. See Table 1.6, which skips the intermediate calculations and just gives the output noise and NF.

At 3 mA the noise with a 1 Ω source is 0.7 dB better, due to slightly lower voltage noise, but with 100 kΩ noise

Table 1.5 The summation of Johnson noise from the source resistance with transistor noise

1	2	3	4	5	6	7	8	9
	Rsource	Rsource	Rsource	Transistor noise	Transistor noise	Noise	Noise	
Rsource	Johnson	Johnson BW	Johnson BW	incl In in Rs	plus Rs Johnson	in BW	in BW	Noise Fig
Ohms	nV/rtHz	nV	dBu	nV/rtHz	nV/rtHz	nV	dBu	dB
1	0.128	19.0	−152.2	0.93	0.94	139.7	−134.9	17.3
10	0.406	60.2	−142.2	0.93	1.02	150.9	−134.2	8.0
100	1.283	190.3	−132.2	0.94	1.59	236.3	−130.3	1.9
1000	4.057	601.8	−122.2	1.73	4.41	654.4	−121.5	0.7
10000	12.830	1903.0	−112.2	14.64	19.46	2886.9	−108.6	3.6
100000	40.573	6017.9	−102.2	146.06	151.59	22484.8	−90.7	11.4

Table 1.6 How input device collector current affects noise output and noise figure

	Ic = 100 uA Ordinary		Ic = 500 uA Ordinary		Ic = 3 mA Ordinary		Ic = 10 mA Ordinary		Ic = 10 mA 2SB737	
Rsource	Noise	NF	Noise	NF	Noise	NF	Noise	NF	Noise	NF
Ohms	dBu	dB	dBu	dB	dBu	dB	dBu	dB	dBu	dB
1	−129.9	22.3	−134.0	18.2	−135.6	16.6	−135.9	16.3	−145.9	6.3
10	−129.7	12.5	−133.4	8.8	−134.8	7.4	−135.1	7.1	−140.9	1.3
100	−127.9	4.29	−130.0	2.21	−130.5	1.7	−130.3	1.9	−131.5	0.7
1000	−121.5	0.72	−121.7	0.53	−120.6	1.6	−118.5	3.7	−118.6	3.6
10 k	−111.6	0.59	−110.0	2.19	−105.3	6.9	−100.7	11.4	−100.7	11.4
100 k	−98.6	3.61	−93.5	8.7	−86.2	16.0	−81.0	21.2	−81.0	21.2

is higher by no less than 9.8 dB as the current noise is much increased.

If we increase Ic to 10 mA, this makes the 100 Ω noise worse again, and we have lost that slender 0.2 dB improvement.

At 1 Ω the noise is 0.3 dB better, which is not exactly a breakthrough, and for the higher source resistances things worse again, the 100 kΩ noise increasing by another 5.2 dB. It therefore appears that a collector current of 3 mA is actually pretty much optimal for noise with our 100 Ω source resistance.

If we now pluck out our "ordinary" transistor and replace it with a specialised low-r_{bb} part like the much-lamented 2SB737, with its a superbly low r_{bb} of 2 Ω, the noise output at 1 Ω plummets by 10 dB, showing just how important low r_{bb} is for moving-coil head amplifiers. The improvement for the 100 Ω source resistance is much less at 1.0 dB.

If we go back to the ordinary transistor and reduce Ic to 100 uA, we get the last two columns in Table 1.6. Compared with Ic = 3 mA, noise with the 1 Ω source worsens by 5.7 dB, and with the 100Ω source by 2.6 dB, but with the 100 kΩ source there is a hefty 12.4 dB improvement, due to reduced current noise. Quiet BJT inputs for high source impedances can be made by using low collector currents, but JFETs usually give better noise performance under these conditions.

Finally we will look at a source impedance of around 10 kΩ, which is a reasonable design target for noise optimisation with most MM cartridges. Our ordinary transistor with Ic = 3 mA (Note no RIAA equalisation is being applied to any of these calculations—see Chapter 9 for that) gives −105.3 dBu, with an NF of 6.9 dB; not quiet. Reducing Ic to 500 uA drops the NF to a more respectable 2.2 dB, and reducing it drastically to 100uA drops it again to 0.6 dB. Obviously you have to have *some* collector current, but it looks as though reducing it even further might be rewarding. Figure 1.7 shows how the calculated noise figure for 10 kΩ source resistance reaches a shallow minimum just below 0.4 dB around collector currents of 20–50 uA. It is clear that MM cartridges running into BJT input devices require low collector currents.

The transistor will probably be the major source of noise in the circuit, but other sources may need to be considered. The transistor may have a collector resistor of high value, to optimise the stage gain, and this naturally introduces its own Johnson noise. Most discrete-transistor amplifiers have multiple stages, to get enough open-loop gain for linearisation by negative feedback, and an important consideration in discrete noise design is that the gain of the first stage should be high enough to make the noise contribution of the second stage negligible. This can complicate matters considerably. Precisely the same situation prevails in an opamp, but here someone else has done the worrying about second-stage noise for you, and the noise is specified for the complete part.

Noise in JFETs

JFETs operate completely differently from bipolar transistors, and noise arises in different ways. The voltage noise in JFETs arises from the Johnson noise produced by the channel resistance, the effective value of which is the inverse of the transconductance (g_m) of the JFET at the operating point we are looking at.

An approximate but widely accepted equation for this noise is :

$$\text{Noise density } e_n = \sqrt{4kT\frac{2}{3g_m}} \text{ in V/rtHz}$$

(usually nV/rtHz) 1.8

Where:
  k is Boltzmann's constant (1.380662×10^{-23})
  T is absolute temperature in °K

FET transconductance goes up proportionally to the square root of drain current I_d. When the transconductance is inserted into Equation 1.8, it is again square-rooted, so the voltage noise is proportional to the fourth root of drain current and varies with it very slowly. There is thus little point in using high drain currents.

The only current noise source in a JFET is the shot noise associated with the gate leakage current. Because the leakage current is normally extremely low, the current noise is very low, which is why JFETs give a good noise performance with high source resistances. However, don't let the JFET get hot, because gate leakage doubles with each 10 °C rise in temperature; this is why JFETs can actually show *increased* noise if the drain current is increased to the point where they heat up.

The g_m of JFETs is rather variable, but at I_d = 1 mA ranges over about 0.5 to 3 mA/V (or mMho) so the voltage noise density varies from 4.7 to 1.9 nV/rtHz.

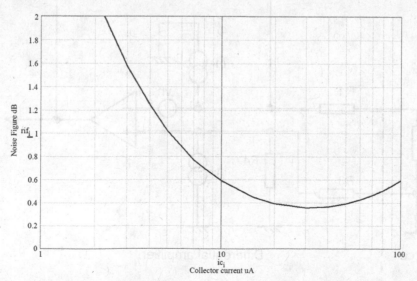

Figure 1.7 Calculated noise figure against collector current Ic with 10 kΩ source resistance

Comparing this with Column 5 in Table 1.5, we can see that the BJTs are much quieter except at high source impedances, where their current noise makes them noisier than JFETs.

However, if you are prepared to use multiple devices, the lowest possible noise may be given by JFETs, because the voltage noise falls faster than the effect of the current noise rises when more devices are added. A low-noise laboratory amplifier design by Samuel Groner achieves a spectacularly low-noise density of 0.39 nV/rtHz by using eight paralleled JFETs.[12]

Noise in Opamps

The noise behaviour of an opamp is very similar to that of a single input amplifier, the difference being that there are now two inputs to consider and usually more associated resistors.

An opamp is driven by the voltage difference between its two inputs, and so the voltage noise can be treated as one voltage v_n connected between them. See Figure 1.8, which shows a differential amplifier.

Opamp current noise is represented by two separate current generators, i_n+ and i_n-, one in parallel with each input. These are assumed to be equal in amplitude and not correlated with each other. It is also assumed that the voltage and current noise sources are likewise uncorrelated, so that rms-addition of their noise components is valid. In reality things are not quite so simple and there is some correlation, and the noise produced can be slightly higher than calculated. In practice the difference is small compared with natural variations in noise performance.

Calculating the noise is somewhat more complex than for the simple amplifier of Figure 1.4. You must:

1) Calculate the voltage noise from the voltage noise density.
2) Calculate the two extra noise voltages resulting from the noise currents flowing through their associated components.
3) Calculate the Johnson noise produced by each resistor.
4) Allow for the noise gain of the circuit when assessing how much each noise source contributes to the output.
5) Add the lot together by rms-addition.

There is no space to go through a complete calculation, but here is a quick example:

Suppose you have an inverting amplifier like that in Figure 1.11a. This is simpler because the noninverting input is grounded, so the effect of i_n+ disappears, as it has no resistance to flow

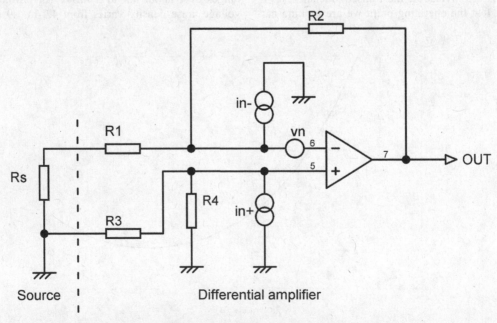

Figure 1.8 The noise sources in an opamp differential amplifier circuit

through and cannot give rise to a noise voltage. This shunt-feedback stage has a "noise gain" that is greater than the signal gain. The input signal is amplified by −1, but the voltage noise source in the opamp is amplified by two times, because the voltage noise generator is amplified as if the circuit was a series-feedback gain stage.

Low-noise Opamp Circuitry

The rest of this chapter deals with designing low-noise opamp circuitry, dealing with opamp selection and the minimisation of circuit impedances. It also shows how adding more stages can actually make the circuitry quieter. This sounds somewhat counterintuitive, but as you will see, it is so.

When you are designing for low noise, it is obviously important to select the right opamp, the great divide being between bipolar and JFET inputs. This chapter concentrates mainly on using the 5532, as it is not only a low-noise opamp with superbly low distortion but also a low-cost opamp, due to its large production quantities. There are opamps with lower noise, such as the AD797 and the LT1028, but these are specialised items and the cost penalties are high. The LT1028 has a bias-cancellation system that increases noise unless the impedances seen at each input are equal, and since audio does not need the resulting DC precision, it is not useful. The new LM4562 is a dual opamp with somewhat lower noise than the 5532, but at present it also is much more expensive.

The AD797 runs its bipolar input transistors at high collector currents (about 1 mA), which reduces voltage noise but increases current noise. The AD797 will therefore only give lower noise for rather low source resistances; these need to be below 1 kΩ to yield benefit for the money spent. There is much more on opamp selection in Chapter 3.

Noise Measurements

There are difficulties in measuring the low-noise levels we are dealing with here. The Audio Precision System 1 test system has a noise floor of −116.4 dBu when its input is terminated with a 47 Ω resistor. When it is terminated in a short circuit, the noise reading only drops to −117.0 dBu, demonstrating that almost all the noise is internal to the AP and the Johnson noise of the 47 Ω resistor is much lower. The significance of 47 Ω is that it is the lowest value of output resistor that will guarantee stability when driving the capacitance

of a reasonable length of screened cable; this value will keep cropping up.

To delve below this noise floor, we can subtract this figure from the noise we measure (on the usual rms basis) and estimate the noise actually coming from the circuit under test. This process is not very accurate when circuit noise is much below that of the test system, because of the subtraction involved, and any figure where the test-gear noise is more than 6 dB above the derived input noise should be regarded with caution. Cross-checking measurements against the theoretical calculations and SPICE results is always wise; in this case it is essential.

We will now look at a number of common circuit scenarios and see how low-noise design can be applied to them.

How to Attenuate Quietly

Attenuating a signal by 6 dB sounds like the easiest electronic task in the world. Two equal-value resistors to make up a potential divider, and *voila*! This knotty problem is solved. Or is it?

To begin with, let us consider the signal going into our divider. Wherever it comes from, the source impedance is not likely to be less than 50 Ω. This is also the lowest output impedance setting for most high-quality signal generators (though it's 40 Ω on my AP SYS-2702). The Johnson noise from 50 Ω is −135.2 dBu, which immediately puts a limit—albeit a very low one—on the performance we can achieve. The maximum signal-handling capability of opamps is about +22 dBu, so we know at once our dynamic range cannot exceed 135 + 22 = 155 dB. This comfortably exceeds the dynamic range of human hearing, which is about 130 dB if you are happy to accept "instantaneous ear damage" as the upper limit.

In the scenario we are examining, there is only one variable—the ohmic value of the two equal resistors. This cannot be too low or the divider will load the previous stage excessively, increasing distortion and possibly reducing headroom. On the other hand, the higher the value, the greater the Johnson noise voltage generated by the divider resistances that will be added to the signal and the greater the susceptibility of the circuit to capacitive crosstalk and general interference pickup. In Table 1.7 the trade-off is examined.

What happens when our signal with its −135.2 dBu noise level encounters our 6 dB attenuator? If it is made up of two 1 kΩ resistors, the noise level at once jumps up to

Table 1.7 Johnson noise from 6 dB resistive divider with different resistor values (bandwidth 22 kHz, temperature 25 °C)

Divider R's value	Divider R_{eff}	Johnson noise	Relative noise
100 Ω	50 Ω	−135.2 dBu	−27.0 dB
500 Ω	250 Ω	−128.2 dBu	−20.0 dB
1 kΩ	500 Ω	−125.2 dBu	−17.0 dB
5 kΩ	2.5 kΩ	−118.2 dBu	−10.0 dB
10 kΩ	5 kΩ	−115.2 dBu	−7.0 dB
50 kΩ	25 kΩ	−108.2 dBu	0 dB reference
100 kΩ	50 kΩ	−105.2 dBu	+3.0 dB

−125.2 dBu, as the effective source resistance from two 1 kΩ resistors effectively in parallel is 500 Ω. We have only deployed two passive components, and 10 dB of signal-to-noise ratio is irretrievably gone already. There will no doubt be more active and passive circuitry downstream, so things can only get worse.

However, a potential divider made from two 1 kΩ resistors in series presents an input impedance of only 2 kΩ, which is too low for most applications. Normally, 10 kΩ is considered the minimum input impedance for a piece of audio equipment in general use, which means we must use two 5 kΩ resistors, and so we get an effective source resistance of 2.5 kΩ. This produces Johnson noise at −118.2 dBu, so the signal-to-noise ratio has been degraded by another 7 dB simply by making the input impedance reasonably high.

In some cases 10 kΩ is not high enough, and a 100 kΩ input impedance is sought. Now the two resistors have

to be 50 kΩ, and the noise is 10 dB higher again, at −108.2 dBu. That is a worrying 27 dB worse than our signal when it arrived.

If we insist on an input impedance of 100 kΩ, how can we improve on our noise level of −108.2 dBu? The answer is by buffering the divider from the outside world. The output noise of a 5532 voltage-follower is about −119 dBu with a 50 Ω input termination. If this is used to drive our attenuator, the two resistors in it can be as low as the opamp can drive. The 5532 has a most convenient combination of low noise and good load-driving ability, and the divider resistors can be reduced to 500 Ω each, giving a load of 1 kΩ and a generous safety margin of drive capability. (Pushing the 5532 to its specified limit of a 500 Ω load tends to degrade its superb linearity by a small but measurable amount.) See Figure 1.9.

The noise from the resistive divider itself has now been lowered to −128.2 dBu, but there is of course the extra −119 dBu of noise from the voltage-follower that drives it. This however is halved by the divider just as the signal is, so the noise at the output will be the rms-sum of −125 dBu and −128.2 dBu, which is −123.3 dBu. A 6 dB attenuator is actually the worst case, as it has the highest possible source impedance for a given total divider resistance. Either more or less attenuation will mean less noise from the divider itself.

So, despite adding active circuitry that intrudes its own noise, the final noise level has been reduced from −108.2 to −123.3 dBu, an improvement of 15.1 dB.

How to Amplify Quietly

OK, we need a low-noise amplifier. Let's assume we have a reasonably low source impedance of 300 Ω, and

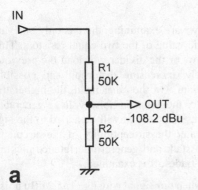

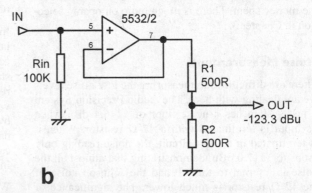

Figure 1.9 Two 6 dB attenuators with a 100 kΩ input impedance: a) Simple attenuator with high resistor values; and b) buffered attenuator with low resistor values. Despite the extra noise from the 5532 voltage-follower this version is 15 dB quieter.

we need a gain of four times (+12 dB). Figure 1.10a shows a very ordinary circuit using half a 5532 with typical values of 3 kΩ and 1 kΩ in the feedback network, and the noise output measures as −105.0 dBu. The Johnson noise generated by the 300 Ω source resistance is −127.4 dBu, and amplifying that by a gain of four gives −115.4 dBu. Compare this with the actual −105.0 dBu we get, and the noise figure is 10.4 dB—in other words the noise from the amplifier is three times the inescapable noise from the source resistance, making the latter essentially negligible. This amplifier stage is clearly somewhat short of noise-free perfection, despite using one of the quieter opamps around.

We need to make things quieter. The obvious thing to do is to reduce the value of the feedback resistances; this will reduce their Johnson noise and also reduce the noise produced in them by the opamp current noise generators. Figure 1.10b shows the feedback network altered to 360 Ω and 120 Ω, adding up to a load of 480 Ω, pushing the limits of the lowest resistance the opamp can drive satisfactorily. This assumes of course that the next stage presents a relatively light load so that almost all of the driving capability can be used to drive

the negative-feedback network; keeping tiny signals free from noise can involve throwing some serious current about. The noise output is reduced to −106.1 dBu, which is only an improvement of 1.1 dB and only brings the noise figure down to 9.3 dB, leaving us still a long way from what is theoretically attainable. However, at least it cost us nothing in extra components.

If we need to make things quieter yet, what can be done? The feedback resistances cannot be reduced further unless the opamp drive capability is increased in some way. An output stage made of discrete transistors could be added, but it would almost certainly compromise the low distortion we get from a 5532 alone. For one answer see the next section on ultra-low noise design.

How to Invert Quietly

Inverting a signal always requires the use of active electronics. (OK, you *could* use a transformer.) Assume that an input impedance of 47 kΩ is required, along with a unity-gain inversion. A straightforward inverting stage as shown in Figure 1.11a will give this input impedance and gain only if both resistors are 47 kΩ. These

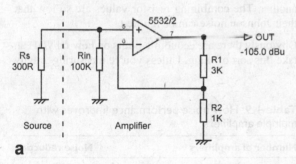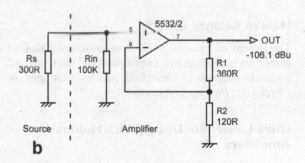

Figure 1.10 4 x amplifier a) with "normal" feedback resistances; b) with low-impedance feedback arm resistances. Noise is only reduced by 1.1 dB.

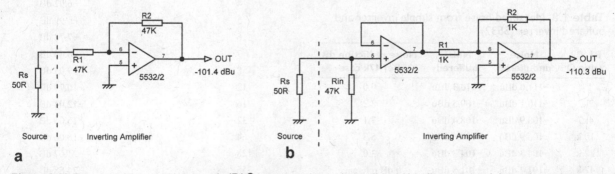

Figure 1.11 The noise from an inverter with 47 kΩ input impedance: a) unbuffered; b) buffered and with low-value resistors

relatively high-value resistors contribute Johnson noise and exacerbate the effect of opamp current noise. Also the opamp is working at a noise gain of two times, so the noise output is high at −101.4 dBu.

The only way to improve this noise level is to add another active stage. It sounds paradoxical—adding more nonsilent circuitry to reduce noise—but that's the way the universe works. If a voltage-follower is added to the circuit give Figure 1.11b, then the resistors around the inverting opamp can be greatly reduced in value without reducing the input impedance, which can now be pretty much as high as we like. The "Noise buffered" column in Table 1.8 shows that if R1 and R2 are reduced to 2.2 kΩ the total noise output is lowered by 8.2 dB, which is a very useful improvement. If R1, R2 are further reduced to 1 kΩ, which is perfectly practical with a 5532's drive capability, the total noise is reduced by 9.0 dB compared with the 47 kΩ case. The "Noise unbuffered" column gives the noise output with specified R value but without the buffer, demonstrating that adding the buffer does degrade the noise slightly, but the overall result is still far quieter than the unbuffered version with 47 kΩ resistors. In each case the circuit input is terminated to ground via 50 Ω.

How to Balance Quietly

The design of low-noise and ultra-low-noise balanced amplifiers using both low impedances and the multipath amplifier technology described here is examined in Chapter 4, "Preamp Architecture".

Ultra-Low-Noise Design With Multipath Amplifiers

Are the circuit structures described earlier the ultimate? Is this as low as noise gets? No. In the search for low noise, a powerful technique is the use of parallel amplifiers with their outputs summed. This is especially useful where source impedances are low and therefore generate little noise compared with the voltage noise of the electronics.

If there are two amplifiers connected, the signal gain increases by 6 dB due to the summation. The noise from the two amplifiers is also summed, but since the two noise sources are completely uncorrelated (coming from physically different components) they partially cancel, and the noise level only increases by 3 dB. Thus there is an improvement in signal-to-noise ratio of 3 dB. This strategy can be repeated by using four amplifiers, in which case the signal-to-noise improvement is 6 dB. Table 1.9 shows how this works for increasing numbers of amplifiers.

In practice the increased signal gain is not useful, and an active summing amplifier would compromise the noise improvement, so the output signals are averaged rather than summed, as shown in Figure 1.12. The amplifier outputs are simply connected together with low-value resistors, so the gain is unchanged but the noise output falls. The amplifier outputs are nominally identical, so very little current should flow from one opamp to another. The combining resistor values are so low that their Johnson noise can be ignored.

Obviously there are economic limits on how far you can take this sort of thing. Unless you're measuring gravity

Table 1.9 How noise performance improves with multiple amplifiers

Number of amplifiers	Noise reduction
1	0 dB ref
2	−3.01 dB
3	−4.77 dB
4	−6.02 dB
5	−6.99 dB
6	−7.78 dB
7	−8.45 dB
8	−9.03 dB
12	−10.79 dB
16	−12.04 dB
32	−15.05 dB
64	−18.06 dB
128	−21.07 dB
256	−24.58 dB

Table 1.8 Measured noise from simple inverter and buffered inverter (5532)

R1, R2 value Ω	Noise unbuffered	Noise buffered	Noise reduction dB Ref 47k case
1 k	−111.0 dBu	−110.3 dBu	9.0
2k2	−110.1 dBu	−109.5 dBu	8.2
4k7	−108.9 dBu	−108.4 dBu	7.1
10 k	−106.9 dBu	−106.6 dBu	5.3
22 k	−104.3 dBu	−104.3 dBu	3.0
47 k	−101.4 dBu	−101.3 dBu	0 dB reference

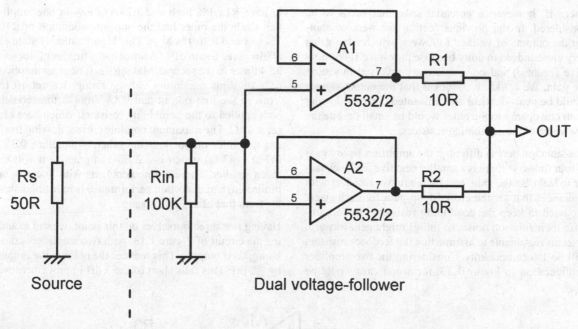

Figure 1.12 A double voltage-follower or buffer. The noise of this simple circuit is below that of the leading test equipment available.

waves or something equally important, 256 parallel amplifiers is probably not a viable choice.

Be aware that this technique does not give any kind of fault redundancy. If one opamp turns up its toes, the low value of the averaging resistors means the whole stage will stop working.

Ultra-Low-Noise Voltage Buffers

The multiple-path philosophy works well even with a minimally simple circuit such as a unity-gain voltage buffer. Table 1.10 gives calculated results for 5532 sections (the noise output is too low to measure reliably even with the best testgear) and shows how the noise output falls as more opamps are added. The distortion performance is not affected.

The 10 Ω output resistors combine the opamp outputs and limit the currents that would flow from output to output as a result of DC offset errors. AC gain errors here will be very small indeed because the opamps have 100% feedback. If the output resistors were raised to 47 Ω they would as usual give HF stability when driving screened cables or other capacitances, but the total output impedance is usefully halved to 23.5 Ω. Another interesting bonus of this technique is that we have

Table 1.10 Noise from parallel-array buffers using 5532 sections

Number of opamps	Calculated noise out
1	−120.4 dBu
2	−123.4 dBu
3	−125.2 dBu
4	−126.4 dBu

doubled the output drive capability; this stage can easily drive 300 Ω. This can be very useful when using low-impedance design to reduce noise in the following stage.

Ultra-Low-Noise Amplifiers

We now return to the problem studied earlier; how to make a really quiet amplifier with a gain of four times. We saw that the minimum noise output using a single 5532 section and a 300 Ω source resistance was −106.1 dBu, with a not particularly impressive noise figure of 9.3 dB. Since almost all the noise is being generated in the amplifier rather than the source resistance, the multiple-path technique should work well here. And it does.

There is, however, a potential snag that needs to be considered. In the previous section we were combining the outputs of voltage followers, which have gains very close indeed to unity because they have 100% negative feedback and no resistors are involved in setting the gain. We could be confident that the output signals would be near-identical and unwanted currents flowing from one opamp to the other would be small despite the low value of the combining resistors.

The situation here is different; the amplifiers have a gain of four times, so there is a smaller negative feedback factor to stabilise the gain, and there are two resistors with tolerances that set the closed-loop gain for each stage. We need to keep the combining resistors low to minimise their Johnson noise, so things might get awkward. It seems reasonable to assume that the feedback resistors will be 1% components. Considering the two-amplifier configuration in Figure 1.13, the worst case would be

to have R1a 1% high and R2a 1% low in one amplifier, while the other had the opposite condition of R1b 1% low and R2b 1% high. This highly unlikely state of affairs gives a gain of 4.06 times in the first amplifier and 3.94 times in the second. Making the further assumption of a 10 Vrms maximum voltage swing, we get 10.15 Vrms at the first output and 9.85 Vrms at the second, both applied to the combining resistors, which here are set at 47 Ω. The maximum possible current flowing from one amplifier output into the other is therefore 0.3V/ (47 Ω + 47 Ω) which is 3.2 mA; in practice it will be much smaller. There are no problems with linearity or headroom, and distortion performance is indistinguishable from that of a single opamp.

Having reassured ourselves on this point, we can examine the circuit of Figure 1.13, with two amplifiers combining their outputs. This reduces the noise at the output by 2.2 dB. This falls short of the 3 dB improvement we

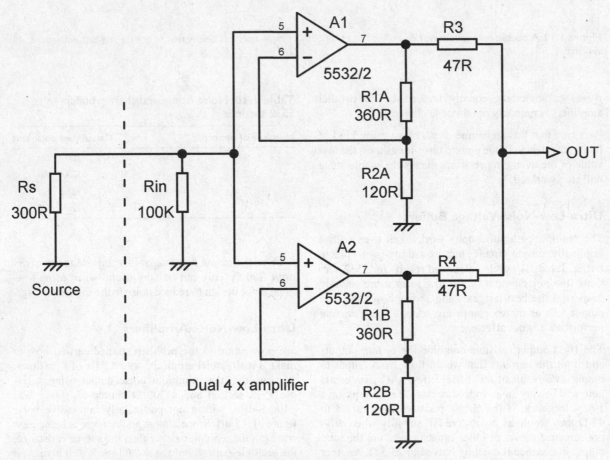

Figure 1.13 A gain of four amplifier using two opamps to reduce noise by approaching 3 dB

Table 1.11 Noise from multiple amplifiers with 4 times gain

Rs Ω	No of opamps	Noise out	Improvement
300	1	−106.1 dBu	0 dB ref
300	2	−108.2 dBu	2.2 dB
300	3	−109.0 dBu	2.9 dB
300	4	−109.6 dBu	3.5 dB
200	1	−106.2 dBu	0 dB ref
200	2	−108.4 dBu	2.2 dB
200	3	−109.3 dBu	3.1 dB
200	4	−110.0 dBu	3.8 dB
100	1	−106.3 dBu	0 dB ref
100	2	−108.7 dBu	2.4 dB
100	3	−109.8 dBu	3.5 dB
100	4	−110.4 dBu	3.9 dB

might hope for because of a significant Johnson noise contribution from source resistance, and doubling the number of amplifier stages again only achieves another 1.3 dB improvement. The improvement is greater with lower source resistances; the measured results with 1, 2, 3, and 4 opamps for three different source resistances are summarised in Table 1.11.

The results for 200 Ω and 100 Ω show that the improvement with multiple amplifiers is greater for lower source resistances, as these resistances generate less Johnson noise of their own.

Multiple Amplifiers for Greater Drive Capability

We have just seen that the use of multiple amplifiers with averaged outputs not only reduces noise but increases the drive capability proportionally to the number of amplifiers employed. This is highly convenient because heavy loads need to be driven when pushing hard the technique of low-impedance design.

Using multiple amplifiers gets difficult when the stage has variable feedback to implement gain control or tone control. In this case the configuration in Figure 1.14 doubles the drive capability in a foolproof manner; I have always called it "mother's little helper". A1 may be enmeshed in as complicated a circuit as you like, but unity-gain buffer A2 will robustly carry it its humble duty of sharing the load. This is unlikely to give any noise advantage, as most of the noise will presumably come from the more complex circuitry around A1.

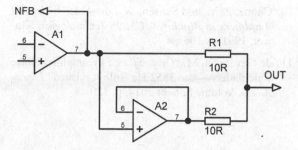

Figure 1.14 Mother's little helper. Using unity-gain buffer A2 to double the drive capability of any opamp stage.

It is assumed that A1 has load-driving capabilities equivalent to those of A2. This approach is more parts-efficient than simply putting a multiple-buffer like that in Figure 1.11 after A1; that would make no use of the drive capability of A1. This technique was used to drive the input of a Baxandall volume control using 1 kΩ pots in the Elektor 2012 preamplifier design.[13]

An interesting point is that any extra distortion contribution from A2 is halved, because its output is averaged with the input signal from A1. Likewise the noise contribution of A2 is halved. Quite a help, really.

References

1. Smith, J. *Modern Operational Circuit Design.* Wiley-Interscience, 1971, p. 129.

2. Intersil Application Note AN9420.1. *Current Feedback Amplifier Theory and Applications.* Apr 1995.

3. Self, D. *Audio Power Amplifier Design.* 6th Edition. Focal Press, 2013, Chapter 3. ISBN 978-0-240-52613-3.

4. Self, D. "The 5532 OpAmplifier" *Elektor*, Oct and Nov 2010.

5. Wikipedia https://en.wikipedia.org/wiki/White_noise Accessed Nov 2016.

6. Wikipedia https://en.wikipedia.org/wiki/Pink_noise Accessed Nov 2016.

7. Wikipedia https://en.wikipedia.org/wiki/Brownian_noise Accessed Nov 2016.

8. Yellott, J. I. Jr. "Spectral Consequences of Photoreceptor Sampling in the Rhesus Retina" *Science*, Volume 221, 1983, pp. 382–385.

9. Johnson, J. "Thermal Agitation of Electricity in Conductors" *Physical Review*, Volume 32, 1928, p. 97.

10. Chong, Z. Y., and Sansen, W. *Low-Noise Wideband Amplifiers in Bipolar & CMOS Technologies*. Kluwer, 1991, p. 106.

11. de Gevel van, M. "Gramophone Preamplifier Noise Calculations—the 3852 Hz Rule Revisited" *Linear Audio*, Volume 8, Sept 2014, p. 129.

12. Groner, S. "A Low-Noise Laboratory-Grade Measurement Preamplifier" *Linear Audio*, Volume 3, Apr 2012, p. 143. ISBN 978-9490929-008.

13. Self, D. "Preamplifier 2012" *Elektor*, Apr, May, June 2012.

Chapter 2

Passive Components

Conductors

It is easy to assume, when wrestling with electronic design, that the active devices will cause most of the trouble. This, like so much in electronics, is subject to Gershwin's law; it ain't necessarily so. Passive components cannot be assumed to be perfect, and while their shortcomings are rarely discussed in polite company, they are all too real. In this chapter I have tried to avoid repeating basic stuff that can be found in many places, to allow room for information that goes deeper.

Normal metallic conductors, such as copper wire, show perfect linearity for our purposes, and as far as I am aware, for everybody's purposes. Ohm's law was founded on metallic conductors, after all, not resistors, which did not exist as we know them at the time. George Simon Ohm published a pamphlet in 1827 titled, "The Galvanic Circuit Investigated Mathematically" while he was a professor of mathematics in Cologne. His work was not warmly received, except by a perceptive few; the Prussian minister of education pronounced that "a professor who preached such heresies was unworthy to teach science". This is the sort of thing that happens when politicians try to involve themselves in science, and in that respect we have progressed little since then.

Although the linearity is generally effectively ideal, metallic conductors will not be perfectly linear in some circumstances. Poorly made connections between oxidised or otherwise contaminated metal parts are capable of generating harmonic distortion at the level of several percent, but this is a property of the contact interface rather than the bulk material and usually means that the connection is about to fail altogether. A more subtle danger is that of magnetic conductors—the soft iron in relay frames causes easily detectable distortion at power amplifier current levels.

From time to time some of the dimmer audio commentators speculate that metallic conductors are actually a kind of "sea of micro-diodes", and that nonlinearity can be found if the test signal levels are made small enough. This is both categorically untrue and physically impossible. There is no threshold effect for metallic conduction. I have myself added to the mountain of evidence on this, by measuring distortion at very low signal levels.[1] Renardsen has some more information online.[2]

One account of distortion in a metal, in this case a binary alloy, is known to me. Takahisa's test[3] subjected a very thin (less than 0.001 mm) nickel-chrome alloy film to 250 volts at 10 kHz. This is the kind of film used in metal film resistors. The distortion measured was only 0.00004% of third harmonic. Other harmonics were found at a much lower level. All of these results indicate that Takahisa was measuring thermal distortion, caused by changes in resistance due to cyclic heating by the test signal and correspond with the results for actual metal film resistors—see later in this chapter. Takahisa would have found much higher distortion at lower frequencies such as 10 Hz. I would emphasise that these results actually relate only to the thin films found in resistors and not wiring or cables, where the metal thickness is far greater and cyclic heating utterly negligible.

Copper and Other Conductive Elements

Copper is the preferred metal for conducting electricity in almost all circumstances. It has the lowest resistance of any metal but silver, is reasonably resistant to corrosion, and can be made mechanically strong; it's wonderful stuff. Being a heavy metal, it is unfortunately not that common in the earth's crust, and so is expensive compared with iron and steel. It is however cheap compared with silver. The price of metals varies all the time due to changing economic and political factors, but at the time of writing silver was 100 times more expensive than copper by weight. Given the same cross-section of conductor, the use of silver would only reduce the resistance of a circuit by 5%. Despite this, silver connection wire has been used in some very expensive hi-fi amplifiers;

output impedance-matching transformers wound with silver wire are not unknown in valve amplifiers. Since the technical advantages are usually negligible, such equipment is marketed on the basis of indefinable subjective improvements. The only exception is the moving-coil step-up transformer, where the use of silver in the primary winding might give a measurable reduction in Johnson noise.

Table 2.1 gives the resistivity of the commonly used conductors, plus some insulators to give it perspective.

Table 2.1 Properties of conductors and nonconductors

Material	Resistivity ρ (Ω – m)	Temperature coefficient per degree C	Electrical usage
Silver	1.59×10^{-8}	0.0061	conductors
Copper	1.72×10^{-8}	0.0068	conductors
Gold	2.2×10^{-8}	0.0041	inert coatings
Aluminium	2.65×10^{-8}	0.00429	conductors
Tungsten	5.6×10^{-8}	0.0045	lamp filaments
Iron	9.71×10^{-8}	0.00651	barreters*
Platinum	10.6×10^{-8}	0.003927	electrodes
Tin	11.0×10^{-8}	0.0042	coatings
Mild steel	15×10^{-8}	0.0066	busbars
Solder (60:40 tin/lead)	15×10^{-8}	0.006	soldering
Lead	22×10^{-8}	0.0039	storage batteries
Manganin (Cu,Mn,Ni)**	48.2×10^{-8}	0.000002	resistances
Constantan (Cu,Ni)**	$49–52 \times 10^{-8}$	±0.00002	resistances
Mercury	98×10^{-8}	0.0009	relays
Nichrome (Ni,Fe,Cr alloy)	100×10^{-8}	0.0004	heating elements
Carbon (as graphite)	$3–60 \times 10^{-5}$	−0.0005	brushes
Glass	$1–10000 \times 10^{9}$	. . .	insulators
Fused quartz	More than 10^{18}	. . .	insulators

* A barreter is an incredibly obsolete device consisting of thin iron wire in an evacuated glass envelope. It was typically used for current regulation of the heaters of RF oscillator valves, to improve frequency stability.

** Constantan and Manganin are resistance alloys with moderate resistivity and a low temperature coefficient. Constantan is preferred as it has a flatter resistance/temperature curve and its corrosion resistance is better.

The difference between copper and quartz is of the order of 10 to the 25, an enormous range that is not found in many other physical properties.

There are several reasonably conductive metals that are lighter than copper, but their higher resistivity means they require larger cross-sections to carry the same current, so copper is always used when space is limited, as in electric motors, solenoids, etc. However, when size is not the primary constraint, the economics work out differently. The largest use of noncopper conductors is probably in the transmission line cables that are strung between pylons. Here minimal weight is more important than minimal diameter, so the cables have a central steel core for strength, surrounded by aluminium conductors.

It is clear that simply spending more money does not automatically bring you a better conductor; gold is a somewhat poorer conductor than copper, and platinum, which is even more expensive, is worse by a factor of six. Another interesting feature of this table is the relatively high resistance of mercury, nearly 60 times that of copper. This often comes as a surprise; people seem to assume that a metal of such high density must be very conductive, but it is not so. There are many reasons for not using mercury-filled hoses as loudspeaker cables, and their conductive inefficiency is just one. The cost and the insidiously poisonous nature of the metal are two more. Nonetheless . . . it is reported that the Hitachi Cable company has experimented with speaker cables made from polythene tubes filled with mercury. There appear to have been no plans to put such a product on the market. Restriction of Hazardous Substances (RoHS) compliance might be a problem.

We also see that the resistivity of solder is high compared with that of copper—nine times higher if you compare copper with the 60/40 tin/lead solder. This is unlikely to be a problem because the thickness of solder the current passes through in a typical joint is very small. There are many formulations of lead-free solder, with varying resistivities, but all are high compared with copper.

The Metallurgy of Copper

Copper is a good conductor because the outermost electrons of its atoms have a large mean free path between collisions. The electrical resistivity of a metal is inversely related to this electron mean free path, which in the case of copper is approximately 100 atomic spacings.

Copper is normally used as a very dilute alloy known as electrolytic tough pitch (ETP) copper, which consists of very high purity metal alloyed with oxygen in the range of 100 to 650 ppm. In view of the wide exposure that the concept of oxygen-free copper has had in the audio business, it is worth underlining that the oxygen is deliberately alloyed with the copper to act as a scavenger for dissolved hydrogen and sulphur, which become water and sulphur dioxide. Microscopic bubbles form in the mass of metal but are completely eliminated during hot rolling. The main use of oxygen-free copper is in conductors exposed to a hydrogen atmosphere at high temperatures. ETP copper is susceptible to hydrogen embrittlement in these circumstances, which arise in the hydrogen-cooled alternators in power stations.

Gold and Its Uses

As stated earlier, gold has a higher resistivity than copper, and there is no incentive to use it as the bulk metal of conductors, not least because of its high cost. However it is very useful as a thin coating on contacts because it is almost immune to corrosion, though it is chemically attacked by fluorine and chlorine. (If there is a significant amount of either gas in the air then your medical problems will be more pressing than your electrical ones.) Other electrical components are sometimes gold-plated simply because the appearance is attractive. A carat (or karat) is a 1/24 part, so 24-carat gold is the pure element, while 18-carat gold contains only 75% of the pure metal. Eighteen-carat gold is the sort usually used for jewellery because it retains the chemical inertness of pure gold but is much harder and more durable; the usual alloying elements are copper and silver.

Eighteen-carat gold is widely used in jewellery and does not tarnish, so it is initially puzzling to find that some electronic parts plated with it have a protective transparent coating which the manufacturer claims to be essential to prevent blackening. The answer is that if gold is plated directly onto copper, the copper diffuses through the gold and tarnishes on its surface. The standard way of preventing this is to plate a layer of nickel onto the copper to prevent diffusion, then plate on the gold. I have examined some transparent-coated gold-plated parts and found no nickel layer; presumably the manufacturer finds the transparent coating is cheaper than another plating process to deposit the nickel. However, it does not look as good as bare gold.

Cable and Wiring Resistance

Electrical cable is very often specified by its cross-sectional area and current-carrying capacity, and the resistance per metre is seldom quoted. This can however be a very important parameter for assessing permissible voltage drops and for predicting the crosstalk that will be introduced between two signals when they unavoidably share a common ground conductor. Given the resistivity of copper from Table 2.1, the resistance R of L metres of cable is simply:

$$R = \frac{resistivity \cdot L}{area} \qquad 2.1$$

Note that the area, which is usually quoted in catalogues in square millimetres, must be expressed here in square metres to match up with the units of resistivity and length. Thus 5 metres of cable with a cross-sectional area of 1.5 mm² will have a resistance of:

$$(1.72 \times 10^{-8}) \times 5 / (0.0000015) = 0.057 \text{ ohms}$$

This gives the resistance of our stretch of cable, and it is then simple to treat this as part of a potential divider to calculate the voltage drop down its length.

PCB Track Resistance

It is also useful to be able to calculate the resistance of a PCB track for the same reasons. This is slightly less straightforward to do; given the smorgasbord of units that are in use in PCB technology, determining the cross-sectional area of the track can present some difficulty.

In the USA and the UK, and probably elsewhere, there is inevitably a mix of metric and imperial units on PCBs, as many important components come in dual-in-line packages which are derived from an inch grid; track widths and lengths are therefore very often in thousandths of an inch, universally (in the UK at least) referred to as "thou". Conversely, the PCB dimensions and fixing-hole locations will almost certainly be metric because they interface with a world of metal fabrication and mechanical CAD that (except in the USA) went metric many years ago. Add to this the UK practice of quoting copper thickness in ounces (the weight of a square foot of copper foil) and all the ingredients for dimensional confusion are in place.

Standard PCB copper foil is known as one-ounce copper, having a thickness of 1.4 thou (= 35 microns). Two-ounce copper is naturally twice as thick; the extra cost of specifying it is small, typically around 5% of the total PCB cost, and this is a very simple way of halving track resistance. It can of course be applied very easily to an existing design without any fear of messing up a satisfactory layout. Four-ounce copper can also be obtained but is more rarely used and is therefore much more expensive. If heavier copper than two-ounce is required, the normal technique is to plate two-ounce up to three-ounce copper. The extra cost of this is surprisingly small, in the region of 10% to 15%.

Given the copper thickness, multiplying by track width gives the cross-sectional area. Since resistivity is always in metric units, it is best to convert to metric at this point, so Table 2.2 gives area in square millimetres. This is then multiplied by the resistivity, not forgetting to convert the area to metres for consistency. This gives the "resistance" column in the table, and it is then simple to treat this as part of a potential divider to calculate the usually unwanted voltage across the track.

For example, if the track in question is the ground return from a 1 kΩ load on an opamp, the load is the top half of a potential divider while the track is the bottom half, and a quick calculation gives the fraction of the input voltage found along the track. This is expressed in the last column of Table 2.2 as attenuation in dB. This shows clearly that circuit sections should not have common return tracks, or the interchannel crosstalk will be poor.

It is very clear from this table that relying on thicker copper on your PCB as means of reducing path resistance is not very effective. In some situations it may be the only recourse, but in many cases a path of much lower resistance can be made by using 32/02 cable soldered between the two relevant points on the PCB.

PCB tracks have a limited current capability because excessive resistive heating will break down the adhesive holding the copper to the board substrate and ultimately melt the copper. This is normally only a problem in power amplifiers and power supplies. It is useful to assess if you are likely to have problems before committing to a

Table 2.3 PCB track current capacity for a permitted temperature rise

Track temp rise	10 °C		20 °C		30 °C	
Copper weight						
Track width thou	1 oz	2 oz	1 oz	2 oz	1 oz	2 oz
10	1.0 A	1.4 A	1.2 A	1.6 A	1.5 A	2.2 A
15	1.2 A	1.6 A	1.3 A	2.4 A	1.6 A	3.0 A
20	1.3 A	2.1 A	1.7 A	3.0 A	2.4 A	3.6 A
25	1.7 A	2.5 A	2.2 A	3.3 A	2.8 A	4.0 A
30	1.9 A	3.0 A	2.5 A	4.0 A	3.2 A	5.0 A
50	2.6 A	4.0 A	3.6 A	6.0 A	4.4 A	7.3 A
75	3.5 A	5.7 A	4.5 A	7.8 A	6.0 A	10.0 A
100	4.2 A	6.9 A	6.0 A	9.9 A	7.5 A	12.5 A
200	7.0 A	11.5 A	10.0 A	16.0 A	13.0 A	20.5 A
250	8.3 A	12.3 A	12.3 A	20.0 A	15.0 A	24.5 A

PCB design, and Table 2.3, based on MIL-standard 275, gives some guidance.

Note that Table 2.3 applies to tracks on the PCB surface only. Internal tracks in a multi-layer PCB experience much less cooling and need to be about three times as thick for the same temperature rise. This factor depends on laminate thickness and so on, and you need to consult your PCB vendor.

Traditionally, overheated tracks could be detected visually because the solder mask on top of them would discolour to brown. I am not sure if this still applies with modern solder mask materials, as in recent years I have been quite successful in avoiding overheated tracking.

Table 2.2 Thickness of copper cladding and the calculation of track resistance

Weight	Thickness	Thickness	Width	Length	Area	Resistance	Atten ref 1 kω
oz	thou	micron	thou	inch	mm²	ohm	dB
1	1.38	35	12	3	0.0107	0.123	−78.2
1	1.38	35	50	3	0.0444	0.029	−90.8
2	2.76	70	12	3	0.0213	0.061	−84.3
2	2.76	70	50	3	0.0889	0.015	−96.5
4	5.52	140	50	3	0.178	0.0074	−102.6

PCB Track-to-Track Crosstalk

The previous section described how to evaluate the amount of crosstalk that can arise because of shared track resistances. Another crosstalk mechanism is caused by capacitance between PCB tracks. This is not very susceptible to calculation, so I did the following experiment to put some figures to the problem.

Figure 2.1 shows the setup; four parallel conductors 1.9 inches long on a standard piece of 0.1 inch pitch proto-type board were used as test tracks. These are perhaps

rather wider than the average PCB track, but one must start somewhere. The test signal was applied to track A, and track C was connected to a virtual-earth summing amplifier A1.

The tracks B and D were initially left floating. The results are shown as Trace 1 in Figure 2.2; the coupling at 10 kHz is −65 dB, which is worryingly high for two tracks 0.2 inch apart. Note that the crosstalk increases steadily at 6 dB per octave, as it results from a very small capacitance driving into what is effectively a short circuit.

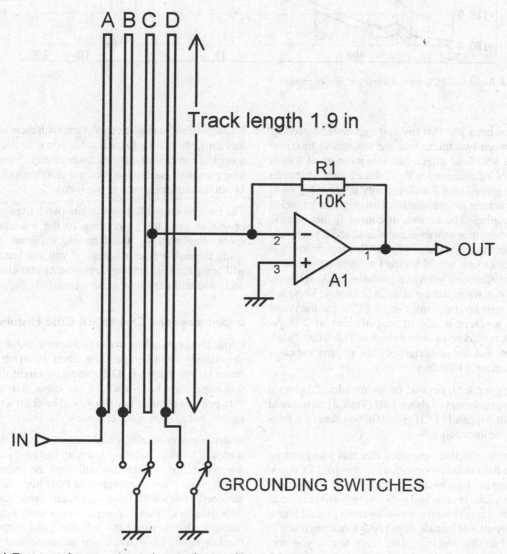

Figure 2.1 Test circuit for measuring track-to-track crosstalk on a PCB

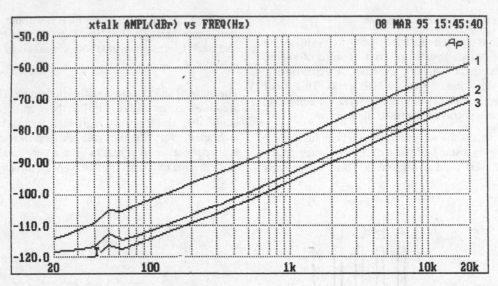

Figure 2.2 Results of PCB track-to-track crosstalk tests

It has often been said that running a grounded screening track between two tracks that are susceptible to crosstalk has a beneficial effect, but how much good does it really do? Grounding track B, to place a screen between A and C, gives Trace 2 and has only improved matters by 9 dB; not the dramatic effect that might be expected from screening. The reason, of course, is that electric fields are very much three-dimensional, and if you could see the electrostatic "lines of force" that appear in physics textbooks you would notice they arch up and over any planar screening such as a grounded track. It is easy to forget this when staring at a CAD display. There are of course two-layer and multi-layer PCBs, but the visual effect on a screen is still of several slices of 2-D. As Mr Spock remarked in one of the Star Trek films, "He's intelligent, but not experienced. His pattern indicates two-dimensional thinking."

Grounding track D, beyond receiving track C, gives a further improvement of about 3 dB (Trace 3); this would clearly not happen if PCB crosstalk was simply a line-of-sight phenomenon.

To get more effective screening than this you must go into three dimensions too; with a double-sided PCB you can put one track on each side, with ground plane opposite. With a four-layer board it should be possible to sandwich critical tracks between two layers of ground plane, where they should be safe from pretty much anything. If you can't do this and things are really tough, you may need to resort to a screened cable between two points on the PCB; this is of course expensive in assembly time.

If components such as electrolytics, with their large surface area, are talking to each other you may need to use a vertical metal wall, but this costs money. A more cunning plan is to use electrolytics not carrying signal, such as rail decouplers, as screening items.

The internal crosstalk between the two halves of a dual opamp is very low, according to the manufacturer's specs. Nevertheless, avoid having different channels going through the same opamp if you can because this will bring the surrounding components into close proximity and will permit capacitive crosstalk.

Impedances and Crosstalk: A Case History

Capacitive crosstalk between two opamp outputs can be surprisingly troublesome. The usual isolating resistor on an opamp output is 47 Ω, and you might think that this impedance is so low that the capacitive crosstalk between two of these outputs would be completely negligible, but . . . you would be wrong.

A stereo power amplifier had balanced input amplifiers with 47 Ω output isolating resistors included to prevent any possibility of instability, although the opamps were driving only a few centimetres of PCB track rather than screened cables with their significant capacitance. Just downstream of these opamps was a switch to enable biamping by driving both left and right outputs with the left input. This switch and its associated tracking brought the left and right signals into close proximity, and the capacitance between them was not negligible.

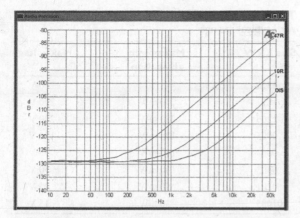

Figure 2.3 Crosstalk between opamp outputs with 47 Ω and 10 Ω output isolating resistors

Crosstalk at low frequencies (below 1 kHz) was pleasingly low, being better than −129 dB up to 70 Hz, which was the difference between the noise floor and the maximum signal level. (The measured noise floor was unusually low at −114 dBu because each input amplifier was a quadruple noise cancelling type as described in Chapter 1, and that figure includes the noise from an AP System 1.) At higher frequencies things were rather less gratifying, being −96 dB at 10 kHz, as shown by the "47R" trace in Figure 2.3. In many applications this would be more than acceptable, but in this case the highest performance possible was being sought.

I therefore decided to reduce the output isolating resistors to 10 Ω so the interchannel capacitance would have less effect. (Checks were done at the time and all through the prototyping and preproduction process to make sure that this would be enough resistance to ensure opamp stability—it was.) This handily reduced the crosstalk to −109 dB at 10 kHz, an improvement of 13 dB at zero cost. This is the ratio between the two resistor values.

The third trace, marked "DIS", shows the result of removing the isolating resistor from the speaking channel, so no signal reached the biamping switch. As usual, this reveals a further crosstalk mechanism, at about −117 dB, for reducing crosstalk is proverbially like peeling onions. There is layer after layer, and even strong men are reduced to tears.

Resistors

In the past there have been many types of resistor, including some interesting ones consisting of jars of liquid, but only a few kinds are likely to be met with now. (Jars of liquid are still used as resistances in high-voltage testing because of their ability to absorb huge amounts of peak power.) These are usually classified by the kind of material used in the resistive element, as this has the most important influence on the fine details of performance. The major materials and types are shown in Table 2.4,

These values are illustrative only, and it would be easy to find exceptions. As always, the official data sheet for the component you have chosen is the essential reference. The voltage coefficient is a measure of linearity (lower is better), and its sinister significance is explained later.

It should be said that you are most unlikely to come across carbon composition resistors in modern signal circuitry, but they frequently appear in vintage valve equipment so they are included here. They also live on in specialised applications such as switch-mode snubbing circuits, where their ability to absorb a high peak power in a mass of material rather than a thin film is very useful.

Carbon film resistors are currently still sometimes used in low-end consumer equipment, but elsewhere have been supplanted by the other three types. Note from Table 2.4 that they have a significant voltage coefficient.

Metal film resistors are now the usual choice when any degree of precision or stability is required. These have no nonlinearity problems at normal signal levels. The voltage coefficient is usually negligible.

Metal oxide resistors are more problematic. Cermet resistors and resistor packages are metal oxide and are made of the same material as thick film SM resistors. thick film resistors can show significant nonlinearity at opamp-type signal levels and should be kept out of high-quality signal paths.

Wirewound resistors are indispensable when serious power needs to be handled. The average wirewound

Table 2.4 Characteristics of resistor types

Type	Resistance tolerance	Temperature coefficient	Voltage coefficient
Carbon composition	±10%	+400 to −900 ppm/°	350 ppm
Carbon film	±5%	−100 to −700 ppm/°C	100 ppm
Metal film	±1%	+100 ppm/°C	1 ppm
Metal oxide	±5%	+300 ppm/°C	variable but too high
Wirewound	±5%	±70% to ±250%	1 ppm

resistor can withstand very large amounts of pulse power for short periods, but in this litigious age component manufacturers are often very reluctant to publish specifications on this capability, and endurance tests have to be done at the design stage; if this part of the system is built first then it will be tested as development proceeds. The voltage coefficient of wirewound resistors is usually negligible.

Resistors for general PCB use come in both through-hole and surface-mount types. Through-hole (TH) resistors can be any of the types in Table 2.4; surface-mount (SM) resistors are always either metal film or metal oxide. There are also many specialised types; for example, high-power wirewound resistors are often constructed inside a metal case that can be bolted down to a heatsink.

Through-Hole Resistors

These are too familiar to require much description; they are available in all the materials mentioned earlier: carbon film, metal film, metal oxide, and wirewound. There are a few other sorts, such as metal foil, but they are restricted to specialised applications. Conventional through-hole resistors are now almost always 250 mW 1% metal film. Carbon film used to be the standard resistor material, with the expensive metal film resistors reserved for critical places in circuitry where low tempco and an absence of excess noise were really important, but as metal film got cheaper so it took over many applications.

TH resistors have the advantage that their power and voltage rating greatly exceed those of surface-mount versions. They also have a very low voltage coefficient, which for our purposes is of the first importance. On the downside, the spiral construction of the resistance element means they have much greater parasitic inductance; this is not a problem in audio work.

Surface-Mount Resistors

Surface-mount resistors come in two main formats, the common chip type and the rarer (and much more expensive) MELF format.

Chip surface-mount (SM) resistors come in a flat tombstone format, which varies over a wide size range; see Table 2.5.

MELF surface-mount resistors have a cylindrical body with metal endcaps, the resistive element is metal film, and the linearity is therefore as good as conventional resistors, with a voltage coefficient of less than 1 ppm.

Table 2.5 The standard surface-mount resistor sizes with typical ratings

Size L x W	Max power dissipation	Max voltage
2512	1 W	200 V
1812	750 mW	200 V
1206	250 mW	200 V
0805	125 mW	150 V
0603	100 mW	75 V
0402	100 mW	50 V
0201	50 mW	25 V
01005	30 mW	15 V

MELF is apparently an acronym for "Metal ELectrode Face-bonded", though most people I know call them "Metal Ended Little Fellows" or something quite close to that.

Surface-mount resistors may have thin film or thick film resistive elements. The latter are cheaper and so more often encountered, but the price differential has been falling in recent years. Both thin film and thick film SM resistors use laser trimming to make fine adjustments of resistance value during the manufacturing process. There are important differences in their behaviour.

Thin film (metal film) SM resistors use a nickel-chromium (Ni-Cr) film as the resistance material. A very thin Ni-Cr film of less than 1 um thickness is deposited on the aluminium oxide substrate by sputtering under vacuum. Ni-Cr is then applied onto the substrate as conducting electrodes. The use of a metal film as the resistance material allows thin film resistors to provide a very low temperature coefficient, much lower current noise and vanishingly small nonlinearity. Thin film resistors need only low laser power for trimming (one-third of that required for thick film resistors) and contain no glass-based material. This prevents possible micro-cracking during laser trimming and maintains the stability of the thin film resistor types.

Thick film resistors normally use ruthenium oxide (RuO_2) as the resistance material, mixed with glass-based material to form a paste for printing on the substrate. The thickness of the printing material is usually 12 um. The heat generated during laser trimming can cause micro-cracks on a thick film resistor containing glass-based materials which can adversely affect stability. Palladium/silver (PdAg) is used for the electrodes.

The most important thing about thick film surface-mount resistors from our point of view is that they do not obey Ohm's law very well. This often comes as a shock to people who are used to TH resistors, which have been the highly linear metal film type for many years. They have much higher voltage coefficients than TH resistors, at between 30 and 100 ppm. The nonlinearity is symmetrical around zero voltage and so gives rise to third-harmonic distortion. Some SM resistor manufacturers do not specify voltage coefficient, which usually means it can vary disturbingly between different batches and different values of the same component, and this can have dire results on the repeatability of design performance.

Chip-type surface-mount resistors come in standard formats with names based on size, such as 1206, 0805, 0603 and 0402. For example, 0805, which used to be something like the "standard" size, is 0.08 in by 0.05 in; see Table 2.5. The smaller 0603 is now more common. Both 0805 and 0603 can be placed manually if you have a steady hand and a good magnifying glass.

The 0402 size is so small that the resistors look rather like grains of pepper; manual placing is not really feasible. They are only used in equipment where small size is critical, such as mobile phones. They have very restricted voltage and power ratings, typically 50V and 100 mW. The voltage rating of TH resistors can usually be ignored, as power dissipation is almost always the limiting factor, but with SM resistors it must be kept firmly in mind.

Recently, even smaller surface-mount resistors have been introduced; for example several vendors offer 0201, and Panasonic and Yageo offer 01005 resistors. The latter are truly tiny, being about 0.4 mm long; a thousand of them weigh less than a twentieth of a gram. They are intended for mobile phones, palmtops, and hearing aids; a full range of values is available from 10 Ω to 1 MΩ (jumper inclusive). Hand placing is really not an option.

Surface-mount resistors have a limited power-dissipation capability compared with their through-hole cousins, because of their small physical size. SM voltage ratings are also restricted, for the same reason. It is therefore sometimes necessary to use two SM resistors in series or parallel to meet these demands, as this is usually more economic than hand-fitting a through-hole component of adequate rating. If the voltage rating is the issue then the SM resistors will obviously have to be connected in series to gain any benefit.

Resistor Tolerances

As noted in Table 2.4, the most common tolerance for metal film resistors today is 1%; there is not likely to be much if any economic incentive to use 2% or even 5% parts. It is perhaps surprising that 5% carbon film resistors are still so freely available; a quick survey of distributors shows that they are not much cheaper than metal film. For some resistances in a phono amplifier 5% would be quite adequate; for example DC drain resistors or output isolation resistors. However in phono amplifiers most resistors need to be accurate, and it is unlikely to be worthwhile keeping two different resistor tolerances in stock, even if DC drain resistors are standardised at 22 kΩ and output isolation resistors at 47 Ω (which is quite feasible).

If you want a closer tolerance than 1%, then the next that is readily available in metal film is 0.1%; a few 0.5% resistor ranges are available, but they seem to be specialised parts with high power ratings and are not relevant to phono amplifiers. While there is considerable variation in the price of 0.1% resistors, roughly speaking they will be from 10 to 15 times more expensive than 1%. Very roughly, at the time of writing they are going to come in at something like 15p each, which I think is really quite reasonable considering their accuracy. Even so, it will usually be best to keep 0.1% for critical components; it helps if every critical resistance can be made the same nominal value, or at least there are only a few values, as this increases purchasing power and eases stock issues. For an example of this see The Devinyliser in Chapter 12, where only two different 0.1% values are used.

If 0.1% is not accurate enough—which I think it always will be for audio use—you can go to 0.05%, but then you are paying three or four GBP for each part. At about 10 GBP each, 0.02% can also be had, 0.01% at around 15 GBP each, and 0.005% at about 25 GBP. Clearly you are going to have to be working at the highest of the high end for this to make any vestige of economic sense. It might be marketing but it's not engineering.

Resistor Selection for Awkward Values

Phono amplifiers are one of the notable fields of electronics where nonpreferred component values come up, due to the need for accurate RIAA equalisation. Awkward values are also likely to occur in subsonic and ultrasonic filters. The other big field for awkward values is active crossover design, where the crossover filters need to be accurate.

Resistors are widely available in the E24 series (twenty-four values per decade) and the E96 series (ninety-six values per decade). There is also the E192 series (you guessed it, 192 values per decade), but this is less freely available. The E3 and E6 series are used for capacitors. E3, E6, E12, E24 and E96 values are listed in Appendix 1. A quirk of this system is that while E3, E6, and E12 are all subsets of E24, and E96 is a subset of E192, E24 is not a subset of E96. Very few of the E24 values appear in E96. If you look for, say, the E24 value of 300 Ω in E96 you will not find it; the nearest values are 294 Ω and 301 Ω. There is an E48 series (every other value from E96), but it seems to get little or no use. I have never come across it in the wild. Appendix 1 also lists pairs of resistors in integer ratios for each series. For example, there are six E24 pairs in a 1:2 ratio, such as 120Ω–240Ω; these a very useful for building 2nd-order Butterworth highpass filters. Similarly, there are two E24 pairs in a 1:4 ratio (300Ω–1200Ω, 750Ω–3000Ω) which occurs in two-stage 3rd-order Butterworth high-pass filters. See Chapter 12.

Using the E96 or, worse, the E192 series means that if, like me, you make many short production runs, to be able to get whatever value required you have to keep an enormous number of different resistor values in stock; when non-E24 values are required it is usually more convenient to use a series or parallel combination of two E24 resistors.

So, faced with what is effectively a random resistance value, what do you do? Here are three ways to address the problem. In Chapter 12 on subsonic filtering, thirty-six effectively random resistor values were dealt with in this way, and the averaged results for accuracy of the nominal value come from there.

1) Use the nearest E96 value and keep your fingers crossed; this is simple, but the way that requires the least thought is rarely the best way. The accuracy will simply be that of the resistor series chosen. Despite the close spacing of the values, at about 2%, E96 resistors are often available at 1% tolerance.
I call this the 1xE96 format. The average absolute error for 1xE96 was 0.805%.
2) Use two E24 1% resistors Ra, Rb in parallel, making them as equal as possible to get the best reduction in effective tolerance. I call this the 2xE24 format. It is often necessary to balance accuracy of nominal value against reduction of effective tolerance. I normally use the criterion that the nominal value should be accurate to better than half of the resistor tolerance; i.e. an error window of ±0.5%. Once

that is achieved reduction in effective tolerance can be pursued. Writing some code that explores all the combinations of two resistors in parallel is straight-forward; you set up a list of the E24 values, input the desired value Rreqd, then step through the list until you find the first resistor Ra that is greater than twice Rreqd. Put another resistor Rb from the E24 list in parallel; evaluate the combination, and keep at it until you have bracketed the required value with one result Rb1 too high and the other result Rb2 too low. If neither answer is within the error window, you know that an answer is impossible with that value of Ra. Increase Ra by one E24 step, then go round the loop again looking for bracketing values of Rb. When an answer is found within the error window, print the resistor values, the error in the nominal value, and the effective tolerance. However, do not stop; reduce Ra to the next lower E24 value and repeat. This will give you a series of bracketing values for Rb so you can choose the best solution. This process was used to generate all the 2xE24 resistor pairs in this book, and inevitably some have a more accurate nominal value than others. I have attempted to explain the algorithm in a good old-fashioned flowchart in Figure 2.4.

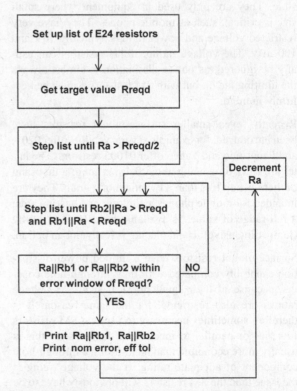

Figure 2.4 Algorithm for selecting optimal resistor pairs

The average absolute error in Chapter 12 for 2xE24 was 0.285%, which is three times better than 1xE96. The average effective tolerance, assuming 1% resistors, was 0.764%, which is not far from 0.707%, the best possible figure (this is explained in detail shortly).

3) Using three E24 1% resistors in parallel not only allows us to get much closer to a desired nominal value, but also gives a better chance of getting near-equal resistors that give most of the potential $1/\sqrt{3}$ (= 0.577) improvement in accuracy, because there are more combinations. The design process is not obvious; I used a Willmann table, which lists, in order of combined value, all combinations of three E24 resistors that give a combined value within a decade. The Willmann process is fully explained in Chapter 7 by means of practical examples in the design of RIAA equalisation networks. This book only makes use of the 3xE24 Willmann table; there are however many more that list E12, E48, and E96 combinations, etc. Gert Willmann intends to make the tables available as free software under the terms of the so-called GNU Lesser General Public License (LGPL); for more details, see www.gnu.org/licenses/. By the time this book is published the tables will be available free of charge either on my website or a site managed by Gert.

The brute-force search used for 2xE24 does not look promising for dealing with three resistors because of the large number of combinations available.

I call this the 3xE24 format. The average absolute error in Chapter 12 for 3xE24 was only 0.025%, more than ten times better than 2xE24. The average effective tolerance, assuming 1% resistors, was 0.659%, which is not too far from 0.577%, the best possible figure.

4) This does not exhaust the possibilities. You could use four E24 resistors (4xE24) to get phenomenally accurate nominal values, but there is not much point unless the component tolerance is upgraded to 0.1% or better. Likewise you could use two E96 values (2xE96) or even three (3xE96) to get very accurate nominal values, but again the component tolerance will be the ultimate limit on the overall accuracy. Four or five paralleled capacitors are often very useful in RIAA networks–see Chapters 7 and 17.

Improving the Effective Tolerance

Using two, three, or more resistors to make up a desired value has a valuable hidden benefit. It will actually increase the average accuracy of the total resistance

value, so it is *better* than the tolerance of the individual resistors; this may sound paradoxical, but it is simply an expression of the fact that random errors tend to partly cancel out if you have a number of them. This also works for capacitors, and indeed any parameter that is subject to random variations. Note that this assumes that the mean (i.e. average) value of the resistors is accurate. It is generally a sound assumption as it is much easier to control a single value such as the mean in a manufacturing process than to control all the variables that lead to scatter about that mean. This is confirmed by measurement.

Component values are usually subject to a Gaussian distribution, also called a normal distribution. It has a familiar peaky shape, not unlike a resonance curve, showing that the majority of the values lie near the central mean and that they get rarer the further away from the mean you look. This is a very common distribution, cropping up wherever there are many independent things going on that affect the value of a given component. The distribution is defined by its mean and its standard deviation, which is the square root of the sum of the squares of the distances from the mean—the rms-sum, in other words. Sigma (σ) is the standard symbol for standard deviation. A Gaussian distribution will have 68.3% of its values within ±1 σ, 95.4% within ±2 σ, 99.7% within ±3 σ, and 99.9% within ±4 σ. This is illustrated in Figure 2.5, where the X-axis is calibrated in numbers of standard deviations on either side of the central mean value.

If we put two equal-value resistors in series, or in parallel (see Figure 2.6a and 2.6b), the total value has proportionally a narrower distribution that of the original components. The standard deviation of summed components is the rms-sum of the individual standard deviations, as shown in Equation 2.2; σ_{sum} is the overall standard deviation, and σ_1 and σ_2 are the standard deviations of the two resistors in series or parallel.

$$\sigma_{sum} = \sqrt{(\sigma_1)^2 + (\sigma_2)^2} \qquad 2.2$$

Equation 2.2 is only correct if there is no correlation between the two resistor values; this is true for two separate resistors but would not hold for two film resistors on the same substrate.

Thus if we have four 100 Ω 1% resistors in series, the standard deviation of the total resistance increases only by the square root of 4, that is two times, while the total resistance has increased by four times; thus we have made a 0.5% close-tolerance 400 Ω resistor for four times the price, whereas a 0.1% resistor would be at least ten and maybe fifteen times the price and may give more

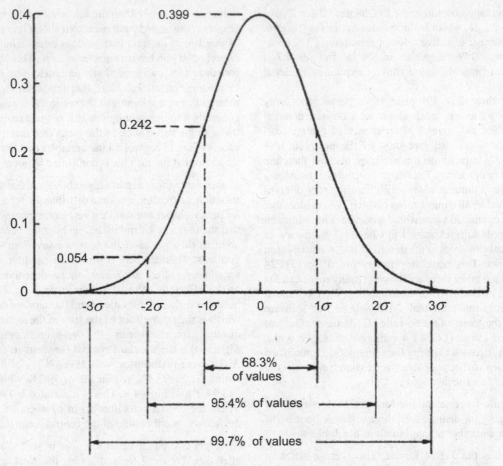

Figure 2.5 A Gaussian (normal) distribution with the X-axis marked in standard deviations on either side of the mean. The apparently strange value for the height of the peak is actually carefully chosen so the area under the curve is one.

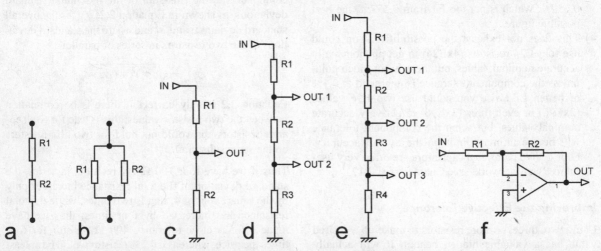

Figure 2.6 Resistor combinations: a) series b) parallel c) one-tap divider d) two-tap divider e) three-tap divider f) inverting amplifier

accuracy than we need. There is a happy analogue here with the use of multiple amplifiers to reduce electrical noise; we are using essentially the same technique of rms-summation to reduce "statistical noise".

You may object that putting four 1% resistors in series means that the worst-case errors can be four times as great. This is obviously true—if they are all 1% low or 1% high, the total error will be 4%. But the probability of this occurring is actually very, very small indeed. The more resistors you combine, the more the values cluster together in the centre of the range.

The mathematics for series resistors is very simple; see Equation 2.2, which also holds for two parallel resistors as in Figure 2.6b, though this is mathematically much less obvious. Other resistor networks get complicated very quickly. I verified it by the use of Monte-Carlo methods.[4] A suitable random number generator is used to select two resistor values, and their combined value is calculated and recorded. This is repeated many times (by computer, obviously), and then the mean and standard deviation of all the accumulated numbers is recorded. This will never give the *exact* answer, but it will get closer and closer as you make more trials. For the series and parallel cases the standard deviation is $1/\sqrt{2}$ of the standard deviation for a single resistor. If you are not wholly satisfied that this apparently magical improvement in average accuracy is genuine, seeing it happen on a spreadsheet makes a convincing demonstration.

In an Excel spreadsheet, random numbers with a uniform distribution are generated by the function RAND(), but random numbers with a Gaussian distribution and specified mean and standard deviation can be generated by the function NORMINV(). Let us assume we want to make an accurate 20 kΩ resistance. We can simulate the use of a single 1% tolerance resistor by generating a column of Gaussian random numbers with a mean of 20 and a standard deviation of 0.2; we need to use a lot of numbers to smooth out the statistical fluctuations, so we generate 400 of them. As a check we calculate the mean and standard deviation of our 400 random numbers using the AVERAGE() and STDEV() functions. The results will be very close to 20 and 0.2, but not identical, and will change every time we hit the F9 recalculate key because this generates a new set of random numbers. The results of five recalculations are shown in Table 2.6, demonstrating that 400 numbers are enough to get us quite close to our targets.

To simulate two 10 kΩ resistors of 1% tolerance in series we generate two columns of 400 Gaussian random numbers with a mean of 10 and a standard deviation of 0.1. We then set up a third column which is the sum of the

two random numbers on the same row, and if we calculate the mean and standard deviation using AVERAGE() and STDEV() again, we find that the mean is still very close to 20 but the standard deviation is reduced on average by the expected factor of $\sqrt{2}$. The result of five trials is shown in Table 2.7. Repeating this experiment with two 40 kΩ resistors in parallel gives the same results.

If we repeat this experiment by making our 20 kΩ resistance from a series combination of four 5 kΩ resistors of 1% tolerance we generate four columns of 400 Gaussian random numbers with a mean of 5 and a standard deviation of 0.05. We sum the four numbers on the same row to get a fifth column and calculate the mean and standard deviation of that. The result of five trials is shown in Table 2.8. The mean is very close to 20, but the standard deviation is now reduced on average by a factor of $\sqrt{4}$, which is 2.

I think this demonstrates quite convincingly that the spread of values is reduced by a factor equal to the

Table 2.6 Mean and standard deviation of five batches of 400 Gaussian random resistor values

Mean kΩ	Standard deviation
20.0017	0.2125
19.9950	0.2083
19.9910	0.1971
19.9955	0.2084
20.0204	0.2040

Table 2.7 Mean and standard deviation of five batches of 400 Gaussian 10 kΩ 1% resistors, two in series

Mean kΩ	Standard deviation
19.9999	0.1434
20.0007	0.1297
19.9963	0.1350
20.0114	0.1439
20.0052	0.1332

Table 2.8 Mean and standard deviation of five batches of 400 Gaussian 5 kΩ 1% resistors, four in series

Mean kΩ	Standard deviation
20.0008	0.1005
19.9956	0.0995
19.9917	0.1015
20.0032	0.1037
20.0020	0.0930

square root of the number of the components used. The principle works equally well for capacitors or indeed any quantity with a Gaussian distribution of values. The downside is the fact that the improvement depends on the square root of the number of equal-value components used, which means that big improvements require a lot of parts and the method quickly gets unwieldy. Table 2.9 demonstrates how this works; the rate of improvement slows down noticeably as the number of parts increases. The largest number of components I have ever used in this way for a production design is five; see Chapter 17.

But what happens if the series resistors used are *not* equal? The overall standard deviation is still the rms-sum of the standard deviations of the two resistors, as shown in Equation 2.2. Since both resistors have the same percentage tolerance, the larger of the two has the greater standard deviation and dominates the total result. The minimum total deviation is thus achieved with equal resistor values.

Figure 2.7 shows how this works for either series or parallel connection. As the resistors move away from a ratio of 1:1 the effective tolerance quickly degrades from 0.707%, but using two resistors in the ratio 2:1 or 3:1 still gives a worthwhile improvement. Much larger ratios, which may be required to get a given nominal value, still show some improvement, but it slowly falls off and asymptotes to 1.0, where one resistor of the pair effectively does not exist at all. The plot is based on a 1% resistor tolerance, but any other tolerance behaves proportionally.

Table 2.9 The improvement in value tolerance with number of equal-value parts

Number of equal-value parts	Tolerance reduction factor
1	1.000
2	0.707
3	0.577
4	0.500
5	0.447
6	0.408
7	0.378
8	0.354
9	0.333
10	0.316

Other Resistor Networks

So far we have looked at serial and parallel combinations of components to make up one value, as in Figure 2.6a,b. Other important networks are the resistive divider in Figure 2.6c, (frequently used as the negative-feedback network for noninverting amplifiers) and the inverting amplifier in Figure 2.6f, where the gain is set by the ratio R2/R1. All resistors are assumed to have the same tolerance about an exact mean value.

I suggest it is not obvious whether the divider ratio of Figure 2.6c, which is R2/(R1 + R2), will be more or less accurate than the resistor tolerance. In the simplest case with R1 = R2 the Monte-Carlo method shows that partial

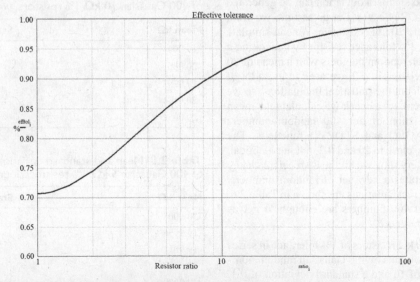

Figure 2.7 The improvement in effective tolerance versus the resistor ratio

cancellation of errors still occurs and the division ratio is improved in accuracy by a factor of √2.

However—this factor actually depends on the divider ratio, as a simple physical argument shows:

- If the top resistor R1 is zero, then the divider ratio is obviously one with complete accuracy, the resistor values are irrelevant, and the output voltage tolerance is zero.
- If the bottom resistor R2 is zero, there is no output and accuracy is meaningless, but if instead R2 is very small compared with R1 then the R1 completely determines the current through R2, and R2 turns this into the output voltage. Therefore the tolerances of R1 and R2 act independently, and so the combined output voltage tolerance is worse by their rms-sum, which is √2.

Some more Monte-Carlo work, with 8000 trials per data point, revealed that there is a linear relationship between accuracy and the "tap position" of the output between R1 and R2, as shown in Figure 2.8. Plotting against division ratio would not give a straight line. With R1 = R2 the tap is at 50% and accuracy improved by a factor of √2, as noted earlier. With a tap at about 30% (R1 = 7 kΩ, R2 = 3 kΩ) the accuracy is the same as the resistors used. This assessment is *not* applicable to potentiometers as the two sections of the pot are not uncorrelated; in linear pots they are very much correlated.

The two-tap divider (Figure 2.6d) and three-tap divider (Figure 2.6e) were also tested with equal resistors. The two-tap divider has an accuracy factor of 0.404 at OUT 1 and 0.809 at OUT 2. These numbers are very close to √2/(2√3) and √2/(√3) respectively. The three-tap divider has an accuracy factor of 0.289 at OUT 1, of 0.500 at OUT 2, and of 0.864 at OUT 3. The middle figure is clearly 1/2 (twice as many resistors as a one-tap divider, so √2 times more accurate), while the first and last numbers are very close to √3/6 and √3/2 respectively. It would be helpful if someone could prove analytically that the factors proposed are actually correct.

For the inverting amplifier of Figure 2.6f, the accuracy of the gain is always √2 worse than the tolerance of the two resistors, assuming the tolerances are equal. The nominal resistor values have no effect on this. We therefore have the interesting situation that a noninverting amplifier will always be equally or more accurate in its gain than an inverting amplifier. So far as I know this is a new result.

Resistor Pair Choice

In many cases the resistance value required is completely determined, and the only decision is how to make up the value with a pair that will meet the requirements for an accurate nominal value and for effective tolerance. However, sometimes there is an extra degree of freedom

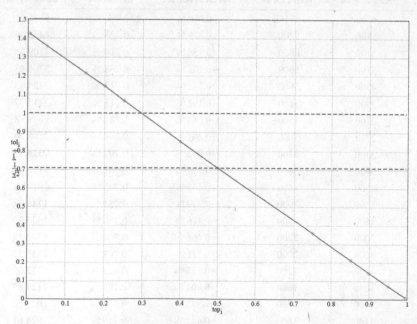

Figure 2.8 The accuracy of the output of a resistive divider made up with components of the same tolerance varies with the divider ratio

that allows more optimisation; this is an important point, and I am going to look at it in detail.

Consider the simple potential divider in Figure 2.6c, and we will assume we want exactly 3 dB of attenuation, which is 0.70795 times. The divider will be driven by an opamp, so we want to keep the loading reasonably light for good distortion performance; this implies the resistance to ground through the divider is not less than 1 kΩ. This suggests high resistor values should be used. The divider is feeding an opamp input, so we want a low value of divider output impedance to keep down Johnson noise and the effects of opamp input-current noise. This suggests low resistor values should be used, and a compromise is required. One more assumption—the top resistor R1 will be a single E24 part. To get an accurate 3dB of attenuation, R2 will therefore have to be a 2xE24 pair. Table 2.10 shows what happens when we pick various E24 values for R1 and then choose a parallel pair R2a, R2b with the criterion that the combined nominal value shall not be in error by more than 0.5% (1% resistors with an accurate mean value are assumed).

Loading and output impedance are calculated from the R2 column. Table 2.10 was arbitrarily started with a value of 1.5 kΩ for R1, which gives an output impedance

of approximately 1 kΩ. The last value of 270 Ω for R1 brings the total resistance of the potential divider below 1 kΩ and so breaks our rules.

This shows using E24 resistors, accepting a minimal load resistance of 1 kΩ loading and a maximum output impedance of 1 kΩ, gives us no fewer than seventeen possible solutions for our potential divider. (If R1 was also made up of a 2xE24 pair, there would be a great many more) Which one we select depends on the circuit design priorities. If an accurate nominal value is the overriding concern, then R1 = 330 Ω gives a nominal error of only +0.004%, combined with an effective tolerance of 0.707 %, which is as good as it gets because the two resistors making up R2 are equal. The output impedance is nicely low at 234 Ω, which will have a Johnson noise of only −128.5 dBu (usual conditions). This looks the best solution out of the seventeen. The loading of the divider is relatively heavy at 1130 Ω, and if on second thoughts we would prefer a higher resistance, R1 = 680 Ω looks promising, with a nominal error of only +0.062 % and once more an effective tolerance of 0.707%. Other factors may come into play—if you have a huge stock of 620 Ω resistors, then R1 = 620 Ω looks more attractive.

Table 2.10 Values for a −3 dB potential divider with different E24 choices for R1, and R2 as 2xE24

R1 Ω	R2 Ω	R2a Ω E24	R2b Ω E24	Nom error R2	Eff tol R2	Loading Ω	Output Z Ω
1500	3635.03	4700	16k	−0.060 %	0.806 %	5135.03	1061.8
1300	3151.46	5100	8200	−0.225 %	0.726 %	4451.46	920.35
1200	2909.76	5100	6800	+0.156 %	0.714 %	4109.76	849.61
1100	2666.43	5100	5600	+0.102 %	0.708 %	3766.43	778.74
1000	2424.03	4300	5600	+0.342 %	0.713 %	3424.03	707.94
910	2205.77	3900	5100	+0.192 %	0.713 %	3115.77	644.22
820	1986.98	2700	7500	−0.085 %	0.781 %	2806.98	580.45
750	1818.04	2700	5600	+0.201 %	0.749 %	2568.04	530.96
680	**1648.97**	**3300**	**3300**	**+0.062 %**	**0.707 %**	**2328.97**	**481.46**
620	1502.91	3000	3000	−0.194 %	0.707 %	2122.91	438.93
560	1357.48	1800	5600	+0.345 %	0.795 %	1917.48	396.45
510	1236.26	1800	3900	−0.379 %	0.754 %	1746.27	361.05
470	1139.29	1600	3900	−0.416 %	0.766 %	1609.29	332.73
430	1042.74	2000	2200	+0.468 %	0.708 %	1472.74	304.45
390	945.37	1800	2000	+0.211 %	0.708 %	1335.37	276.10
360	872.65	1100	4300	+0.375 %	0.822 %	1232.65	254.86
330	**799.97**	**1600**	**1600**	**+0.004 %**	**0.707 %**	**1129.87**	**233.62**
300	727.20	1000	2700	+0.348 %	0.778 %	1027.20	212.38
270	654.61	1100	1600	−0.421 %	0.719 %	924.61	191.16

Many other circumstances arise in circuit design, where there is a degree of freedom in choosing one component that allows the best option for another component to be selected.

Resistor Value Distributions

At this point you may be complaining that this will only work if the resistor values have a Gaussian (also known as normal) distribution with the familiar peak around the mean (average) value. Actually, it is a happy fact this effect does *not* assume that the component values have a Gaussian distribution; even a batch of resistors with a uniform distribution gives better accuracy when two of them are combined. This is easily demonstrated by Monte-Carlo. An excellent account of how to handle statistical variations to enhance accuracy is in W. J. Smith's *Modern Optical Engineering*.[5] This deals with the addition of mechanical tolerances in optical instruments, but the principles are just the same.

You sometimes hear that this sort of thing is inherently flawed, because, for example, 1% resistors are selected from production runs of 5% resistors. If you were using the 5% resistors, then you would find there was a hole in the middle of the distribution; if you were trying to select 1% resistors from them, you would be in for a very frustrating time as they have already been selected out, and you wouldn't find a single one. If instead you were using the 1% components obtained by selection from the complete 5% population, you would find that the distribution would be much flatter than Gaussian and the accuracy improvement obtained by combining them would be reduced, although there would still be a definite improvement.

However, don't worry. In general this is not the way that components are manufactured nowadays, though it may have been so in the past. A rare contemporary exception is the manufacture of carbon composition resistors,[6] where making accurate values is difficult and selection from production runs, typically with a 10% tolerance, is the only practical way to get more accurate values. Carbon composition resistors have no place in audio circuitry because of their large temperature and voltage coefficients and high excess noise, but they live on in specialised applications such as switch-mode snubbing circuits, where their ability to absorb high peak power in bulk material rather than a thin film is useful, and in RF circuitry, where the inductance of spiral-format film resistors is unacceptable.

So, having laid that fear to rest, what is the actual distribution of resistor values like? It is not easy to find out,

as manufacturers are not exactly forthcoming with this sort of sensitive information, and measuring thousands of resistors with an accurate DVM is not a pastime that appeals to all of us. Any nugget of information in this area is therefore very welcome.

Hugo Kroeze[7] reported the result of measuring 211 metal film resistors from the same batch with a nominal value of 10 kΩ and 1% tolerance. He concluded that:

1) The mean value was 9.995 k Ω (0.05% low).
2) The standard deviation was about 10 Ω, i.e. only 0.1%. This spread in value is surprisingly small (the resistors were all from the same batch, and the spread across batches widely separated in manufacture date might have been less impressive).
3) All resistors were within the 1% tolerance range.
4) The distribution appeared to be Gaussian, with no evidence that it was a subset from a larger distribution.

I decided to add my own morsel of data to this. I measured 100 ordinary metal film 1kΩ resistors of 1% tolerance from Yageo, a Chinese manufacturer, and very tedious it was too. I used a recently calibrated 4.5 digit meter.

1) The mean value was 997.66 ohms (0.23% low).
2) The standard deviation was 2.10 ohms, i.e. 0.21%.
3) All resistors were within the 1% tolerance range. All but one was within 0.5%, with the outlier at 0.7%.
4) The distribution appeared to be Gaussian, with no evidence that it was a subset from a larger distribution.

These are only two reports, but from this and other evidence there seems to be no reason to doubt that the mean value is very well controlled, and the spread is under good control as well. The distribution of resistance values appears to be Gaussian, with nothing selected out.

The Uniform Distribution

As I mentioned earlier, improving average accuracy by combining resistors does not depend on the resistance value having a Gaussian distribution. A uniform distribution of component values would seem to be very unlikely, but the result of combining two or more of them is highly instructive; two combined give a triangular-shaped distribution, combining more gives a shape that gets peakier and eventually is indistinguishable from the Gaussian distribution.

Resistor Imperfections

Ohm's law strictly is a statement about metallic conductors only. It is dangerous to assume that it also invariably applies to "resistors" simply because they have a fixed value of resistance marked on them; in fact resistors— whose main *raison d'etre* is packing a lot of controlled resistance in a small space—do not always adhere to Ohm's law very closely. This is a distinct difficulty when trying to make low-distortion circuitry.

It is well known that resistors have inductance and capacitance and vary somewhat in resistance with temperature. Unfortunately there are other less obvious imperfections, such as excess noise and nonlinearity; these can get forgotten because parameters describing how bad they are often omitted from component manufacturer's data sheets.

Being components in the real world, resistors are not perfect examples of resistance and nothing else. Their length is not infinitely small, and so they have series inductance; this is particularly true for the many kinds that use a spiral resistive element. Likewise, they exhibit stray capacitance between each end and also between the various parts of the resistive element. Both effects can be significant at high frequencies, but can usually be ignored below 100 kHz unless you are using very high or low resistance values.

It is a sad fact that resistors change their value with temperature. Table 2.4 shows some typical temperature coefficients. This is not likely to be a problem in small-signal audio applications, where the ambient temperature range is usually small, and extreme precision is not required unless you are designing measurement equipment. Carbon film resistors are markedly inferior to metal film in this area.

The fact that resistors have non-zero temperature coefficients has however a worrying implication; if significant cyclic changes in temperature occur due to big low-frequency signals, there will be corresponding cyclic changes in resistance that in some circuit positions will cause nonlinear distortion. This is dealt with in more detail later, but in short it is unlikely to reach measurable proportions in metal film resistors unless the signal is bigger than 25 Vrms and the frequency below 100 Hz.

Many resistors also change their value slightly with the voltage placed across them; this is measured by the voltage coefficient and can cause frequency-insensitive nonlinear distortion. This is also dealt with in more detail later.

Resistor Excess Noise

All resistors, no matter what their resistive material or mode of construction, generate Johnson noise. This is white noise, its level being determined solely by the resistance value, the absolute temperature, and the bandwidth over which the noise is being measured. It is based on fundamental physics and is not subject to negotiation. Some cases it places the limit on how quiet a circuit can be, though the noise from active devices is often more significant. Johnson noise is covered in Chapter 1.

Excess resistor noise refers to the fact that some kinds of resistor, with a constant voltage drop imposed across them, generate excess noise in addition to its inherent Johnson noise. According to classical physics, passing a current through a resistor should have no effect on its noise behaviour; it should generate the same Johnson noise as a resistor with no steady current flow. In reality, some types of resistors do generate excess noise when they have a DC voltage across them. It is a very variable quantity, but is essentially proportional to the DC voltage across the component, a typical spec being "1 uV/V" and it has a $1/f$ frequency distribution. Typically it could be a problem in biasing networks at the input of amplifier stages. It is usually only of interest if you are using carbon or thick film resistors- metal film, and wirewound types should have very little excess noise. It is known $1/f$ noise does not have a Gaussian amplitude distribution, which makes it difficult to assess reliably from a small set of data points. A rough guide to the likely specs is given in Table 2.11.

(Wirewound resistors are normally considered to be completely free of excess noise.)

The level of excess resistor noise changes with resistor type, size, and value in ohms; here are the relevant factors:

Thin film resistors are markedly quieter than thick film resistors; this is due to the homogeneous nature of thin film resistive materials, which are metal alloys such as nickel-chromium deposited on a substrate. The thick film resistive material is a mixture of metal (often ruthenium) oxides and

Table 2.11 Resistor excess noise

Type	Noise uV/V
Metal film TH	0
Carbon film TH	0.2–3
Metal oxide TH	0.1–1
Thin film SM	0.05–0.4
Bulk metal foil TH	0.01
Wirewound TH	0

glass particles; the glass is fused into a matrix for the metal particles by high temperature firing. The higher excess noise levels associated with thick film resistors are a consequence of their heterogeneous structure, due to the particulate nature of the resistive material. The same applies to carbon film resistors where the resistive medium is finely-divided carbon dispersed in a polymer binder.

A physically large resistor has lower excess noise than a small resistor, because there is more resistive material in parallel, so to speak. In the same resistor range, the highest wattage versions have the lowest noise. See Figure 2.9.

A low ohmic value resistor has lower excess noise than a high ohmic value. Noise in uV per V rises approximately with the square root of resistance. See Figure 2.9 again.

A low value of excess noise is associated with uniform constriction-free current flow; this condition is not well met in composite thick film materials. However, there are great variations among different thick film resistors. The most readily apparent relationship is between noise level and the amount of conductive material present. Everything else being equal, compositions with lower resistivity have lower noise levels.

Higher resistance values give higher excess noise since it is a statistical phenomenon related to the total number of charge carriers available within the resistive element; the fewer the total number of carriers present, the greater will be the statistical fluctuation.

Traditionally at this point in the discussion of excess resistor noise, the reader is warned against using carbon composition resistors because of their very bad excess noise characteristics. Carbon composition resistors are still made—their construction makes them good at handling pulse loads—but are not likely to be encountered in audio circuitry.

One of the great benefits of dual-rail opamp circuitry is that is noticeably free of resistors with large DC voltages across them. The offset voltages and bias currents are far too low to cause trouble with resistor excess noise. However, if you are getting into low-noise hybrid discrete/opamp stages, such as the MC head amplifier in Chapter 12, you might have to consider it.

To get a feel for the magnitude of excess resistor noise, consider a 100 kΩ 1/4 W carbon film resistor with 10V across it. This, from the graph above, has an excess noise parameter of about 0.7 uV/V and so the excess noise will be of the order of 7 uV, which is −101 dBu. This definitely could be a problem in a low-noise preamplifier stage.

Resistor Nonlinearity: Voltage Coefficient

This form of resistor nonlinearity is normally quoted by manufacturers as a voltage coefficient, usually the number of parts per million (ppm) that the resistor value changes when one volt is applied. The measurement standard for resistor nonlinearity is IEC 6040.

Through-hole metal film resistors usually show perfect linearity at the levels of performance considered here, as

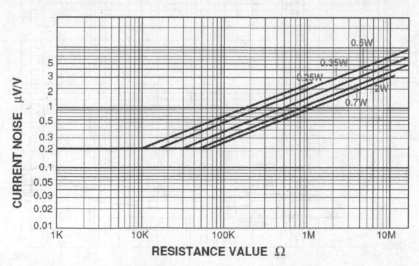

Figure 2.9 The typical variation of excess resistor noise with ohmic value and physical size; this is for a range of carbon film resistors. The flat part of the plot represents the measurement floor, not a change in noise mechanism.

do wirewound types. The voltage coefficient is less than 1 ppm. Carbon film resistors are quoted at less than 100 ppm; 100 ppm is however enough to completely dominate the distortion produced by active devices, if it is used in a critical part of the circuitry. Carbon composition resistors, probably of historical interest only, come in at about 350 ppm, a point that might be pondered by connoisseurs of antique equipment. The greatest area of concern over nonlinearity is thick film surface-mount resistors, which have high and rather variable voltage coefficients; more on this later.

Table 2.12 (calculated with SPICE) gives the THD in the current flowing through the resistor for various voltage

Table 2.12 Resistor voltage coefficients and the resulting distortion at +15 and +20 dBu.

Voltage Coefficient	THD at +15 dBu	THD at +20 dBu
1 ppm	0.00011 %	0.00019 %
3 ppm	0.00032 %	0.00056%
10 ppm	0.0016 %	0.0019 %
30 ppm	0.0032 %	0.0056 %
100 ppm	0.011 %	0.019 %
320 ppm	0.034 %	0.060 %
1000 ppm	0.11 %	0.19 %
3000 ppm	0.32 %	0.58 %

coefficients when a pure sine voltage is applied. If the voltage coefficient is significant this can be a serious source of nonlinearity.

A voltage coefficient model generates all the odd-order harmonics, at a decreasing level as order increases. No even-order harmonics can occur because the model is symmetrical. This is covered in much more detail in my power amplifier book.[8]

My own test setup is shown in Figure 2.10. The resistors are usually of equal value, to give 6 dB attenuation. A very low-distortion oscillator that can give a large output voltage is necessary; the results in Figure 2.11 were taken at a 10 Vrms (+22 dBu) input level. Here thick film SM and through-hole resistors are compared. The gen-mon trace at the bottom is the record of the analyser

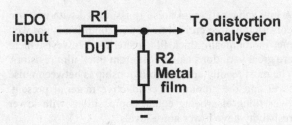

Figure 2.10 Test circuit for measuring resistor nonlinearity. The not-under-test resistor R2 in the potential divider must be a metal film type with negligible voltage coefficient.

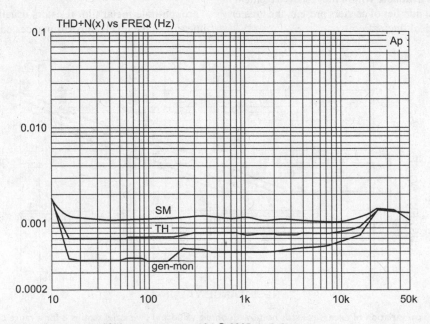

Figure 2.11 SM resistor distortion at 10 Vrms input, using 10 kΩ 0805 thick film resistors

reading the oscillator output and is the measurement floor of the AP System 1 used. The TH plot is higher than this floor, but this is not due to distortion. It simply reflects the extra Johnson noise generated by two 10 kΩ resistors. Their parallel combination is 5 kΩ, and so this noise is at −115.2 dBu. The SM plot, however, is higher again, and the difference is the distortion generated by the thick film component.

For both thin film and thick film SM resistors nonlinearity increases with resistor value and also increases as the physical size (and hence power rating) of the resistor shrinks. The thin film versions are much more linear; see Figures 2.12 and 2.13. Sometimes it is appropriate to reduce the nonlinearity by using multiple resistors in series. If one resistor is replaced by two with the same voltage coefficient in series, the THD in the current flowing is halved. Similarly, using three resistors reduces THD to a third of the original value. There are

obvious economic limits to this sort of thing, but it can be useful in specific cases, especially where the voltage rating of the resistor is a limitation.

Modelling voltage coefficient distortion in SPICE simulation is straightforward. An analogue behavioural model is used to create a resistance whose value varies proportionally with the instantaneous voltage across it; a 1st-order voltage coefficient nonlinearity. Only odd harmonics are generated, and their percentage increases proportionally with the signal level. It is not much more complex to model a resistance whose value varies proportionally with the square of the instantaneous voltage; a 2nd-order voltage coefficient nonlinearity. Once more only odd harmonics appear, but in this case the third harmonic increases as the square of signal level, the fifth as the fourth power, and so on. The SPICE modelling of voltage coefficient nonlinearity is dealt with in much more detail in my book *Audio Power Amplifier Design*,[8] but make sure you get the 6th edition or later.

Resistor Nonlinearity: Temperature Coefficient

Since resistors have nonzero temperature coefficients it is obvious that if cyclic changes in temperature occur, there will be corresponding cyclic changes in resistance. If the resistor concerned is involved in setting the gain of the stage (for example as part of a voltage divider providing negative feedback), the cyclic changes in gain will cause nonlinear distortion. This is often called thermal distortion; there is much talk in the hi-fi world of "thermal distortion" in circumstances where it does not exist, but this example is the real thing. The lower the frequency, the greater the distortion, as the resistor has more time to change temperature. The distortion level rises very slowly as frequency falls, taking three octaves to double in amplitude, and as far as I am aware this behaviour is unique to thermal distortion and provides a good test for its existence. The distortion is reasonably pure third harmonic.

Thermal distortion in carbon film resistors is relatively easy to detect and measure; in a simple 6 dB voltage divider where only one part is carbon film, the other being metal film, it will be around 0.001% at 10 Hz and 20 Vrms. You can argue a) that's not a lot of distortion, and b) 20 Vrms is not going to be encountered in small-signal design, except in a balanced output stage running at maximum level. The counterargument is that many resistors in a practical system, all showing that sort of distortion, are likely to give a rather unhappy overall result.

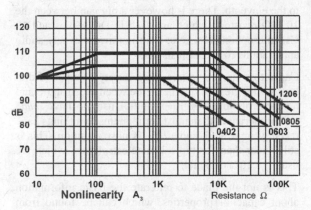

Figure 2.12 Nonlinearity of thin film surface-mount resistors of different sizes. THD is in dB rather than percent.

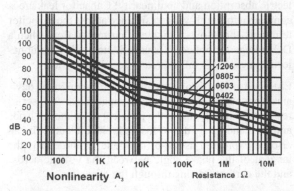

Figure 2.13 Nonlinearity of thick film surface-mount resistors of different sizes

Generally, in small-signal design you can just specify metal film resistors and forget about the issue; an exception being a balanced line output at maximum level. It is a very real issue in power amplifiers of more than modest output, and this is a good way to demonstrate the effect with amounts of distortion that are easily measured. The feedback network will be a voltage divider with the upper resistor having almost all the output voltage across it; at 100W/8Ω this will be about 27 Vrms, and at this level even a high-quality metal film resistor has been found to generate 0.0008% THD+N at 10 Hz. You obviously need to use a really good and clean power amplifier, or other distortion mechanisms will mask the effect. I used one of my Blameless power amplifiers. Figure 2.14 shows the results at 100W/8Ω, where thermal distortion causes the gentle rise in THD +N below 100 Hz. The original feedback resistor had a temperature coefficient specified as 100 ppm/°C, and replacing it with a different type (from the same reputable manufacturer) spec'd at 50 ppm/°C gave significantly reduced distortion.

If you do run into thermal distortion, you might fear that a nonlinearity built into a passive component is going to be hard to cure. Not so. First the distortion is low, and reducing it by a factor of three or four times will usually push it below the noise floor. In dealing with voltage coefficient distortion we can split the signal voltage across two or more resistors in series. For thermal distortion we need to split the power, and either parallel or series connections can be used.

It is quite easy to model voltage coefficient distortion in SPICE simulation, but temperature coefficient distortion is a harder nut to crack. It is straightforward to calculate the instantaneous heat dissipation and combine that with the effective thermal inertia to calculate the cyclic resistance change. The problem is, the thermal inertia of what? It is the temperature of actual resistive element—the helical metal film—that matters, but it is intimately attached to the body of the resistor, which much affects the thermal inertia. The thermal variations inside the body will depend on the distance from the surface, and modelling the effects of this is probably going to require some sort of network of RC time-constants.

Capacitors

Capacitors are diverse components. In the audio business their capacitance ranges from 10 pF to 100,000 uF, a ratio of 10 to the tenth power. In this they handily out-do resistors, which usually vary from 0.1 Ω to 10 MΩ, a ratio of only 10 to the eighth. However, if you include the 10 GΩ bias resistors used in capacitor microphone head amplifiers, this range increases to 10 to the eleventh. There is however a big gap between the 10 MΩ resistors, which are used in DC servos and 10 GΩ microphone resistors; I am not aware of any audio applications for 1 GΩ resistors.

Capacitors also come in a wide variety of types of dielectric, the great divide being between electrolytic and nonelectrolytic types. Electrolytics used to have *much* wider tolerances than most components, but things have recently improved, and ±20% is now common. This is still wider than for typical nonelectrolytics, which are usually ±10% or better.

This is not the place to reiterate the basic information about capacitor properties, which can be found from many sources. I will simply note that real capacitors fall short of the ideal circuit element in several ways, notably leakage, equivalent series resistance (ESR), dielectric absorption and nonlinearity: Capacitor leakage is equivalent to a high-value resistance across the capacitor terminals and allows a trickle of current to flow when a DC voltage is applied. It is usually negligible for nonelectrolytics, but is much greater for electrolytics.

ESR is a measure of how much the component deviates from a mathematically pure capacitance. The series resistance is partly due to the physical resistance of leads and foils and partly due to losses in the dielectric. It can also be expressed as tan-δ (tan-delta). Tan-delta is the tangent of the phase angle between the voltage across and the current flowing through the capacitor.

Dielectric absorption is a well-known phenomenon; take a large electrolytic, charge it up, and then make sure it

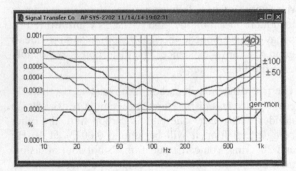

Figure 2.14 Thermal distortion in a metal film resistor. THD of Compact Class-B CFP power amp 100W/8Ω. Single 2k2 750 mW ±100 ppm/°C MF, Single PR 2k2 500mW ±50 ppm/°C MF resistors; "gen-mon" is testgear output. Measurement bandwidth 10–22 kHz.

is fully discharged. Use a 10 Ω WW resistor across the terminals rather than a screwdriver unless you're not too worried about either the screwdriver or the capacitor. Wait a few minutes, and the charge will partially reappear, as if from nowhere. This "memory effect" also occurs in nonelectrolytics to a lesser degree; it is a property of the dielectric and is minimised by using polystyrene, polypropylene, NPO ceramic, or PTFE dielectrics. Dielectric absorption is invariably simulated by a linear model composed of extra resistors and capacitances; nevertheless, dielectric absorption and distortion correlate across the different dielectrics.

Capacitor nonlinearity is undoubtedly the least known of these shortcomings. A typical RC lowpass filter can be made with a series resistor and a shunt capacitor, and if you examine the output with a distortion analyser, you may find to your consternation that the circuit is not linear. If the capacitor is a nonelectrolytic type with a dielectric such as polyester, then the distortion is relatively pure third harmonic, showing that the effect is symmetrical. For a 10 Vrms input, the THD level may be 0.001% or more. This may not sound like much, but it is substantially greater than the mid-band distortion of a good opamp. Capacitor nonlinearity is dealt with at greater length later.

Capacitors are used in audio circuitry for three main functions, where their possible nonlinearity has varying consequences:

1) Coupling or DC-blocking capacitors. These are usually electrolytics, and if properly sized have a negligible signal voltage across them at the lowest frequencies of interest. Likewise, the detailed properties of the capacitor are unimportant unless they have power amplifier current levels through them; power amplifier output capacitors can generate considerable mid-band distortion.[9] Much nonsense has been talked about mysterious coupling capacitor properties, but it *is* all absolute and total nonsense. For small-signal use, as long as the signal voltage across the capacitor is kept low, nonlinearity is not normally detectable. The capacitance value must be noncritical, given the wide tolerances of electrolytics.
2) Supply filtering or decoupling capacitors. These are electrolytics if you are filtering out supply rail ripple, etc., and non-electrolytics, usually around 100 nF, when the task is to keep the supply impedance low at high frequencies and so keep opamps stable. The capacitance value is again noncritical.
3) Setting time-constants, for example the capacitors in the feedback network of an RIAA amplifier. This is

a much more demanding application than the other two. First, the actual value is now crucially important as it defines the accuracy of the frequency response. Second, there is by definition significant signal voltage across the capacitor, as it is involved in filtering, and so nonlinearity can be a serious problem. Nonelectrolytics are normally used; sometimes an electrolytic is used to define the lower end of the bandwidth, but this is a bad practice likely to introduce distortion at the bottom of the frequency range. Small value ceramic capacitors are used for compensation purposes.

In Subjectivist circles it is frequently asserted that electrolytic coupling capacitors (if they are permitted at all) should be bypassed by small nonelectrolytics. There is no sense in this; if the main coupling capacitor has no signal voltage across it, the extra capacitor can have no effect.

Capacitor Selection for Awkward Values

There is much discussion in this book about the best way to obtain awkward resistance values; this is relatively easy because resistors are available in the E24 series. Capacitors are different; they are frequently only obtainable in the E3, E6, or E12 series. This makes the mathematical problem quite different—one approach is to select the nearest capacitor that is smaller than the desired value, then parallel smaller and smaller capacitors until you get as close as desired. A better way is to try to break the capacitance into two or three equal values, plus an extra capacitor, as this gives more improvement in the effective tolerance.

Another factor is the need for a close tolerance on the capacitor value. As explained in more detail in Chapter 7, often the most economical way to get a 1% tolerance is to use polystyrene capacitors, but in many ranges 10 nF is the largest available value, and for phono amplifier use several may need to be paralleled. This improves the effective tolerance in the same way as for resistors. There are many examples of this in Chapter 7. Most of the capacitor combinations in this book were produced manually, but since then Gert Willmann has produced a table for four E6 capacitors in parallel. It is 3360 lines long, but this is perfectly easy to handle and search in any decent text editor.

I used this new table to check all the values for C2 in Chapter 7, where three or four capacitors are used in parallel, and was rather surprised that only two out of ten results could be improved. These are shown in Tables 2.13 and 2.14. In the first table the use of four

Table 2.13 Improvements on manual É6 capacitor selection using the Willmann table (Table 7.30)

Desired value	Actual value	Parallel part A	Parallel part B	Parallel part C	Parallel part D	Nominal error	Effective tolerance
10.371 nF	10.4 nF	4.7 nF	4.7 nF	1 nF	–	+0.28%	0.65%
10.371 nF	10.37 nF	3.3 nF	3.3 nF	3.3 nF	0.47 nF	−0.01%	0.55%

Table 2.14 Improvements on manual E6 capacitor selection using the Willmann table (Table 7.34)

Desired value	Actual value	Parallel part A	Parallel part B	Parallel part C	Parallel part D	Nominal error	Effective tolerance
11.408 nF	11.37 nF	4.7 nF	4.7 nF	1.5 nF	0.47 nF	−0.33%	0.60%
11.408 nF	11.40 nF	4.7 nF	4.7 nF	1 nF	1 nF	−0.07%	0.60%

capacitors rather than three allows the nominal error to be reduced from +0.28% to a negligible 0.01%, with a helpful improvement in the effective tolerance.

In Table 2.14 four capacitors are used in each case

These new values have been used in Chapter 7.

Capacitor Nonlinearity Examined

When attempting the design of linear circuitry, everyone knows that inductors and transformers with ferromagnetic core material can be a source of nonlinearity. It is however less obvious that capacitors and even resistors can show nonlinearity and generate some unexpected and very unwelcome distortion. Resistor nonlinearity has been dealt with earlier in this chapter; let us examine the shortcomings of capacitors.

The definitive work on capacitor distortion is a magnificent series of articles by Cyril Bateman in *Electronics World*.[10] The authority of this work is underpinned by Cyril's background in capacitor manufacture. (The series is long because it includes the development of a low-distortion THD test set in the first two parts.)

Capacitors generate distortion when they are actually implementing a time-constant—in other words, when there is a signal voltage across them. The normal coupling or DC-blocking capacitors have no significant signal voltage across them, as they are intended to pass all the information through, not to filter it or define the system bandwidth. Capacitors with no signal across them do not generally produce distortion at small-signal current levels. This was confirmed for all the capacitors tested here. However, electrolytic types may do so at power amplifier levels where the current through them

is considerable, such as in the output coupling capacitor of a power amplifier.[9]

Nonelectrolytic Capacitor Nonlinearity

It has often been assumed that nonelectrolytic capacitors, which generally approach an ideal component more closely than electrolytics and have dielectrics constructed in a totally different way, are free from distortion. It is not so. Some nonelectrolytics show distortion at levels that are easily measured and can exceed the distortion from the opamps in the circuit. Nonelectrolytic capacitor distortion is primarily third harmonic, because the nonpolarised dielectric technology is basically symmetrical. The problem is serious, because nonelectrolytic capacitors are commonly used to define time-constants and frequency responses (in RIAA equalisation networks, for example) rather than simply for DC blocking.

Very small capacitances present no great problem. Simply make sure you are using the COG (NP0) type, and so long as you choose a reputable supplier, there will be no distortion. I say "reputable supplier" because I did once encounter some allegedly COG capacitors from China that showed significant nonlinearity.[11]

Middle-range capacitors, from 1 nF to 1 uF, present more of a problem. Capacitors with a variety of dielectrics are available, including polyester, polystyrene, polypropylene, polycarbonate, and polyphenylene sulphide, of which the first three are the most common. (Note that what is commonly called "polyester" is actually polyethylene terephthalate, PET.)

Figure 2.15 shows a simple lowpass filter circuit which, in conjunction with a good THD analyser, can be used

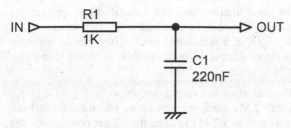

Figure 2.15 Simple lowpass test circuit for nonelectrolytic capacitor distortion

to get some insight into the distortion problem; it is intended to be representative of a real bit of audio circuitry. The values shown give a pole frequency, or −3 dB roll-off point, at 710 Hz. Since it might be expected that different dielectrics give different results (and they definitely do), we will start off with polyester, the smallest, most economical, and therefore the most common type for capacitors of this size.

The THD results for a microbox 220 nF 100 V capacitor with a polyester dielectric are shown in Figure 2.16, for input voltages of 10, 15, and 20 Vrms. They are unsettling.

The distortion is all third harmonic and peaks at around 300 to 400 Hz, well below the pole frequency, and even with input limited to 10 Vrms, will exceed the nonlinearity introduced by opamps such as the 5532 and the LM4562. Interestingly, the peak frequency changes with applied level. Below the peak, the voltage across the capacitor is constant, but distortion falls as frequency is reduced, because the increasing impedance of the capacitor means it has less effect on a circuit node at a

1 kΩ impedance. Above the peak, distortion falls with increasing frequency because the lowpass circuit action causes the voltage across the capacitor to fall.

The level of distortion varies with different samples of the same type of capacitor; six of this type were measured, and the THD at 10 Vrms and 400 Hz varied from 0.00128% to 0.00206%. This puts paid to any plans for reducing the distortion by some sort of cancellation method. The distortion can be seen in Figure 2.16 to be a strong function of level, roughly tripling as the input level doubles. Third-harmonic distortion normally quadruples for doubled level, so there may well be an unanswered question here. It is however clear that reducing the voltage across the capacitor reduces the distortion. This suggests that if cost is not the primary consideration, it might be useful to put two capacitors in series to halve the voltage and the capacitance and then double up this series combination to restore the original capacitance, giving the series-parallel arrangement in Figure 2.17. The results are shown in Table 2.15, and once more it can be seen that halving the level has reduced distortion by a factor of three rather than four. The series-parallel arrangement obviously has limitations in terms of cost and PCB area occupied but might be useful in some cases.

Clearly polyester gives significant distortion, despite its extensive use in audio circuitry of all kinds.

An unexpected complication was that every time a sample was remeasured, the distortion was lower than before. I found a steady reduction in distortion over time if a test signal was left applied; 9 Vrms at 1 kHz halved the THD over 11 hours. This is a semi-permanent

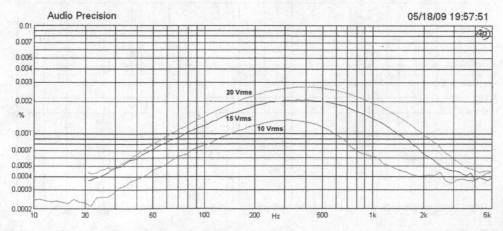

Figure 2.16 Third-harmonic distortion from a 220 nF 100 V polyester capacitor, at 10, 15, and 20 Vrms input level, showing peaking around 400 Hz

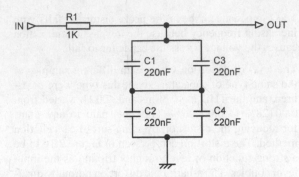

Figure 2.17 Reducing capacitor distortion by series-parallel connection

Table 2.15 The reduction of polyester capacitor distortion by series-parallel connection

Input level Vrms	Single capacitor	Series-parallel capacitors
10	0.0016 %	0.00048 %
15	0.0023 %	0.00098 %
20	0.0034 %	0.0013%

change, as some of the distortion returns over time when the signal is removed. This effect may be of very little practical use, but if nothing else it does demonstrate that polyester capacitors are more complicated than they look. For much more on this see the aforementioned article by Bateman.[12]

The next dielectric we will try is polystyrene. Capacitors with a polystyrene dielectric are extremely useful for some filtering and RIAA equalisation applications because they can be obtained at a 1% tolerance at up to 10 nF at a reasonable price. They can be obtained in larger sizes at an unreasonable, or at any rate much higher, price.

The distortion test results are shown in Figure 2.18 for a 4n7 2.5% capacitor; the series resistor R1 has been increased to 4.7 kΩ to keep the −3 dB point inside the audio band, and it is now at 7200 Hz. Note that the THD scale has been extended down to a subterranean 0.0001%, and if it was plotted on the same scale as Figure 2.16 it would be bumping along the bottom of the graph. Figure 2.18 in fact shows no distortion at all, just the measurement noise floor, and the apparent rise at the HF end is simply due to the fact that the output level is decreasing because of the lowpass action, and so the noise floor is relatively increasing. This is at an input level of 10 Vrms, which is about as high as might be expected to occur in normal opamp circuitry. The test was repeated at 20 Vrms, which might be encountered in discrete circuitry, and the results were the same. No measurable distortion.

The tests were done with four samples of 10nF 1% polystyrene from LCR at 10 Vrms and 20 Vrms, with the same results for each sample. This shows that polystyrene capacitors can be used with confidence; this is in complete agreement with Cyril Bateman's findings.[13]

Having settled the issue of capacitor distortion below 10 nF, we need now to tackle it capacitor values greater than 10 nF. Polyester having proven unsatisfactory, the next most common dielectric is polypropylene, and I might

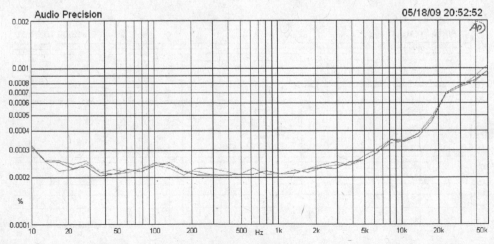

Figure 2.18 The THD plot with three samples of 4n7 2.5% polystyrene capacitors, at 10 Vrms input level. The reading is entirely noise.

as well say at once that it was with considerable relief that I found these were effectively distortion free in values up to 220 nF. Figure 2.19 shows the results for four samples of a 220 nF 250 V 5% polypropylene capacitor from RIFA. Once more the plot shows no distortion at all, just the noise floor, the apparent rise at the HF end being increasing relative noise due to the lowpass roll-off. This is also in agreement with Cyril Bateman's findings. Rerunning the tests at 20 Vrms gave the same result—no distortion. This is very pleasing, but there is a downside. Polypropylene capacitors of this value and voltage rating are physically much larger than the commonly used 63 or 100V polyester capacitor, and more expensive.

It was therefore important to find out if the good distortion performance was a result of the 250 V rating, and so I tested a series of polypropylene capacitors with lower voltage ratings from different manufacturers. Axial 47 nF 160 V 5% polypropylene capacitors from Vishay proved to be THD-free at both 10 Vrms and 20 Vrms. Likewise, microbox polypropylene capacitors from 10 nF to 47 nF with ratings of 63 V and 160 V from Vishay and Wima proved to generate no measurable distortion, so the voltage rating appears not to be an issue. This finding is particularly important because the Vishay range has a 1% tolerance, making them very suitable for precision filters and equalisation networks. The 1% tolerance is naturally reflected in the price.

The only remaining issue with polypropylene capacitors is that the higher values (above 100 nF) appear to be currently only available with 250 V or 400 V ratings, and that means a physically big component. For example, the EPCOS 330 nF 400 V 5% part has a footprint of 26 mm by 6.5 mm, with a height of 15 mm. One way of dealing with this is to use a smaller capacitor in a capacitance multiplication configuration, so a 100 nF 1% component could be made to emulate 470 nF. It has to be said that the circuitry for this is only straightforward if one end of the capacitor is connected to ground.

When I first started looking at capacitor distortion, I thought that the distortion would probably be lowest for the capacitors with the highest voltage rating. I therefore tested some RF-suppression X2 capacitors rated at 275 Vrms, which translates into a peak or DC rating of 389 V. These are designed to be connected directly across the mains and therefore have a thick and tough dielectric layer. For some reason manufacturers seem to be very coy about saying exactly what the dielectric material is, normally describing the layers simply as "film capacitors". A problem that surfaced immediately is that the tolerance is 10 or 20%, not exactly ideal for precision filtering or equalisation. A more serious problem, however, is that they are far from distortion free. Four samples of a 470 nF X2 capacitor showed THD between 0.002% and 0.003% at 10 Vrms. Clearly a high-voltage rating alone does not mean low distortion.

Electrolytic Capacitor Nonlinearity

Cyril Bateman's series in *Electronics World*[14] included two articles on electrolytic capacitor distortion. It proved to be a complex subject, and many long-held assumptions (such as "DC biasing always reduces distortion") were shown to be quite wrong. Distortion was

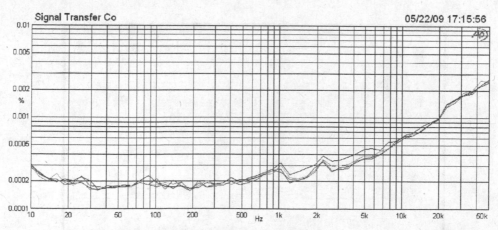

Figure 2.19 The THD plot with four samples of 220 nF 250 V 5% polypropylene capacitors, at 10 Vrms input level. The reading is again entirely noise.

in general a good deal higher than for nonelectrolytic capacitors.

My view is that electrolytics should never, ever, under any circumstances, be used to set time-constants in audio. There should be a time-constant early in the signal path, based on a nonelectrolytic capacitor, that determines the lower limit of the bandwidth, and all the electrolytic-based time-constants should be much longer so that the electrolytic capacitors can never have significant signal voltages across them and so never generate measurable distortion. There is of course also the point that electrolytics have large tolerances and cannot be used to set accurate time-constants anyway.

However, even if you obey this rule, you can still get into deep trouble. Figure 2.20 shows a simple highpass test circuit designed to represent an electrolytic capacitor in use for coupling or DC blocking. The load of 1 KΩ is the sort of value that can easily be encountered if you are using low-impedance design principles. The

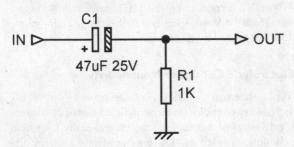

Figure 2.20 Highpass test circuit for examining electrolytic capacitor distortion

calculated −3 dB roll-off point is 3.38 Hz, so the attenuation at 10 Hz, at the very bottom of the audio band, will be only 0.47 dB; at 20 Hz it will be only 0.12 dB, which is surely a negligible loss. As far as frequency response goes, we are doing fine. But . . . examine Figure 2.21, which shows the measured distortion of this arrangement. Even if we limit ourselves to a 10 Vrms level, the distortion at 50 Hz is 0.001%, already above that of a good opamp. At 20 Hz it has risen to 0.01%, and by 10 Hz a most unwelcome 0.05%. The THD is increasing by a ratio of 4.8 times for each octave fall in frequency, in other words increasing faster than a square law. The distortion residual is visually a mixture of second and third harmonic, and the levels proved surprisingly consistent for a large number of 47 uF 25 V capacitors of different ages and from different manufacturers.

Figure 2.21 also shows that the distortion rises rapidly with level; at 50 Hz going from an input of 10 Vrms to 15 Vrms almost doubles the THD reading. To underline the point, consider Figure 2.22, which shows the measured frequency response of the circuit with 47 uF and 1 KΩ; note the effect of the capacitor tolerance on the real versus calculated response. The roll-off that does the damage, by allowing an AC voltage to exist across the capacitor, is very modest indeed, less than 0.2 dB at 20 Hz.

Having demonstrated how insidious this problem is, how do we fix it? Changing capacitor manufacturer is no help. Using 47 uF capacitors of higher voltage does not work—tests showed there is very little difference in the amount of distortion generated. An exception was the subminiature style of electrolytic, which was markedly worse.

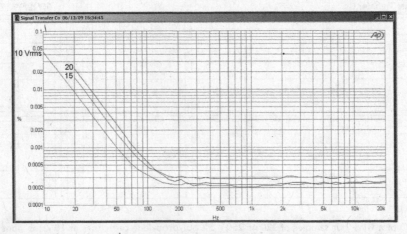

Figure 2.21 Electrolytic capacitor distortion from the circuit in Figure 2.20. Input level 10, 15, and 20 Vrms.

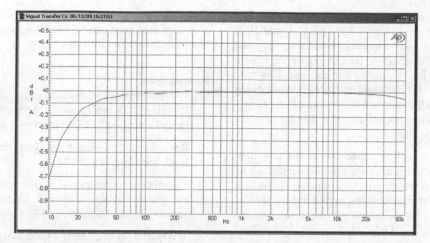

Figure 2.22 The measured roll-off of the highpass test circuit for examining electrolytic capacitor distortion

The answer is simple—just make the capacitor bigger in value. This reduces the voltage across it in the audio band, and since we have shown that the distortion is a strong function of the voltage across the capacitor, the amount produced drops more than proportionally. The result is seen in Figure 2.23, for increasing capacitor values with a 10 Vrms input.

Replacing C1 with a 100uF 25V capacitor drops the distortion at 20 Hz from 0.0080% to 0.0017%, an improvement of 4.7 times; the voltage across the capacitor at 20 Hz has been reduced from 1.66 Vrms to 790 mV rms. A 220 uF 25 V capacitor reduces the voltage across itself to 360 mV and gives another very welcome reduction to

0.0005% at 20 Hz, but it is necessary to go to 1000 uF 25 V to obtain the bottom trace, which is indistinguishable from the noise floor of the AP-2702 test system. The voltage across the capacitor at 20 Hz is now only 80 mV. From this data, it appears that the AC voltage across an electrolytic capacitor should be limited to below 80 mV rms if you want to avoid distortion. I would emphasise that these are ordinary 85 °C rated electrolytic capacitors and in no sense special or premium types.

This technique can be seen to be highly effective, but it naturally calls for larger and somewhat more expensive capacitors, and larger footprints on a PCB. This can be to some extent countered by using capacitors of lower

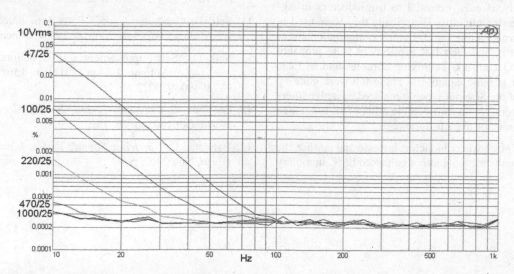

Figure 2.23 Reducing electrolytic capacitor distortion by increasing the capacitor value. Input 10 Vrms.

voltage, which helps to bring back down the CV product and hence the can size. I tested 1000 uF 16 V and 1000 uF 6V3 capacitors, and both types gave exactly the same results as the 1000 uF 25 V part in Figure 2.23, with useful reductions in CV product and can size. This does of course assume that the capacitor is, as is usual, being used to block small voltages from opamp offsets to prevent switch clicks and pot noises rather than for stopping a substantial DC voltage.

The use of large coupling capacitors in this way does require a little care, because we are introducing a long time-constant into the circuit. Most opamp circuitry is pretty much free of big DC voltages, but if there are any, the settling time after switch-on may become undesirably long.

More information on capacitor distortion in subsonic filter applications can be found in Chapter 12.

Inductors

For several reasons, inductors are unpopular with circuit designers. They are relatively expensive, often because they need to be custom-made. Unless they are air-cored (which limits their inductance to low values) the core material is a likely source of nonlinearity. Some types produce substantial external magnetic fields, which can cause crosstalk if they are placed close together, and similarly they can be subject to the induction of interference from other external fields. Because of their series resistance they deviate from being an ideal circuit element much more than resistors or capacitors.

It is rarely, if ever, essential to use inductors in signal-processing circuitry. Historically they were used in tone controls, before the Baxandall configuration swept all before it, and their last applications were probably in mid-EQ controls for mixing consoles and in LCR filters for graphic equalisers. These too were gone by the end of the Seventies, being replaced by active filters and gyrators, to the considerable relief of all concerned (except inductor manufacturers).

The only place where inductors are essential is when the need for galvanic isolation or enhanced EMC immunity makes input and output transformers desirable, and even then they need careful handling; see Chapter 14 on line-in and line-out circuitry.

References

1. Douglas, S. "Ultra-Low-Noise Amplifiers & Granularity Distortion" *JAES*, Nov 1987, pp. 907–915.

2. Renardsen, M. www.angelfire.com/ab3/mjramp/wire. html Accessed Feb 2017.

3. Takahisa. *Nonlinearity of Resistors and Its Geometric Factor*. Electronics and Communications in Japan, Volume 56-C, #6, 1973.

4. Wikipedia https://en.wikipedia.org/wiki/Monte_Carlo_method Accessed Feb 2017.

5. Smith, W. J. *Modern Optical Engineering*. McGraw-Hill, 1990, p. 484. ISBN0-07-059174-1.

6. Johnson, H. www.edn.com/electronics-blogs/signal-integrity/4363408/7-solution Accessed Feb 2017.

7. Kroeze, H. www.rfglobalnet.com/forums/Default. aspx?g=posts&m=61096 Accessed Mar 2002.

8. Self, D. *Audio Power Amplifier Design*. 6th edition. Newnes, 2013, pp. 39–45 (voltage coeff). ISBN 978-0-240-52613-3.

9. Self, D. *Audio Power Amplifier Design*. 6th edition. Newnes, 2013, p. 107 (amp output cap).

10. Bateman, C. *Capacitor Sound?* Parts 1–6. *Electronics World*, July 2002–Mar 2003.

11. Self, D. *Capacitor Sound?* Parts 1–6. *Electronics World*, July 2002–Mar 2003, p. 299 (COG cap).

12. Self, D. "Self-Improvement for Capacitors" *Linear Audio*, Volume 1, Apr 2011, p. 156. ISBN 9–789490–929022.

13. Bateman, C. *Capacitor Sound?* Part 3. *Electronics World*, Oct 2002, p. 16, 18.

14. Bateman, C. *Capacitor Sound?* Part 4. *Electronics World*, Nov 2002, p. 47.

Chapter 3

Opamps and Their Properties

Introduction

Audio design has for many years relied on a very small number of opamp types; the TL072 and the 5532 dominated the audio small-signal scene for many years. The TL072, with its JFET inputs, was used wherever its negligible input bias currents and low cost were important. For a long time the 5534/5532 was much more expensive than the TL072, so the latter was used wherever feasible in an audio system, despite its inferior noise, distortion, and load-driving capabilities. The 5534 was reserved for critical parts of the circuitry. Although it took many years, the price of the 5534 is now down to the point where you need a very good reason to choose any other type of opamp for audio work.

The TL072 and the 5532 are dual opamps; the single equivalents are TL071 and 5534. Dual opamps are used almost universally, as the package containing two is usually cheaper than the package containing one, simply because it is more popular.

There are however other opamps, some of which can be useful, and a selected range is covered here.

Opamp Properties: Noise

There is no point in regurgitating manufacturer's data sheets, especially since they are readily available on the internet. Here I have simply ranked the opamps most commonly used for audio in order of voltage noise (Table 3.1).

The great divide is between JFET input opamps and BJT input opamps. The JFET opamps have more voltage noise but less current noise than bipolar input opamps, the TL072 being particularly noisy. If you want the lowest voltage noise, it has to be a bipolar input. The difference however between a modern JFET input opamp such as the OPA2134 and the old faithful 5532 is only 4 dB; but the JFET part is a good deal more costly. The

Table 3.1 Opamps ranked by voltage noise density (typical)

Opamp	e_n nV/rtHz	i_n pA/rtHz	Input device type	Bias cancel?
LM741	20	??	BJT	No
TL072	18	0.01	FET	No
OPA604	11	0.004	FET	No
NJM4556	8	Not spec'd	BJT	No
OPA2134	8	0.003	FET	No
OP275	6	1.5	BJT+FET	No
OPA627	5.2	0.0025	FET	No
5532A	5	0.7	BJT	No
LM833	4.5	0.7	BJT	No
MC33078	4.5	0.5	BJT	No
5534A	3.5	0.4	BJT	No
OP270	3.2	0.6	BJT	No
OP27	3	0.4	BJT	Yes
LM4562	2.7	1.6	BJT	No
AD797	0.9	2	BJT	No
LT1028	0.85	1	BJT	Yes

bipolar AD797 seems to be out on its own here, but it is a specialised and expensive part. The LT1028 is not suitable for audio use because its bias-current cancellation system makes it noisy in most circumstances. The LM741 has no noise specs on its data sheets, and the 20 nV/rtHz is from measurements.

Both voltage and current noise increase at 6 dB/octave below the $1/f$ corner frequency, which is usually around 100 Hz. The only way to minimise this effect is to choose an appropriate opamp type

Opamps with bias-cancellation circuitry are normally unsuitable for audio use due to the extra noise this creates. The amount depends on circuit impedances and is not taken into account in Table 3.1. The general noise behaviour of opamps in circuits is dealt with in Chapter 1.

Opamp Properties: Slew Rate

Slew rates vary more than most parameters; a range of 100:1 is shown here in Table 3.2. The slowest is the LM741, which is the only type not fast enough to give full output over the audio band. There are faster ways to handle a signal, such as current-feedback architectures, but they usually fall down on linearity. In any case, a maximum slew rate greatly in excess of what is required appears to confer no benefits whatever.

The 5532 slew rate is typically ±9 V/us. This version is internally compensated for unity-gain stability, not least because there are no spare pins for compensation when you put two opamps in an 8-pin dual package. The single-amp version, the 5534, can afford a couple of compensation pins, and so is made to be stable only for gains of 3x or more. The basic slew rate is therefore higher at ±13 V/us.

Compared with power amplifier specs, which often quote 100 V/us or more, these opamp speeds may appear rather sluggish. In fact they are not; even ±9 V/us is more than fast enough. Assume you are running your opamp from ±18V rails and that it can give a ±17V swing on its output. For most opamps this is distinctly optimistic, but never mind. To produce a full-amplitude 20 kHz

Table 3.2 Opamps ranked by slew rate (typical)

Opamp	V/us
LM741	0.5
OP270	2.4
OP27	2.8
NJM4556	3
MC33078	7
LM833	7
5532A	9
LT1028	11
TL072	13
5534A	13
OPA2134	20
LM4562	20
AD797	20
OP275	22
OPA604	25
OPA627	55

sine wave you only need 2.1 V/us, so even in the worst case there is a safety margin of at least four times. Such signals do not of course occur in actual use, as opposed to testing. More information on slew-limiting is given in the section on opamp distortion.

Opamp Properties: Common-Mode Range

This is simply the range over which the inputs can be expected to work as proper differential inputs. It usually covers most of the range between the rail voltages, with one notable exception. The data sheet for the TL072 shows a common-mode (CM) range that looks a bit curtailed at −12V. This bland figure hides the deadly trap this IC contains for the unwary. Most opamps, when they hit their CM limits, simply show some sort of clipping. The TL072, however, when it hits its negative limit, promptly inverts its phase, so your circuit either latches up or shows nightmare clipping behaviour with the output bouncing between the two supply rails. The positive CM limit is in contrast trouble-free. This behaviour can be especially troublesome when TL072s are used in highpass Sallen and Key filters.

Opamp Properties: Input Offset Voltage

A perfect opamp would have its output at 0 V when the two inputs were exactly at the same voltage. Real opamps are not perfect, and a small voltage difference—usually a few millivolts—is required to zero the output. These voltages are large enough to cause switches to click and pots to rustle, and DC blocking is often required to keep them in their place.

The typical offset voltage for the 5532A is ±0.5 mV typical, ±4 mV maximum at 25 °C; the 5534A has the same typical spec but a lower maximum at ±2 mV. The input offset voltage of the new LM4562 is only ±0.1 mV typical, ±4 mV maximum at 25 °C.

Opamp Properties: Bias Current

Bipolar-input opamps not only have larger noise currents than their JFET equivalents, they also have much larger bias currents. These are the base currents taken by the input transistors. This current is much larger than the input offset current, which is the difference between the bias current for the two inputs. For example, the 5532A has a typical bias current of 200 nA, compared with a much smaller input offset current of 10 nA. The LM4562 has a lower bias current of 10 nA typical, 72 nA maximum. In the case of the 5532/4 the bias current flows into the input pins as the input transistors are NPN.

Bias currents are a considerable nuisance when they flow through variable resistors; they make them noisy when moved. They will also cause significant DC offsets when they flow through high-value resistors.

It is often recommended that the effect of bias currents can be cancelled out by making the resistance seen by each opamp input equal. Figure 3.1a shows a shunt-feedback stage with a 22 kΩ feedback resistor. When 200 nA flows through this it will generate a DC offset of 4.4 mV, which is rather more than we would expect from the input offset voltage error.

If an extra resistance, *Rcompen*, of the same value as the feedback resistor, is inserted into the noninverting input circuit then the offset will be cancelled. This strategy works well and is done almost automatically by many designers. However, there is a snag. The resistance *Rcompen* generates extra Johnson noise, and to prevent this it is necessary to shunt the resistance with a capacitor, as in Figure 3.1b. This extra component costs money and takes up PCB space, so it is questionable if this technique is actually very useful for audio work. It is usually more economical to allow offsets to accumulate in a chain of opamps and then remove the DC voltage with a single output blocking capacitor. This assumes that there are no stages with a large DC gain and that the offsets are not large enough to significantly reduce the available voltage swing. Care must also be taken if controls are involved, because even a small DC voltage across a potentiometer will cause it become crackly, especially as it wears.

FET-input opamps have very low bias current at room temperature; however it doubles for every 10 degree Centigrade rise. This is pretty unlikely to cause trouble in most audio applications, but a combination of high internal temperatures and high-value pots could lead to some unexpected crackling noises.

Opamp Properties: Cost

While it may not appear on the data sheet, the price of an opamp is obviously a major factor in deciding whether or not to use it. Table 3.3 was derived from the averaged prices for 1+ and 25+ quantities across a number of UK distributors. At the time of writing the cheapest popular opamps are the TL072 and the 5532, and these happened to come out at exactly the same price for 25+, so their price is taken as unity and used as the basis for the price ratios given.

Table 3.3 was compiled using prices for DIL packaging and the cheapest variant of each type. Price is per package and not per opamp section. It is obviously only a

Table 3.3 Opamps ranked by price (2009) relative to 5532 and TL072

Type	Format	Price ratio 1+	Type	Price ratio 25+
LM833	Dual	1.45	5532	**1.00**
5532	Dual	1.64	TL072	**1.00**
MC33078	Dual	1.97	LM833	1.12
TL072	Dual	2.45	MC33078	1.27
OPA604	Single	5.09	TL052	2.55
OPA2134PA	Dual	5.55	OP275GP	3.42
TL052	Dual	5.76	OPA2134PA	4.45
OP275GP	Dual	7.18	OPA604	5.03
OP27	Single	8.67	OP27	6.76
LM4562	Dual	12.45	LM4562	9.06
AD797	Single	25.73	AD797	13.09
OP270	Dual	29.85	LT1028	17.88
LT1028	Single	30.00	OPA270	24.42
OPA627	Single	51.91	OPA627	48.42

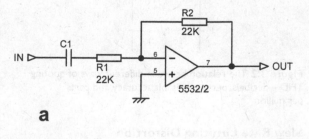

a

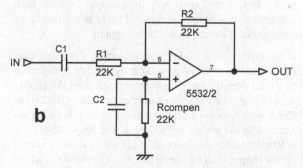

b

Figure 3.1 Compensating for bias-current errors in a shunt-feedback stage. The compensating resistor must be bypassed by a capacitor, C2, to prevent it adding Johnson noise to the stage.

rough guide. Purchasing in large quantities or in different countries may change the rankings somewhat (even going from 1+ to 25+ causes some changes) but the basic look of things will not alter too much. One thing is obvious—the 5532 is one of the great opamp bargains of all time.

Opamp Properties: Distortion

Relatively few discussions of opamp behaviour deal with nonlinear distortion, perhaps because it is a complex business. Opamp "accuracy" is closely related, but the term is often applied only to DC operation. Accuracy here is often specified in terms of bits, so "20-bit accuracy" means errors not exceeding one part in 2 to the 20, which is −120 dB or 0.0001%. Audio signal distortion is of course a dynamic phenomenon, very sensitive to frequency, and DC specs are of no use at all in estimating it.

Distortion is always expressed as a ratio and can be quoted as a percentage, as number of decibels, or in parts per million. With the rise of digital processing, treating distortion as the quantisation error arising from the use of a given number of bits has become more popular. Figure 3.2 hopefully provides a way of keeping perspective when dealing with these different metrics.

There are several different causes of distortion in opamps. We will now examine them.

Opamp Internal Distortion

This is what might be called the basic distortion produced by the opamp you have selected. Even if you scrupulously avoid clipping, slew-limiting, and common-mode issues, opamps are not distortion free, though some types such as the 5532 and the LM4562 have very low levels. If distortion appears when the opamp is run with shunt feedback, to prevent common-mode voltages on the inputs, and with very light output loading, then it is probably wholly internal and there is nothing to be done about it except pick a better opamp.

If the distortion is higher than expected, the cause may be internal instability provoked by putting a capacitive load directly on the output or neglecting the supply decoupling. The classic example of the latter effect is the 5532, which shows high distortion if there is not a capacitor across the supply rails close to the package; 100 nF is usually adequate. No actual HF oscillation is visible on the output with a general purpose oscilloscope, so the problem may be instability in one of the intermediate gain stages.

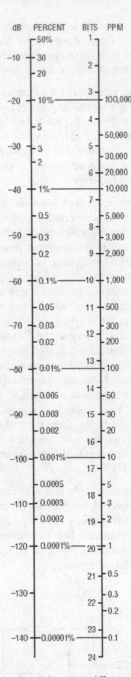

Figure 3.2 The relation between different ways of quoting THD—decibels, percentages, bit accuracy, and parts per million

Slew Rate Limiting Distortion

While this is essentially an overload condition, it is wholly the designer's responsibility. If users whack up the gain until the signal is within a hair of clipping, they should still be able to assume that slew-limiting will never occur, even with aggressive material full of high frequencies.

Arranging this is not too much of a problem. If the rails are set at the usual maximum voltage, i.e. ±18V, then the maximum possible signal amplitude is 12.7 Vrms, ignoring the saturation voltages of the output stage. To reproduce this level cleanly at 20 kHz requires a minimum slew rate of only 2.3 V/usec. Most opamps can do much better than this, though with the OP27 (2.8 V/usec) you are sailing rather close to the wind. The old LM741 looks as though it would be quite unusable, as its very limited 0.5 V/usec slew rate allows a full output swing only up to 4.4 kHz.

Horrific as it may now appear, audio paths full of LM741s were quite common in the early 1970s. Entire mixers were built with no other active devices, and what complaints there were tended to be about noise rather than distortion. The reason for this is that full-level signals at 20 kHz simply do not occur in reality; the energy at the HF end of the audio spectrum is well known to be much lower than that at the bass end.

This assumes that slew-limiting has an abrupt onset as level increases, rather like clipping. This is in general the case. As the input frequency rises and an opamp gets closer to slew-limiting, the input stage is working harder to supply the demands of the compensation capacitance. There is an absolute limit to the amount of current this stage can supply, and when you hit it the distortion shoots up, much as it does when you hit the supply rails and induce voltage clipping. Before you reach this point, the linearity may be degraded, but usually only slightly until you get close to the limit. It is not normally necessary to keep big margins of safety when dealing with slew-limiting. If you are employing the Usual Suspects in the audio opamp world—the TL072, the 5532, and the LM4562, with maximal slew rates of 13, 9, and 20 V/usec respectively—you are most unlikely to suffer any slew rate nonlinearity.

Distortion Due to Loading

Output stage distortion is always worse with heavy output loading because the increased currents flowing exacerbate the gain changes in the Class-B output stage. These output stages are not individually trimmed for optimal quiescent conditions (as are audio power amplifiers), and so the crossover distortion produced by opamps tends to be both higher and can be more variable between different specimens of the same chip. Distortion increases with loading in different ways for different opamps. It may rise only at the high-frequency end, (e.g. the OP2277), or there may be a general rise at all frequencies. Often both effects occur, as in the TL072.

The lowest load that a given opamp can be allowed to drive is an important design decision. It will typically be a compromise between the distortion performance required and opposing factors such as number of opamps in the circuit, cost of load-capable opamps, and so on. It even affects noise performance, for the lower the load resistance an amplifier can drive, the lower the resistance values in the negative feedback can be, and hence the lower the Johnson noise they generate. There are limits to what can be done in noise reduction by this method, because Johnson noise is proportional to the square root of circuit resistance and so improves only slowly as opamp loading is increased.

Thermal Distortion

Thermal distortion is that caused by cyclic variation of the properties of the amplifier components due to the periodic release of heat in the output stage. The result is a rapid rise in distortion at low frequencies, which gets worse as the loading becomes heavier.

Those who have read my work on audio power amplifiers will be aware that I am highly sceptical—in fact totally sceptical—about the existence of thermal distortion in amplifiers built from discrete components.[1] The power devices are too massive to experience per-cycle parameter variations, and there is no direct thermal path from the output stage to the input devices. There is no rise, rapid or otherwise, in distortion at low frequencies in a properly designed discrete power amplifier.

The situation is quite different in opamps, where the output transistors have much less thermal inertia and are also on the same substrate as the input devices. Nonetheless, opamps do not normally suffer from thermal distortion; there is generally no rise in low-frequency distortion, even with heavy output loading. Integrated-circuit power amplifiers are another matter, and the much greater amounts of heat liberated on the substrate do appear to cause serious thermal distortion, rising at 12 dB/octave below 50 Hz. I have never seen anything resembling this in any normal opamp.

Common-Mode Distortion

This is the general term for extra distortion that appears when there is a large signal voltage on both the opamp inputs. The voltage difference between these two inputs will be very small, assuming the opamp is in its linear region, but the common-mode (CM) voltage can be a large proportion of the available swing between the rails.

It appears to be by far the least understood mechanism, and gets little or no attention in opamp textbooks, but it is actually one of the most important influences on opamp distortion. It is simple to separate this effect from the basic forward-path distortion by comparing THD performance in series and shunt-feedback modes; this should be done at the same noise gain. The distortion is usually a good deal lower for the shunt-feedback case where there is no common-mode voltage. Bipolar and JFET input opamps show different behaviour, and they are treated separately. See Figure 3.3 for the various configurations examined.

Common-Mode Distortion: Bipolar Input Opamps

Figure 3.4 shows the distortion from a 5532 working in shunt mode with low-value resistors of 1 kΩ and 2k2 setting a gain of 2.2 times, at an output level of 5 Vrms. This is the circuit of Figure 3.3a with R_s set to zero; there is no CM voltage. The distortion is well below 0.0005% up to 20 kHz; this underlines what a superlative bargain the 5532 is.

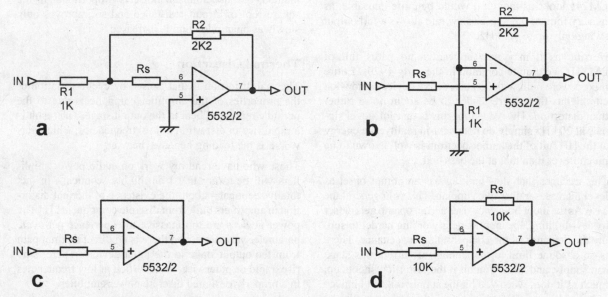

Figure 3.3 Opamp test circuits with added source resistance R_s: a) shunt; b) series; c) voltage-follower; d) voltage-follower with cancellation resistor in feedback path

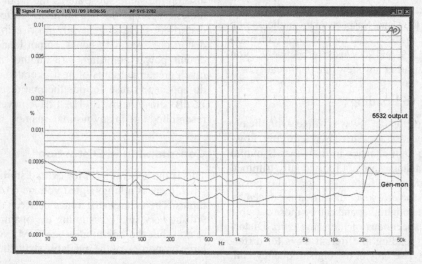

Figure 3.4 5532 distortion in a shunt-feedback circuit at 5 Vrms out. This shows the AP SYS-2702 output (lower trace) and the opamp output (upper trace). Supply ±18V.

Figure 3.5 shows the same situation but with the output increased to 10 Vrms (the clipping level on ±18V rails is about 12 Vrms) and there is now significant distortion above 10 kHz, though it only exceeds 0.001% at 18 kHz.

This remains the case when R_s in Figure 3.3a is increased to 10 kΩ and 47 kΩ—the noise floor is higher, but there is no real change in the distortion behaviour. The significance of this will be seen in a moment.

We will now connect the 5532 in the series-feedback configuration, as in Figure 3.3b; note that the stage gain is greater at 3.2 times but the opamp is working at the same noise gain. The CM voltage is 3.1 Vrms. With a 10 Vrms output we can see in Figure 3.6 that even with no added source resistance the distortion starts to rise from 2 kHz, though it does not exceed 0.001% until 12 kHz. But when we add some source resistance R_s, the picture is radically worse, with serious mid-band distortion rising at 6 dB/octave and roughly proportional to the amount of resistance added. We will note it is 0.0085% at 10 kHz with $R_s = 47$ kΩ.

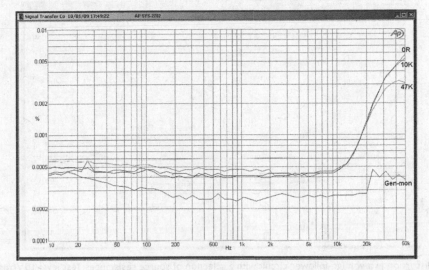

Figure 3.5 5532 distortion in the shunt-feedback circuit of Figure 3.3b. Adding extra resistances of 10 kΩ and 47 kΩ in series with the inverting input does not degrade the distortion at all but does bring up the noise floor a bit. Test level 10 Vrms out, supply ±18V.

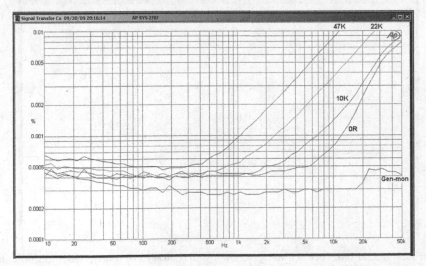

Figure 3.6 5532 distortion in a series-feedback stage with 2k2 and 1kΩ feedback resistors and varying source resistances. 10 Vrms output.

The worst case for CM distortion is the voltage-follower configuration, as in Figure 3.3c, where the CM voltage is equal to the output voltage. Figure 3.7 shows that even with a CM voltage of 10 Vrms, the distortion is no greater than for the shunt mode. However, when source resistance is inserted in series with the input, the distortion mixture of second, third, and other low-order harmonics increases markedly. It increases with output level, approximately quadrupling as the level doubles. The THD is now 0.018% at 10 kHz with R_s = 47 kΩ,

more than twice that of the series-feedback amplifier, due to the increased CM voltage.

It would be highly inconvenient to have to stick to the shunt-feedback mode because of the phase inversion and relatively low input impedance that comes with it, so we need to find out how much source resistance we can live with. Figure 3.8 zooms in on the situation with resistance of 10 kΩ and below; when the source resistance is below 2k2, the distortion is barely distinguishable from

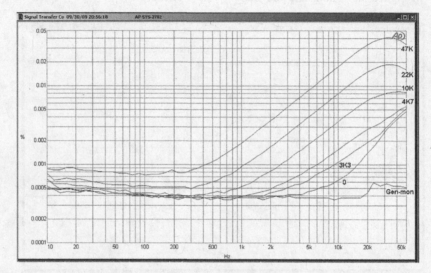

Figure 3.7 5532 distortion in a voltage-follower circuit with a selection of source resistances. Test level 10 Vrms, supply ±18V. The lowest trace is the analyser output measured directly, as a reference.

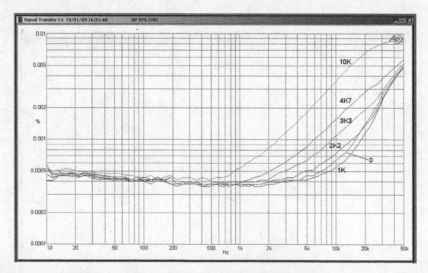

Figure 3.8 A closer look at 5532 distortion in a voltage-follower with relatively low source resistances; note that a 1 kΩ source resistance actually gives less distortion than none. Test level 10 Vrms, supply ±18V.

the zero source resistance trace. This is why the lowpass Sallen and Key filters in Chapter 13 have been given series resistors that do not in total exceed this figure.

Close examination reveals the intriguing fact that a 1 kΩ source actually gives *less* distortion than no source resistance at all, reducing THD from 0.00065% to 0.00055% at 10 kHz. Minor resistance variations around 1 kΩ make no difference. This must be due to the cancellation of distortion from two different mechanisms. It is hard to say whether it is repeatable enough to be exploited in practice; I wouldn't want to rely on it.

So, what's going on here? Is it simply due to nonlinear currents being drawn by the opamp inputs? Audio power amplifiers have discrete input stages which are very simple compared with those of most opamps and draw relatively large input currents. These currents show appreciable nonlinearity even when the output voltage of the amplifier is virtually distortion free, and if they flow through significant source resistances will introduce added distortion.[2]

If this was the case with the 5532 then the extra distortion would manifest itself whenever the opamp was fed from a significant source resistance, no matter what the circuit configuration. But we have just seen that it only occurs in series-feedback situations; increasing the source resistance in a shunt feedback does not perceptibly increase distortion. The effect may be present, but if so it is very small, no doubt because opamp

signal input currents are also very small and it is lost in the noise.

The only difference is that the series circuit has a CM voltage of about 3 Vrms while the shunt circuit does not, and the conclusion is that with a bipolar input opamp, you must have *both* a CM voltage and a significant source resistance to see extra distortion. The input stage of a 5532 is a straightforward long-tailed pair with a simple tail current source and no fancy cascoding, and I suspect that Early effect operates on it when there is a large CM voltage, modulating the quite high input bias currents, and this is what causes the distortion. The signal input currents are much smaller, due to the high open-loop gain of the opamp, and as we have seen appear to have a negligible effect.

Common-mode Distortion: JFET Opamps

FET-input opamps behave differently from bipolar input opamps. Take a look at Figure 3.9, taken from a TL072 working in shunt and in series configuration with a 5 Vrms output. The circuits are as in Figure 3.3a, except that the resistor values have to be scaled up to 10 kΩ and 22 kΩ because the TL072 is nothing like so good at driving loads as the 5532. This unfortunately means that the inverting input is seeing a source resistance of 10k∥22k – 6.9k, which introduces a lot of common-mode (CM) distortion—five times as much at 20 kHz as for the shunt case. Adding a similar resistance in the input

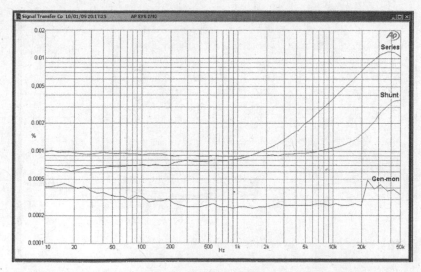

Figure 3.9 A TL072 shunt-feedback stage using 10 kΩ and 22 kΩ resistors shows low distortion. The series version is much worse due to the impedance of the NFB network, but it can be made the same as the shunt case by adding cancellation source resistance in the input path. No external loading, test level 5 Vrms, supply ±18V.

path cancels out this distortion, and the trace then is the same as the "Shunt" trace in Figure 3.9. Disconcertingly, the value that achieved this was not 6.9k but 9.1k. That means adding −113 dBu of Johnson noise, so it's not always appropriate.

It's worth mentioning that the flat part of the shunt trace below 10 kHz is not noise, as it would be for the 5532; it is distortion.

A voltage-follower has no inconvenient medium-impedance feedback network, but it does have a much larger CM voltage. Figure 3.10 shows a voltage-follower working at 5 Vrms. With no source resistance the distortion is quite low, due to the 100% NFB, but as soon as a 10 kΩ source resistance is added we are looking at 0.015% at 10 kHz.

Once again, this can be cured by inserting an equal resistance in the feedback path of the voltage-follower, as in Figure 3.3d. This gives the "Cancel" trace in Figure 3.10. Adding resistances for distortion cancellation in this way has the obvious disadvantage that they introduce extra Johnson noise into the circuit. Another point is that stages of this kind are often driven from pot wipers, so the source impedance is variable, ranging between zero and one-quarter of the pot track resistance. Setting a balancing impedance in the other opamp input to a mid-value, i.e. one-eighth of the track resistance, should reduce the average amount of input distortion, but it is inevitably a compromise.

With JFET inputs the problem is not the operating currents of the input devices themselves, which are negligible, but the currents drawn by the nonlinear junction capacitances inherent in field-effect devices. These capacitances are effectively connected to one of the supply rails. For P-channel JFETs, as used in the input stages of most JFET opamps, the important capacitances are between the input JFETs and the substrate, which is normally connected to the V-rail. See Jung.[3]

According to the Burr-Brown data sheet for the OPA2134, "The P-channel JFETs in the input stage exhibit a varying input capacitance with applied CM voltage". It goes on to recommend that the input impedances should be matched if they are above 2 kΩ.

Common-mode distortion can be minimised by running the opamp off the highest supply rails permitted, though the improvements are not large. In one test on a TL072, going from ±15V to ±18V rails reduced the distortion from 0.0045% to 0.0035% at 10 kHz.

Opamps Surveyed: BJT Input Types

The rest of this chapter looks at some opamp types and examines their performance, with the 5534A the usual basis for comparison. The parts shown here are not necessarily intended as audio opamps, though some, such as the OPA2134, were specifically designed as such. They have however all seen use, in varying numbers, in audio applications. Bipolar input opamps are dealt with first.

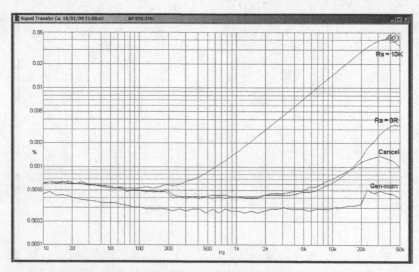

Figure 3.10 A TL072 voltage-follower working at 5 Vrms with a low source resistance produces little distortion (R_s = 0R), but adding a 10 kΩ source resistance makes things much worse (R_s = 10K). Putting a 10 kΩ resistance in the feedback path as well gives complete cancellation of this extra distortion (Cancel). Supply ±18V.

The NE5532/5534 Opamp

The 5532 is a low-noise, low-distortion bipolar dual opamp, with internal compensation for unity-gain stability. The 5534 is a single version internally compensated for gains down to three times, and an external compensation capacitor can be added for unity-gain stability; 22 pF is the usual value. The 5532 achieves unity-gain stability by having degeneration resistors in the emitter circuits of the input transistors, to reduce the open-loop gain, and this is why it is noisier than the 5534.

The common-mode range of the inputs is a healthy ±13V, with no nasty phase inversion problems if this is exceeded; there is more on the CM behaviour of the 5532/4 in the earlier section on common-mode distortion. It has a distinctly higher power consumption than the TL072, drawing approx. 4 mA per opamp section when quiescent. The DIL version runs perceptibly warm when quiescent on ±17 V rails.

The 5534/5532 has bipolar transistor input devices. This means it gives low noise with low source resistances but draws a relatively high bias current through the input pins. The input devices are NPN, so the bias currents flow into the chip from the positive rail. If an input is fed through a significant resistance, then the input pin will be more negative than ground due to the voltage drop caused by the bias current. The inputs are connected together with back-to-back diodes for reverse-voltage protection and should not be forcibly pulled to different

voltages. The 5532 is intended for linear operation, and using it as a comparator is not recommended.

As can be seen from Figure 3.11, the 5532 has low distortion, even when driving the maximum 500 ohm load. The internal circuitry of the 5532 has never been officially explained but appears to consist of nested Miller loops that permit high levels of internal negative feedback. The 5532 is the dual of the 5534 and is much more commonly used than the single as it is cheaper per opamp and does not require an external compensation capacitor when used at unity gain.

The 5532/5534 is made by several companies, but they are not all created equal. Those by Fairchild, JRC, and ON-Semi have significantly lower THD at 20 kHz and above, and we're talking about a factor of two or three here.

The 5532 and 5534 type opamps require adequate supply decoupling if they are to remain stable; otherwise they appear to be subject to some sort of internal oscillation that degrades linearity without being visible on a normal oscilloscope. The essential requirement is that the +ve and −ve rails should be decoupled with a 100 nF capacitor between them, at a distance of not more than a few millimetres from the opamp; normally one such capacitor is fitted per package as close to it as possible. It is *not* necessary, and often not desirable, to have two capacitors going to ground; every capacitor between a supply rail and ground carries the risk of injecting rail noise into the ground.

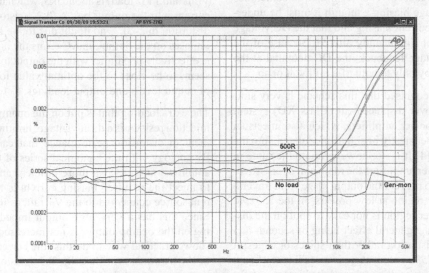

Figure 3.11 Distortion is very low from the 5532, though loading makes a detectable difference. Here it is working in series-feedback mode at the high level of 10 Vrms with 500 Ω, 1 kΩ loads, and no load. The gen-mon trace is the output of the distortion analyser measured directly. Gain of 3.2 times. Supply ±18V.

The 5534, and particularly the 5534A, have voltage and current noise parameters that are well suited to giving a good noise performance with MM cartridges. Using a standard cartridge with a series resistance of 610 Ω and a series inductance of 470 mH, (derived from the popular Shure M75ED Mk2), the calculated equivalent input noise is −122.5 dBu, so the output noise of a phono amplifier with +30 dB of gain (1 kHz) is −92.5 dBu. The effect of all circuit resistances is included with $R_{in} = 47$ kΩ and $R_0 = 220$ Ω. Naturally this is with RIAA equalisation. A completely noiseless amplifier with all of the noise-generating resistances still present would have an EIN of −124.9 dBu with the same cartridge, which is only 2.4 dB better. This shows that using discrete devices instead of opamps can offer only small advantages. The subject of MM noise is dealt with in detail in Chapter 9.

Reducing 5532 Distortion by Output Stage Biasing

There is a useful, though relatively little-known (and where it is known, almost universally misunderstood and misapplied) technique for reducing the distortion of the 5532 opamp. While the method may be applicable to some other opamps, here I concentrate on the 5532, and it must not be assumed that the results will be emulated by any other opamp.

If a biasing current of the right polarity is injected into the opamp output, then the output stage distortion can be significantly reduced. This technique is sometimes called "output stage biasing", though it must be understood that this is current biasing and that the DC voltage conditions are not significantly altered. Because of the high level of voltage feedback, the DC potential at the output is shifted by only a tenth of a millivolt or so.

You may have recognised that this scheme is very similar to the crossover displacement (Class XD) system I introduced for power amplifiers, which also injects an extra current, either steady or signal-modulated, into the amplifier output.[4] It is not however quite the same in operation. In power amplifiers the main aim of crossover displacement is to prevent the output stage from traversing the crossover region at low powers. In the 5532, at least, the crossover region is not easy to spot on the distortion residual, the general effect being of second- and third-harmonic distortion rather than spikes or edges; it appears that the 5532 output stage is more linear when it is pulling rather than pulling up, and the biasing current is compensating for this.

For the 5532, the current *must* be injected from the positive rail; currents from the negative rail make the distortion emphatically worse. This confirms that the output stage of the 5532 is in some way asymmetrical in operation, for if it was simply a question of suppressing crossover distortion by crossover displacement, a bias current of either polarity would be equally effective. The continued presence of the crossover region, albeit displaced, would mean that the voltage range of reduced distortion would be quite small and centred on 0 V. It is rather the case that there is a general reduction in distortion across the whole of the 5532 output range, which seems to indicate that the 5532 output stage is better at sinking current than sourcing it, and therefore injecting a positive current is effective at helping out.

Figure 3.12a shows a 5532 running in shunt-feedback mode with a moderate output load of 1 kΩ; the use of shunt feedback makes it easier to see what's going on by eliminating the possibility of common-mode distortion. With normal operation we get the upper trace in Figure 3.13, labelled "No bias". If we then connect a current-injection resistor between the output and to the V+ rail, we find that the LF distortion (the flat bit) drops almost magically, giving the trace labelled "3K3", which is only just above the gen-mon trace. Since noise makes a significant contribution to the THD residual at these levels, the actual reduction in distortion is greater than it appears.

The optimum resistor value for the conditions shown (5 Vrms and 1 kΩ load) is about 3k3, which injects a 5.4 mA current into the output pin. A 2k2 resistor gives greater distortion than 3k3, no doubt due to the extra loading it imposes on the output; in AC terms the injection resistor is effectively in parallel with the output load. In fact, 3k3 seems to be close to the optimal value for a wide range of output levels and output loadings.

The extra loading that is put on the opamp output by the injection resistor is a disadvantage, limiting the improvement in distortion performance that can be obtained. By analogy with the canonical series of Class-A power amplifier outputs,[5] a more efficient and elegant way to inject the required biasing current is by using a current source connected to the V+ rail, as in Figure 3.12b. Since this has a very high output impedance the loading on the opamp output is not increased. Figure 3.12c shows practical way to do this; the current source is set to the same current that the 3k3 resistor injects when the output is at 0 V, (5.4 mA), but the improvement in distortion is greater. There is nothing magical about

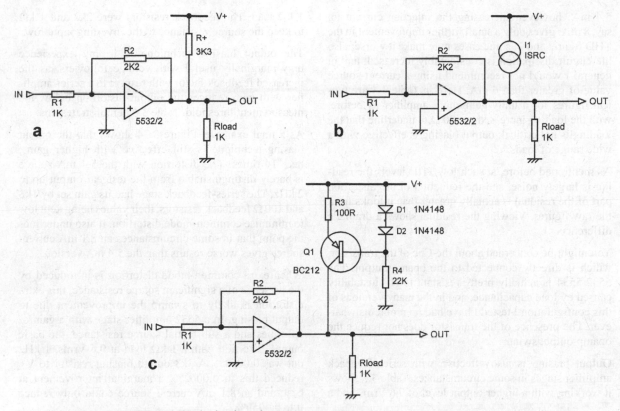

Figure 3.12 Reducing 5532 distortion in the shunt-feedback mode by biasing the output stage with a current injected through a resistor R+ or a current source

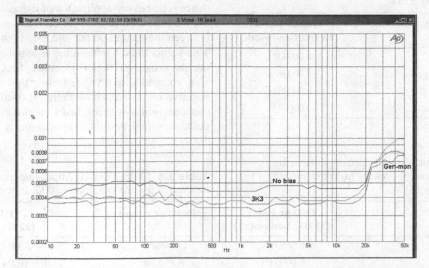

Figure 3.13 The effect of output biasing, with a 3k3 resistor to V+, on a unity-gain shunt-feedback 5532 stage. Output load 1 kΩ, input and feedback resistors are 2k2, noise gain 2.0 times. Output 5 Vrms, supply ±18V.

5.4 mA; however, increasing the injection current to, say, 8 mA, gives only a small further improvement in the THD figure, and in some cases may make it worse; also the circuit dissipation is considerably increased, and in general I would not recommend using a current-source value of greater than 6 mA. Here in Table 3.4 are typical figures for a unity-gain shunt amplifier as before, with the loading increased to 680 Ω to underline that the loading is not critical; output biasing is effective with a wide range of loads.

As mentioned before, at such low THD levels the reading is largely noise, and the reduction of the distortion part of the residual is actually greater than it looks from the raw figures. Viewing the residual shows a dramatic difference.

You might be concerned about the Cbc of the transistor, which is directly connected to the opamp output. The 5532/5534 is actually pretty resistant to HF instability caused by load capacitance, and in the many versions of this configuration I tested I have had no problems whatever. The presence of the transistor does not reduce the opamp output swing.

Output biasing is also effective with series-feedback amplifier stages in some circumstances. Table 3.5 shows it working with a higher output level of 9.6 Vrms and a 1 kΩ load. The feedback resistors were 2k2 and 1 kΩ to keep the source resistance to the inverting input low.

The output biasing technique is in my experience only marginally useful with voltage followers, as the increased feedback factor with respect to a series amplifier with gain reduces the output distortion below the measurement threshold. Table 3.6 demonstrates this.

As a final example, Figure 3.14 shows that the output biasing technique is still effective with higher gains, here 14 times. The distortion with the 5.4 mA source is barely distinguishable from the testgear output up to 2 kHz. The series-feedback stage had its gain set by 1k3 and 100 Ω feedback resistors, their values being kept low to minimise common-mode distortion. It also underlines the point that in some circumstances an 8.1 mA current source gives worse results than the 5.4 mA version.

When extra common-mode distortion is introduced by the presence of a significant source resistance, this extra distortion is likely to swamp the improvement due to output biasing. In a 5532 amplifier stage with a gain of 3.2 times and a substantial source resistance, the basic output distortion with a 1 kΩ load at 9.6 Vrms, 1 kHz out was 0.0064%. A 3k3 output biasing resistor to V+ reduced this to 0.0062%, a marginal improvement at best, and an 8.1 mA current source could only reduce it to 0.0059%.

Earlier I said that the practice of output stage biasing appears to be pretty much universally misunderstood, judging by how it is discussed on the Internet. The evidence is that every application of it that my research has exposed shows a resistor (or current source) connected between the opamp output and the *negative* supply rail. This seems to be based on the assumption that displacing the crossover region in either direction is a good idea, coupled with a vague feeling that a resistor to the negative rail is somehow more "natural", though how that conclusion was reached I cannot guess. However, the assumption that the output stage is symmetrical is

Table 3.4 Output biasing improvements with unity-gain shunt feedback, 5 Vrms out, load 680 Ω, supply ±18V

Injection method	THD at 1 kHz (22 kHz bandwidth)
None	0.00034%
3k3 resistor	0.00026%
5.4 mA current source	0.00023%
8.1 mA current source	0.00021%

Table 3.5 Output biasing improvements with 3.2 times gain, series feedback, 9.6 Vrms out, load 1 KΩ, supply ±18V

Injection method	THD at 1 kHz (22 kHz bandwidth)
None	0.00037%
3k3 resistor	0.00033%
5.4 mA current source	0.00027%
8.1 mA current source	0.00022%

Table 3.6 Output biasing improvements for voltage-follower, 9.6 Vrms out, load 680 Ω, supply ±18V

Injection method	THD at 1 kHz (22 kHz bandwidth)
None	0.00018% (almost all noise)
3k3 resistor	0.00015% (all noise)
5.4 mA current source	0.00015% (all noise)
8.1 mA current source	0.00015% (all noise)

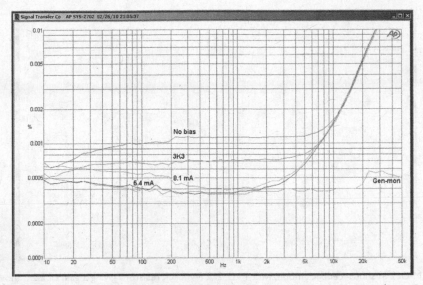

Figure 3.14 Reducing 5532 distortion with series feedback by biasing the output stage with a 3k3 resistor, or 5.2 or 8.1 mA current sources. Gain 14 times, no external load. Test level 5 Vrms out, supply ±18V.

usually incorrect; as we have seen, it is certainly not true for the 5532/5534. For the 5532—which surely must be the most popular audio opamp by a long way—a pulldown resistor would be completely inappropriate, as it *increases* rather than decreases the output stage distortion.

You may be thinking that this is an ingenious method of reducing distortion, but rather clumsy compared with simply using a more linear opamp like the LM4562. This is true, but on the other hand, if the improvement from output biasing is adequate, it will be much cheaper than switching to a more advanced opamp that costs ten times as much.

The LM4562 Opamp

The LM4562 is a new opamp, which first become freely available at the beginning of 2007. It is a National Semiconductor product. It is a dual opamp—there is no single or quad version. It costs about ten times as much as a 5532.

The input noise voltage is typically 2.7 nV/√Hz, which is substantially lower than the 4 nV/√Hz of the 5532. For suitable applications with low source impedances this translates into a useful noise advantage of 3.4 dB. However, with MM cartridges the greater current noise means that the overall noise is higher than that of the 5532 or 5534.

The bias current is 10 nA typical, which is very low and would normally imply that bias cancellation, with its attendant noise problems, was being used. However in my testing I have seen no sign of excess noise, and the data sheet is silent on the subject. No details of the internal circuitry have been released so far, and quite probably never will be.

It is not fussy about decoupling, and as with the 5532, 100 nF across the supply rails close to the package should ensure HF stability. The slew rate is typically ±20 V/us, more than twice as quick as the 5532.

The first THD plot in Figure 3.15 shows the LM4562 working at a closed-loop gain of 2.2x in shunt-feedback mode, at a high level of 10 Vrms. The top of the THD scale is 0.001%, something you will see with no other opamp in this survey. The no-load trace is barely distinguishable from the AP SYS-2702 output, and even with a heavy 500 Ω load driven at 10 Vrms there is only a very small amount of extra THD, reaching 0.0007% at 20 kHz.

Figure 3.16 shows the LM4562 working at a gain of 3.2x in series-feedback mode, both modes having a noise gain of 3.2x. There is little extra distortion from 500 Ω.

For Figures 3.15 and 3.16 the feedback resistances were 2k2 and 1 kΩ, so the minimum source resistance presented to the inverting input is 687 Ω. In Figure 3.17 extra source resistances were then put in series with the

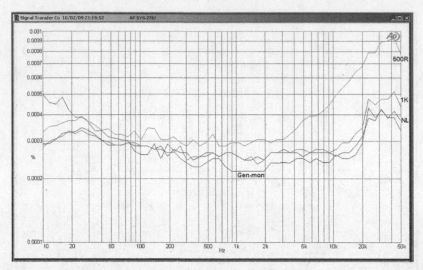

Figure 3.15 The LM4562 in shunt-feedback mode, with 1 kΩ, 2k2 feedback resistors giving a gain of 2.2x. Shown for no load (NL) and 1 kΩ, 500 Ω loads. Note the vertical scale ends at 0.001% this time. Output level is 10 Vrms. ±18V supply rails.

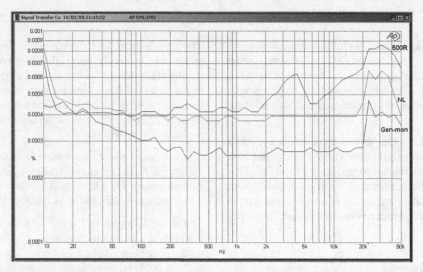

Figure 3.16 The LM4562 in series-feedback mode, with 1 kΩ, 2k2 feedback resistors giving a gain of 3.2x. No load (NL) and 500 Ω load. 10 Vrms output. ±18V supply rails.

input path, (as was done with the 5532 in the earlier section on common-mode distortion), and this revealed a remarkable property of the LM4562—it is much more resistant to common-mode distortion than the 5532. At 10 Vrms and 10 kHz, with a 10 kΩ source resistance, the 5532 generates 0.0014% THD (see Figure 3.6), but the LM4562 gives only 0.00046% under the same conditions. I strongly suspect that the LM4562 has a more sophisticated input stage than the 5532, probably incorporating cascoding to minimise the effects of common-mode voltages.

Note that only the rising curves to the right represent actual distortion. The raised levels of the horizontal traces at the LF end are due to Johnson noise from the extra series resistance.

It has taken an unbelievably long time—nearly thirty years—for a better audio opamp than the 5532 to come along, but at last it has happened. At present it also has a much higher price, but hopefully that will change. The LM4562 is superior in just about every parameter, except for its higher current noise. This leads to an EIN

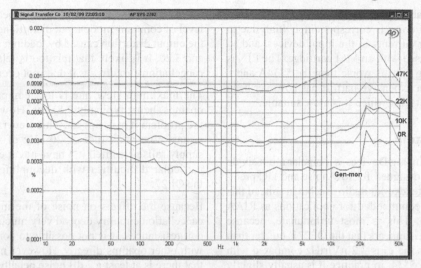

Figure 3.17 The LM4562 in series-feedback mode, gain 3.2x, with varying extra source resistance in the input path. The extra distortion is much lower than for the 5532. 10Vrms out, ±18V supply rails.

with the standard cartridge of −117.9 dB, a significant 4.6 dB noisier than the humble 5534A.

The AD797 Opamp

The AD797 (Analog Devices) is a single opamp with very low voltage noise and distortion. It appears to have been developed primarily for the cost-no-object application of submarine sonar, but it works very effectively with normal audio—if you can afford to use it. The cost is something like twenty times that of a 5532. No dual

version is available, so the cost ratio per opamp section is forty times.

Early versions appeared to be rather difficult to stabilise at HF, but the current product is no harder to apply than the 5532. Possibly there has been a design tweak, or on the other hand my impression may be wholly mistaken.

The AD797 incorporates an ingenious feature for internal distortion cancellation. This is described on the manufacturer's data sheet. Figure 3.18 shows that it works effectively.

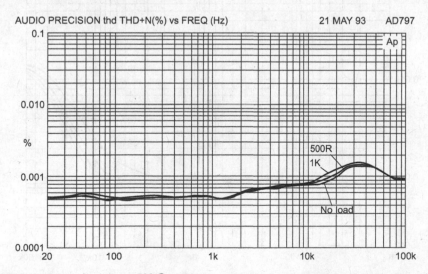

Figure 3.18 AD797 THD into loads down to 500 Ω, at 7.75 Vrms. Output is virtually indistinguishable from input. Series feedback, but no CM problems. Gain=3.2x.

This is a remarkably quiet device in terms of voltage noise, but current noise is correspondingly high due to the high collector currents in the input devices, and so it is noisy when used with an MM cartridge. The EIN is −116.6 dBu, a big 6.0 dB worse than the 5534A and a poor return for an expensive part. It has however been used for MC inputs, for which it is more suitable.

The OP27 Opamp

The OP27 from Analog Devices is a bipolar input single opamp primarily designed for low noise and DC precision. It was not intended for audio use, but in spite of this it is frequently recommended for applications as RIAA phono amplifiers. This is most unfortunate, because while at first sight it appears that the OP27 is quieter than the 5534/5532, as the e_n is 3.2 nV/rtHz compared with 4 nV/rtHz for the 5534, in practice it is usually slightly noisier. This is because the OP27 is in fact optimised for DC characteristics and so has input bias-current cancellation circuitry that generates common-mode noise. When the impedances on the two inputs are very different—which is the case in RIAA preamps—the CM noise does not cancel, and this appears to degrade the overall noise performance significantly. This opamp may be useful in other parts of the circuitry, but is not recommended for MM input amplifiers.

For a bipolar input opamp, there appears to be a high level of common-mode input distortion, enough to bury the output distortion caused by loading; see Figures 3.19 and 3.20. It is likely that this too is related to the bias-cancellation circuitry, as it does not occur in the 5532.

The maximum slew rate is low compared with other opamps, being typically 2.8V/us. However, this is not the problem it may appear. This slew rate would allow a maximum amplitude at 20 kHz of 16 Vrms, if the supply rails permitted it. I have never encountered any particular difficulties with decoupling or stability of the OP27.

Because the effects on noise of the input bias-current cancellation circuitry depend very much on external circuit resistances, it is not possible to compare the OP27 with other opamps directly. However my experience is that there is at least a 2 dB noise penalty compared with the 5534A.

Opamps Surveyed: JFET Input Types

Opamps with JFET inputs tend to have higher voltage noise and lower current noise than BJT input types and are therefore give a better noise performance with high source resistances. Their very low bias currents often allow circuitry to be simplified.

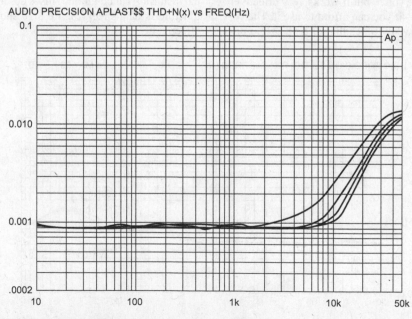

AUDIO PRECISION APLAST$$ THD+N(x) vs FREQ(Hz)

Figure 3.19 OP27 THD in shunt-feedback mode with varying loads. This opamp accepts even heavy (1 kΩ) loading gracefully.

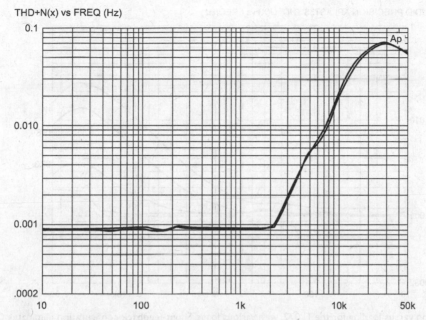

THD+N(x) vs FREQ (Hz)

Figure 3.20 OP27 THD in series-feedback mode. The common-mode input distortion completely obscures the output distortion.

The TL072 Opamp

The TL072 is one of the most popular opamps, having very high-impedance inputs with effectively zero bias and offset currents. The JFET input devices give their best noise performance at medium impedances, in the range 1 kΩ–10 kΩ. It has a modest power consumption, at typically 1.4 mA per opamp section, which is significantly less than the 5532. The slew rate is higher than for the 5532, at 13 V/us against 9 V/us. The TL072 is a dual opamp. There is a single version called the TL071 which has offset null pins.

However, the TL072 is not THD-free in the way the 5532 is. In audio usage, distortion depends primarily upon how heavily the output is loaded. The maximum loading is a trade-off between quality and circuit economy, and I would put 2 kΩ as the lower limit. This opamp is not the first choice for audio use unless the near-zero bias currents (which allow circuit economies by making blocking capacitors unnecessary), the low price, or the modest power consumption are dominant factors.

It is an unhappy quirk of this device that the input common-mode range does not extend all the way between the rails. If the common-mode voltage gets to within a couple of volts of the V-rail, the opamp suffers phase reversal and the inputs swap their polarities. There may

be really horrible clipping, where the output hits the bottom rail and then shoots up to hit the top one, or the stage may simply latch up until the power is turned off. There is more on the CM behaviour of the TL072 in the earlier section on common-mode distortion.

TL072s are relatively relaxed about supply rail decoupling, though they will sometimes show very visible oscillation if they are at the end of long, thin supply tracks. One or two rail-to-rail decoupling capacitors (e.g. 100 nF) per few centimetres is usually sufficient to deal with this, but the usual practice is to not take chances and allow one capacitor per package, as with other opamps.

Because of common-mode distortion, a TL072 in shunt configuration is always more linear. In particular compare the results for a 3k3 load in Figures 3.21 and 3.22. At heavier loadings the difference is barely visible because most of the distortion is coming from the output stage.

Distortion always gets worse as the loading increases. This factor together with the closed-loop NFB factor determines the THD.

TL072/71 opamps are prone to HF oscillation if faced with significant capacitance to ground on the output pin; this is particularly likely when they are used as

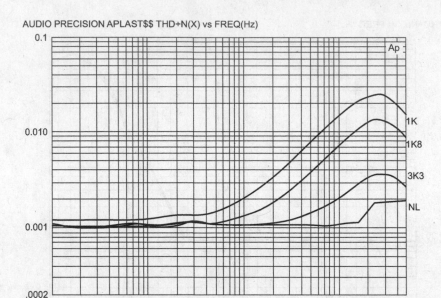

AUDIO PRECISION APLAST$$ THD+N(X) vs FREQ(Hz)

Figure 3.21 Distortion versus loading for the TL072, with various loads. Shunt-feedback configuration eliminates CM input distortion. Output level 3 Vrms, gain 3.2x, rails ±15V. No output load except for the feedback resistor. The no-load plot is indistinguishable from that of the testgear alone.

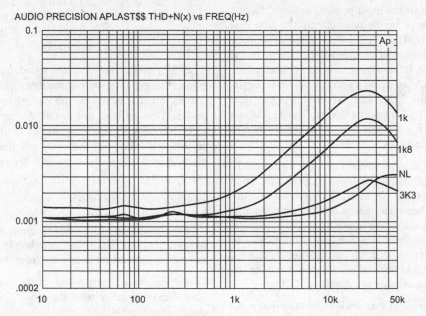

AUDIO PRECISION APLAST$$ THD+N(x) vs FREQ(Hz)

Figure 3.22 Distortion versus loading for the TL072, with various loads. Series-feedback configuration, Output level 3 Vrms, gain 3.2x, rails ±15V. Distortion at 10 kHz is with no load is 0.0015% compared with 0.0010% for the shunt configuration. This is due to the 1 Vrms CM signal on the inputs.

unity-gain buffers with 100% feedback. A few inches of track can sometimes be enough. This can be cured by an isolating resistor, in the 47–75 Ω range, in series with the output, placed at the opamp end of the track.

The TL072 has low current noise, because of its JFET inputs, but the voltage noise is high at 18nV/√Hz, leading to an EIN with the standard cartridge of −113.4 dBu, 9.1 dB worse than the 5534A. It is the worst noise result so far.

The OPA2134 Opamp

The OPA2134 is a Burr-Brown product, the dual version of the OPA134. The manufacturer claims it has superior sound quality due to its JFET input stage. Regrettably, but not surprisingly, no evidence is given to back up this assertion. The slew rate is typically ±20 V/us, which is ample. It does not appear to be optimised for DC precision, the typical offset voltage being ±1 mV, but this is usually good enough for audio work. I have used it many times as a DC servo in power amplifiers, the low bias currents allowing high resistor values and correspondingly small capacitors.

The OPA2134 does not show phase reversal anywhere in the common-mode range, which immediately marks it as superior to the TL072.

The two THD plots in Figures 3.23 and 3.24 show the device working at a gain of 3x in both shunt and series-feedback modes. It is obvious that a problem emerges in the series plot, where the THD is higher by about three times at 5 Vrms and 10 kHz. This distortion increases with level, which immediately suggests common-mode distortion in the input stage. Distortion increases with even moderate loading; see Figure 3.25.

This is a relatively modern and sophisticated opamp. When you need JFET inputs (usually because significant input bias currents would be a problem), this definitely beats the TL072; it is however four to five times more expensive.

The input noise voltage is 8 nV/√Hz, more than twice that of the 5532, and despite the low current noise this leads to an EIN of −116.6 dBu with the standard cartridge; this is 5.9 dB noisier than the 5534A.

Other Opamps

This chapter has only space to cover the opamps most likely to be useful in electronics for vinyl. Other opamps are examined in *Small Signal Audio Design*;[5] the LM741, OP270, and OP275 with BJT input devices, and the TL052, OPA604, and OP627 with JFET input

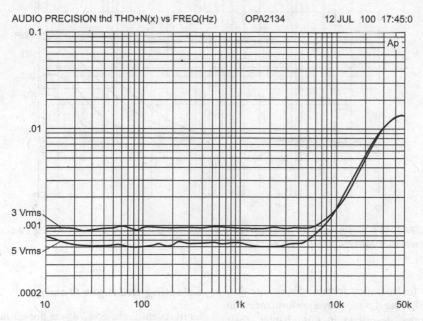

Figure 3.23 The OPA2134 working in shunt-feedback mode. The THD is below the noise until frequency reaches 10 kHz; it appears to be lower at 5 Vrms simply because the noise floor is relatively lower.

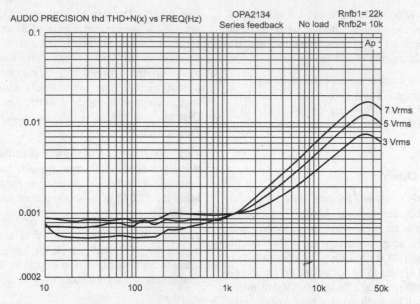

Figure 3.24 The OPA2134 in series-feedback mode. Note much higher distortion at HF.

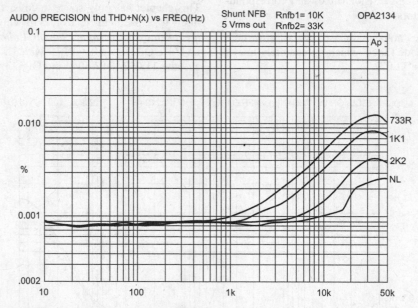

Figure 3.25 The OPA2134 in shunt-feedback mode (to remove input CM distortion), and with varying loads on the output. As usual, more loading makes linearity worse. 5 Vrms out, Gain = 3.3x.

devices. Opamps for operation from a +5V rail are also analysed. None of these devices have performance that exceeds the opamps described in this chapter; their noise performance can however be found in Chapter 9 of this book.

Selecting the Right Opamp

Until recently, the 5532/4 was pre-eminent in almost all audio electronics. It is found in almost every mixing console and in a large number of preamplifiers. Distortion is almost very low, even when driving 600 Ω loads. Noise is very low, and the balance of voltage and current noise in the input stage is well-matched to moving-magnet phono cartridges; using exotic discrete devices cannot give more than a dB or two advantage. Large-quantity production has brought the price down to a point where a powerful reason is required to pick any other device. The lowest noise version, and the best noise match to an MM cartridge, is the 5534A, but this comes in a single package and so costs more per opamp.

The 5532 is not, however, perfect. It suffers common-mode distortion. It has high bias and offset currents at the inputs as an inevitable result of using a bipolar input stage (for low noise) without any sort of bias-cancellation circuitry. The 5532 is not in the forefront for DC accuracy, though it's not actually that bad. The offset voltage spec is 0.5 mV typical, 4 mV max, compared with 3 mV typical, 6 mV max for the popular TL072. I have actually used 5532s to replace TL072s when offset voltage was a problem, but the increased bias current was acceptable.

With horrible inevitability, the very popularity and excellent technical performance of the 5532 has led to it being criticised by Subjectivists who have contrived to convince themselves that they can tell opamps apart by listening to music played through them. This always makes me laugh like a drain, because there is probably no music on the planet that has not passed through a hundred or more 5532s on its way to the consumer.

There are two distinct roles for opamps in phono amplifiers. The critical first stage driven by an MM cartridge requires not only low voltage noise but also low current noise for the lowest noise output, and the 5534A wins this handily. It is the first choice, though the 5532 is not far behind and saves money because of its dual package.

The other role is in what might be called general purpose signal processing; subsonic and ultrasonic filters, flat gain stages, and so on. Here the LM4562, with its lower voltage noise and very low distortion, represents a real advance on the 5532/4. It is however still a good deal more expensive, and is not perfect—it appears to be more easily damaged by excess common-mode voltages, and there is some evidence it is more susceptible to RF demodulation when used as a voltage-follower.

References

1. Self, D. *Audio Power Amplifier Design*. 6th edition, Focal Press, 2013, p. 120. ISBN 978-0-240-52613-3, 978-0-240-52614-0.

2. Self, D. *Audio Power Amplifier Design*. 6th edition, Focal Press, 2013, p. 119.

3. Jung, W. ed. *Op-amp Applications Handbook*. Newnes, 2006, Chapter 5, p. 399.

4. *Audio Power Amplifier Design*, 6th edition, chapter 18.

5. *Audio Power Amplifier Design*, 6th edition, p. 425.

6. Self, D. *Small Signal Audio Design*. 2nd edition. Focal Press, 2015, Chapter 4. ISBN: 978-0-415-70974-3 (hbk) ISBN: 978-0-415-70973-6 (pbk) ISBN: 978-1-315-88537-7 (ebk).

Preamp Architecture

The purpose of this chapter is to look at how phono amplifier stages fit into the architecture of a complete preamplifier that can handle a variety of input source types.

Passive Preamplifiers

Some sort of preamplifier or control unit is required in all hi-fi systems, even if its only function is to select the source and set the volume. You could even argue that the source-selection switch could be done away with, if you are prepared to plug and unplug connectors, leaving a "preamplifier" that basically consists solely of a volume-control potentiometer in a box.

I am assuming here that a selector switch will be required, and that gives us the "passive preamplifier" (oxymoron alert!) in Figure 4.1a.

There is of course no such thing as a passive phono pre-amplifier; even with high output MM cartridges at least 35 dB of gain is required to get a signal level that can be reasonably applied to a power amplifier. If a passive preamplifier is used then the phono amplifier must be an external active unit with its own power supply. All the amplification is external to the passive preamp, so the outboard phono unit must have a high enough output to drive a power amplifier fully without any further help. This implies that the phono preamplifier must have variable or switchable gain, or its overload margin will be impractically small.

While a passive preamplifier may have only one component, it does not follow that it is easy to design, even though the only parameter to decide is the resistance of the volume pot. Any bit of gear that embodies its internal contradictions in its very name needs to be treated with caution. The pot resistance of a "passive preamplifier" cannot be too high because the output impedance, maximal at one-quarter the track resistance when volume is set to −6 dB, will cause an HF roll-off in conjunction

with the connecting cable capacitance. It also makes life difficult for those designing RF filters on the inputs of the equipment being driven.

On the other hand, if the volume pot resistance is too low the source equipment will suffer excessive loading, and this includes an external phono unit.

If, however, we can assume that our source equipment has a reasonable drive capability, we can use a 10 kΩ pot. Its maximum output impedance (at −6 dB) will then be 2.5 kΩ. The capacitance of most audio cable is 50–150 pF/metre, so with a 2.5 kΩ source impedance and 100 pF/metre cable, a maximum length of 5 metres is permissible before the HF loss hits the magic figure of −0.1 dB at 20 kHz. A very rapid survey of current "passive preamplifiers" confirms that 10 kΩ seems to be the most popular value for the pot. This value will cause no trouble at all for any competently designed opamp-based phono unit but might embarrass a discrete design if the output stage is ill-conceived. It is very unwise to design for an external load that is lighter than 2 kΩ. It is always necessary to avoid unbuffered filter stages at the output of the phono unit, as even a 10 kΩ load is likely to severely degrade their accuracy.

Transformer Preamplifiers

At the time of writing there are at least nine passive pre-amplifiers on the market that control volume by changing the taps on the secondary of a transformer. These are sometimes called "passive magnetic preamplifiers". Examples include the Luxman AT3000, the Audio Tekne 9701, and the DaVinci Audio Labs Grandezza. While such amplifiers can give voltage gain by using the transformer to step up the signal voltage, they are still regarded as passive preamplifiers because they have no active electronics and do not require a power supply.

The load on the external phono unit is the power amplifier input impedance reflected from the secondary winding

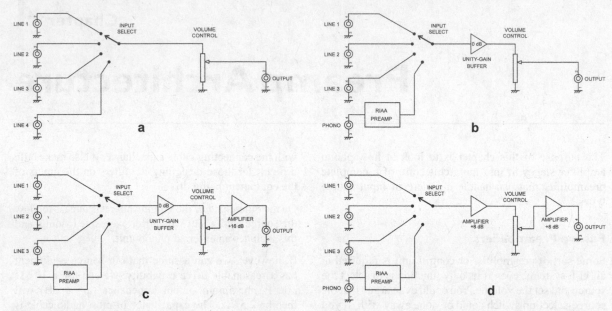

Figure 4.1 Preamplifier evolution: a) passive preamplifier; b) input buffer and phono amplifier added; c) amplification after the volume control added; d) amplification split into two stages, before and after volume control

of the transformer to the primary. If the gain is set to unity, i.e. a 1:1 transformer ratio, the input impedance of the power amplifier being driven appears unaltered at the preamplifier input and will load the phono unit. If however the transformer preamplifier is set for a gain of +6 dB, the input impedance will be a quarter of that and may be approaching our figure of 2 kΩ. It is important to remember that while the voltage gain is proportional to the turns ratio, the impedance transformation is proportional to the turns ratio squared, as both voltage and current are transformed. In general preamplifiers are more often set to attenuate rather than amplify, and in this case the preamplifier input impedance will be higher than the power amplifier input impedance, and it is unlikely that an excessive load will be placed on the external phono unit.

There are several potential problems with the transformer approach—they are well known to fall much further short of being an ideal component than most electronic parts do. They can introduce frequency response irregularities, LF distortion, and hum. They are relatively heavy and expensive, and the need for a large number of taps on the secondary puts the price up further. The multiway switch to select the desired tap will be expensive if there are a reasonable number of steps.

There are however some advantages. The output impedance of a resistive potential divider varies according to

its setting, being a maximum at −6 dB. Assuming it is fed from a low impedance, a transformer volume control has a low impedance at every tap, greater than zero only by the resistance of the windings, and handily lower than even a low-resistance pot. This gives lower Johnson noise and minimises the effect of the current noise of the following stage. The much-respected Sowter transformer company makes a number of different volume-control transformers, of which the most representative is probably the 9335 model. This is basically a 1:1 transformer with taps on the secondary that give attenuation from 0 to −50 dB in 26 steps of 2 dB each. It comes in a mumetal can 45 mm in diameter and 52 mm high, so it is not a cumbersome component. The DC resistance of both primary and secondary is 310 Ω. The total resistance of the windings is therefore 620 Ω, which is much lower than the value of the volume pots normally used. Note, however that it is not much lower than the 1 kΩ pots I used in the Elektor preamplifier.[1] Driving from a low impedance is recommended, and the point is forcibly made that DC must be kept out of the transformer. At the time of writing 9335 transformers cost £153 each. You naturally need two for stereo, and an expensive 2-pole multiway switch.

Transformers can of course provide balanced inputs without any added electronics.

Active Preamplifiers

Once we permit ourselves active electronics, we can design a much more flexible and effective preamplifier. Even a simple unity-gain buffer gives us more flexibility.

If a unity-gain buffer stage is added after the selector switch, as in Figure 4.1b, the volume pot resistance can be reduced to much less than 10 kΩ, while presenting a high impedance to the sources. If a 5532 is used for the buffer there is no technical reason why the pot could not be as low as 1 kΩ, which will give a much more useable maximum output impedance of only 250 Ω and also reduce Johnson noise by 10 dB. Note that an internal phono preamplifier has also been added. Now we've paid for a power supply, it might as well supply something else. Alternatively the phono amplifier can be external; since there is no gain in the preamplifier there will still be potential overload margin problems.

A unity-gain buffer still leaves us with a "preamplifier" that has a maximum gain of only one. Normally only CD players and other digital sources with an output of 2 Vrms can fully drive a power amplifier without additional gain, and there are some high-power amplifiers that require more than this for full output. iPods appear to have a maximum output of 1.2 Vrms. Output levels for tuners, phono amps, and so on vary but may be as low as 150 mV rms, while power amplifiers rarely have sensitivities lower than 500 mV. Clearly some gain would be good thing, so one option is adding a gain stage after the volume control as in Figure 4.1c. The output level can be increased and the output impedance kept down to 100 Ω or lower.

This amplifier stage introduces its own difficulties. If its nominal output level with the volume control fully up is taken as 1 Vrms for 150 mV in, which will let us drive most power amps to full output from most sources most of the time, we will need a gain of 6.7 times or 16.5 dB. If we decide to increase the nominal output level to 2 Vrms, to be sure of driving most if not all exotica to its limits, we need 22.5 dB. The problem is that the gain stage is amplifying its own noise at all volume settings and amplifying a proportion of the Johnson noise of the pot whenever the wiper is off the zero stop. The noise performance will therefore deteriorate markedly at low volume levels, which are the ones most used.

Adding amplification makes it easier to design the phono amplifier as it does not need to be able to drive a power amplifier directly and so requires less gain, and the overload margin is better.

Balanced Line Inputs

Balanced inputs are now common on preamplifiers with aspirations to quality. They ignore (to a first approximation) noise and hum currents in ground connections and allow hum loops to be rendered harmless. The only drawback seems to be the need for slightly more expensive cables and connectors, and of course you have to pay for the balanced input amplifier that converts the balanced signal back to single ended and at the same time cancels out ground noise and hum. But . . . there is another disadvantage which is rarely discussed in polite circles, and this has implications for the design of external phono amplifiers.

Balanced inputs are inherently noisier than unbalanced inputs by a large margin, in terms of the noise generated by the input circuitry itself rather than external noise. This may appear paradoxical, but it is all too true. Many people feel that this is the wrong way round. Surely the balanced input, with its professional XLR connector and its much-vaunted rejection of ground noise, should be completely superior? Well, it is—except as regards the internal noise generated by a balanced input amplifier.

If we assume that the unbalanced input stage is a 5532 voltage-follower, then with its input terminated by 50 Ω to ground the output noise is a very low −119.0 dBu over the usual 22–22 kHz bandwidth. This is because there are no series input resistors and no feedback resistors, so the noise seen is only the unamplified voltage noise of the opamp. If however the unbalanced input stage is configured to give gain, its noise output will inevitably be greater.

A balanced input stage is by comparison regrettably noisy. If it is built with 10 kΩ resistors and a 5532 section, as in the standard differential circuit of Figure 4.2a, the noise output is −104.8 dBu with both inputs similarly terminated. This is a 14 dB discrepancy which is both clearly audible and hard to explain away to suspicious potential customers.

The extra noise is due to the relatively high resistor values around the opamp which generate Johnson noise, and also the effect of opamp current noise flowing in these resistors. Their value cannot be reduced in Figure 4.2a without reducing the input impedances to below what is acceptable. If however two input buffers are added as in Figure 4.2b, the input impedances are defined solely by R5 and R6, and therefore resistors R1–R4 can be much reduced in value, here to 820 Ω. This reduces the output noise to −110.2 dB, which is 5.4 dB

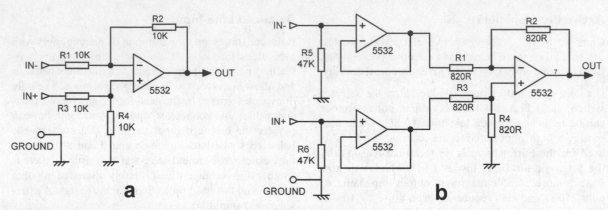

Figure 4.2 Simple (a) and buffered (b) balanced input stages. Unity gain.

quieter than Figure 4.2a, but still noisier than the unbalanced voltage-follower.

This technique can be carried much further; the use of multiple input buffers and multiple amplifiers, whose noncorrelated noise partially cancels, allows us to get to within 2 dB of the unbalanced input without using anything more exotic than 5532s; see Table 4.1. However this requires quadruple input buffers and quadruple differential amplifiers, so the extra complexity is not negligible. Getting closer than that requires more expensive opamps such as the LM4562 or the AD797; both of these are a mixed blessing because they have lower voltage noise but higher current noise than the 5532. The technique, using double buffers and quad differential amplifiers, was used in the Cambridge Audio 840W power amplifier, a design of mine which, I might modestly mention in passing, won a CES Innovation Award in January 2008. There is much more on this method in *Small Signal Audio Design*.[2]

The important conclusion is that unless you are prepared to deploy a lot of hardware, the balanced input will be noisier than the unbalanced input. This has implication for the design of external phono amplifiers. If there are ground current problems, then using a balanced link will fix them, and it is not necessary for the sending equipment to have balanced output; connecting the cold (out-of-phase) balanced input to the output ground of the external source will give the full noise rejection that the balanced input is capable of; this is limited by its common-mode rejection ratio (CMRR). However, if the phono amplifier does have a balanced output, then the signal level in the connecting cable is effectively doubled, and the signal/noise ratio of the balanced input is improved by 6 dB. This means we can use much less hardware and still have a balanced input that is as quiet as the quietest of unbalanced inputs. Table 4.1 shows this

can be achieved with single 5532 buffers and dual 5532 diff amps, requiring just two 5532 packages per channel.

I therefore recommend that external phono amplifiers should always have balanced outputs if they are expected to drive balanced inputs, solely to reduce the effect of noise in the balanced input. The extra cost is not great; see Chapter 14 for more on this.

Table 4.1 A summary of the noise improvements made to a balanced input stage

Buffer type	Differential amplifier	Noise output	Improvement over 4 × 10 kΩ diff amp dB	Noisier than unbal input by dB
	5532 voltage-follower	−119.0		0.0 dB ref
None	Standard diff amp 10k 5532	−104.8	0.0 dB ref	14.2
Single 5532	Single diff amp 820R 5532	−110.2	5.4	8.8
Single 5532	Dual diff amp 820R 5532	−112.5	7.4	6.5
Single 5532	Quad diff amp 820R 5532	−114.0	9.2	5.0
Dual 5532	Quad diff amp 820R 5532	−116.2	11.4	2.8
Quad 5532	Quad diff amp 820R 5532	−117.0	12.2	2.0
Quad 5532	Quad diff amp 820R LM4562	−118.9	14.1	0.1

A ground-cancelling output on an external phono amplifier, driving an unbalanced input on the preamplifier, would give the same rejection of ground noise but lower electronic noise in the link; it is also more economical on components. See Chapter 14 for all about ground-cancelling outputs.

Balanced Line Input Selection

If you have more than one balanced input, there are two ways to implement input selection. Figure 4.3a shows separate balanced input amplifiers for each balanced input; if you have a lot of balanced inputs this puts the cost up.

Alternatively you can have just one balanced input amplifier and use a 4-pole select switch as in Figure 4.3b, so that the XLR connectors are the only part added for each extra input. This saves electronics cost but increases the cost of the select switch, and this is likely to dominate. Another drawback is that the unbalanced inputs have to go through a relatively noisy balanced input amplifier. Note that the cold input of this amplifier is grounded when unbalanced inputs are in use. I used this technique in the *Linear Audio* low-noise preamplifier,[3] where the gain of the balanced input stage was variable over a limited range to implement an active balance control.

Amplification and the Gain-Distribution Problem

One answer to the noise/headroom issue in preamplifiers is to take the total gain and split it so there is some before and some after the volume control, so there is less gain amplifying the noise at low volume settings. One version of this is shown in Figure 4.1d. The question is—how much gain before, and how much after? This is inevitably a compromise, and it might be called the gain-distribution problem. Putting more of the total

gain before the volume control reduces the headroom because there is no way to reduce the signal level, while putting more after increases the noise output at low volume settings. The first amplifier is sometimes called the normalisation amplifier, because after it the signals are at a standard nominal level. This may involve it having different gain for different inputs, which complicates the input selection switch.

If you are exclusively using sources with a predictable output, of which the 2 Vrms from a CD player will be the maximum, the overload situation is well defined, and if we assume that the pre-volume gain stage is capable of at least 8 Vrms out, so long as the pre-volume-control gain is less than four times there will never be a clipping problem. However, phono cartridges, particularly moving-coil ones, which have a very wide range of sensitivities, produce much less predictable outputs after fixed-gain preamplification, and it is a judgement call as to how much safety margin is desirable.

As an aside, it's worth bearing in mind that even putting a unity-gain buffer before the volume control, which we did as the first step in preamp evolution, does place a constraint on the signal levels that can be handled, albeit at rather a high level of 8–10 Vrms depending on the supply rails in use. There is also the ultimate constraint that a volume-control pot can only handle so much power, and the manufacturers ratings are surprisingly low, sometimes only 50 mW. This means that a 10 kΩ pot would be limited to 22 Vrms across it, and if you are planning to use lower resistance pots than this to reduce noise, their power rating needs to be kept very much in mind.

Whenever a compromise appears in engineering, you can bet that someone will try to find a way round it and get the best of both worlds. What can be done about the gain-distribution dilemma?

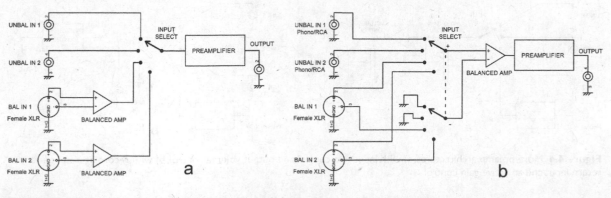

Figure 4.3 a) Balanced input amp for each input; b) single balanced input amp switched between inputs

One possibility is the use of a special low-noise amplifier after the volume control, combined with a low-resistance volume pot as suggested earlier. This could be done either by a discrete device and opamp hybrid stage or by using a multiple opamp array, as described in Chapter 1. It is doubtful if it is possible to obtain more than a 10 dB noise improvement by these means, but it would be an interesting project.

Another possible solution is the use of double gain controls. There is an input gain control before any amplification stage which is used to set the internal level appropriately, thus avoiding overload, and after the active stages there is an output volume control, which gives the much-desired silence at zero volume. See Figure 4.4a. The input gain controls can be separate for each channel, so they double as a balance facility; this approach was used on the Radford HD250 amplifier and also in one of my early preamplifier designs.[4] This helps to offset the cost of the extra pot. However, having two gain controls is operationally rather awkward, and however attenuation and fixed amplification are arranged, there are always going to be some trade-offs between noise and headroom. It could also be argued that this scheme does not make a lot of sense unless some means of metering the signal level after the input gain control is provided, so it can be set appropriately.

If the input and output gain controls are ganged together, to improve ease of operation at the expense of flexibility, this is sometimes called a distributed gain control.

Active Gain Controls

The noise/headroom compromise is completely avoided by replacing the combination of volume-control-and-amplifier with an active gain control, i.e. an amplifier stage whose gain is variable from near zero to the required maximum; see Figure 4.4b. We get lower noise at gain settings below maximum, and we can increase that maximum gain so even the least sensitive power amplifiers can be fully driven, without impairing the noise performance at lower settings. We also get the ability to generate a quasi-logarithmic law from a linear pot, which gives excellent channel balance as it depends only on mechanical alignment. The only snags are that:

a) most active gain controls phase-invert, though this can be corrected by suitable connection of a balanced input or balanced output stage or a Baxandall tone control.
b) the noise out is very low but not zero at zero volume as it would be with a passive pot, since the noise gain does not fall below unity.

Balance Controls

I assert that any preamplifier, be it passive, active, or based on quantum entanglement, needs a balance control to be usable. We do not all have precisely symmetrical listening spaces. A channel gain imbalance of 10 dB is quite enough to shift the stereo image wholly to one side, and there is no need to fade out one channel totally. A passive balance control introduces the same noise/headroom compromise as a volume control, and an active-gain solution is preferred. Either a balanced input stage or a tone control can have its gain made variable over a limited range.

Tone Controls

Let us now consider adding tone controls. They have been unfashionable for a while, but this is definitely changing now. I think they are absolutely necessary, and it is a startling situation when, as frequently happens, anxious inquirers to hi-fi advice columns are advised to

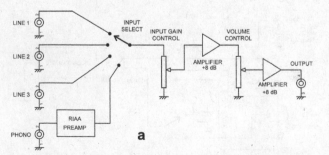
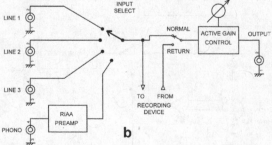

Figure 4.4 More preamp architectures: a) with input gain control and output volume control; b) with recording output and return input and an active gain control

change their loudspeakers to correct excess or lack of bass or treble. This is an extremely expensive alternative to tone controls.

There are many possible types, but one thing most of them have in common is that they must be fed from a low-impedance source to give the correct boost/cut figures and predictable EQ curves. Likewise most of them, and certainly all the really useful types, including the famous Baxandall configuration, give a phase inversion. Since there is now pretty much a consensus that all audio equipment should maintain absolute phase polarity for all input and outputs, this can be highly inconvenient. Adding another inverting stage to do nothing but correct the phase is not an attractive prospect.

However, as noted earlier, this phase inversion can very neatly be undone by the use of an active gain control, which also uses shunt feedback and so also phase-inverts.

The tone control can be placed before or after the active gain control in Figure 4.4b, but if placed afterwards it generates noise that cannot be turned down. Putting it before the active gain control reduces headroom if boost is in use, but if we assume the maximum boost used is +10 dB, the tone control will not clip until an input of 3V rms is applied, and domestic equipment rarely generates such levels. It therefore seems best to put the tone control before the active gain control, and this is exactly what I did in most of my preamplifier designs. See [1], [3], [5], and [6].

Phono Amplifier Integration

Having looked briefly at some of the issues in preamplifier design, we can turn again to the question of how best to integrate a phono amplifier into the rest of the preamplifier. It is assumed that the preamp is based on opamp technology, and so the maximum signal that can be handled is about 10 Vrms.

First we will assume that the phono amplifier has a fixed gain. If the output of the phono amplifier is simply treated as just another line input, then the output level will need to be high to match line inputs from digital sources; otherwise there will be annoying level changes on switching sources. This means the nominal output from the phono amp will have to be 2 Vrms for a 5 mV rms (1 kHz) input; this limits the headroom (more often called the overload margin in phono amplifiers) to $10/2 = 5$ times or 14 dB. This is a very small safety margin considering the unpredictability of vinyl velocities and the wide range of MC cartridge sensitivities; 30 dB or more overload margin is

expected in quality equipment. Line inputs fed by digital equipment do not suffer this problem because their maximum output is rigidly defined.

So how do we deal with this? There are essentially three ways:

1) The preamplifier has a dedicated low-level line input. This may use a separate amplifier stage so the phono amp can have a lower nominal output and a correspondingly greater overload margin, though there is then of course the problem of clipping in the second amplifier. This is not very feasible unless both bits of equipment are made by the same company.
2) Another option is to switch the gain of the normalisation amplifier. This doubles the complexity of the source-select switch and will probably require precautions to avoid changes in DC offset, which will cause thumps and bumps.
3) A better solution is for the phono amplifier to have the variable gain. Making the gain continuously variable requires a dual-gang pot and will introduce stereo tracking errors, unless something like the Baxandall active volume stage is used, which allows linear pots to be used and cancels out the effects of track resistance changes. Continuously variable gain is not particularly useful, and switched gain, say with 5 dB steps, is much to be preferred.

The issues involved in designing variable-gain phono stages are dealt with in Chapter 5 and Chapter 6.

References

1. Self, D. "Preamplifier 2012" *Elektor*, Apr, May, June 2012.

2. Self, D. *Small Signal Audio Design*. 2nd edition. Focal Press, 2015. ISBN: 978-0-415-70974-3 (hbk) ISBN: 978-0-415-70973-6 (pbk) ISBN: 978-1-315-88537-7 (ebk).

3. Self, D. "A Low Noise Preamplifier With Variable-Frequency Tone Controls" *Linear Audio*, Volume 5, pub. Jan Didden, pp. 141–162.

4. Self, D. "An Advanced Preamplifier" *Wireless World*, Nov 1976.

5. Self, D. "A Precision Preamplifier" *Wireless World*, Oct 1983.

6. Self, D. "Precision Preamplifier 96" *Electronics World*, July/Aug and Sept 1996.

Chapter 5

Moving-Magnet Inputs

Phono Amp Architecture

Moving-magnet (MM) Phono Amplifiers

The previous chapter treated the phono amplifier as a black box, possibly with variable gain, while examining how it can be integrated into a preamplifier. This chapter dives inside the black box to look at the various ways in which a phono amplifier can be implemented, taking the desirability of variable gain as axiomatic.

It is explained in Chapter 12 that a highpass filter operating below 20 Hz is highly desirable to remove subsonic disturbances produced by the vinyl surface, and this is considered an integral part of the phono amplifier. There are thus a minimum of three stages to couple together; an input stage giving RIAA equalisation with a gain of around +30 dB at 1 kHz, a unity-gain subsonic filter, and a stage switched from 0 to +20 dB of gain in 5 dB steps. The choice of these specific numbers is explained in this and the following chapters. Figure 5.1 shows various approaches to this. In Figure 5.1a the gain stage raises the signal to the final nominal level before it passes through the subsonic filter, minimising the noise contribution of the latter but perhaps increasing its distortion contribution.

In Figure 5.1b the subsonic filter comes before the gain stage, which may make its noise contribution more significant, but subsonic disturbances have been removed so they cannot cause intermodulation or clipping in the gain stage.

The choice here depends on the noise and distortion characteristics of the subsonic filter and the gain stage. Subsonic filters are usually quiet; in Chapter 12 even a sophisticated 6th-order unity-gain elliptical subsonic filter has a noise output at a very low −111.4 dBu. If the RIAA stage has a gain of +30 dB at 1 kHz, then Chapter 9 shows that even a completely noiseless amplifier would have a noise output of −94.9 dBu (see Case 3 in

Table 9.2) derived from the MM cartridge. Adding these rms-fashion, we find that the combined noise only rises by 0.1 dB, an imperceptible amount. In a more realistic scenario with a 5534A as the amplifier, the RIAA stage output noise will be −92.5 dBu, and the combined noise only increases by a negligible 0.06 dB. For these reasons I usually adopt the arrangement of Figure 5.1b in my designs.

Figure 5.1c offers another alternative; it is possible to build the subsonic filter into the RIAA stage, and this rejects subsonic disturbances before they are amplified at all, as well as saving an opamp or two. See Chapter 12.

Vinyl produces ultrasonic disturbances as well as subsonic ones, and a fully equipped phono amplifier will have some means of removing this; Chapter 13 gives much information on this. Ultrasonic stuff is much more likely than subsonic stuff to cause intermodulation in amplifiers, so Figure 5.1d shows the ultrasonic filter placed just after the RIAA stage and before the subsonic filter. Even if no ultrasonic filter is fitted, the HF correction pole (see Chapter 7) has a response going down at 6dB/octave forever and so gives some protection if placed as usual immediately after the RIAA stage.

Active RIAA Equalisation

The complex issue of active RIAA equalisation is dealt with in great detail in Chapter 7, but I will tell you here and now that a series-feedback stage that performs the equalisation in one go is by far the best way, and the only downside is the difficulty of designing an RIAA network with interacting time-constants; see Figure 5.2a. The shunt-feedback version in Figure 5.2b has the crippling disadvantage that the input resistor must be 47 kΩ for correct cartridge loading, leading to a 14 dB noise disadvantage; more on this in Chapter 6.

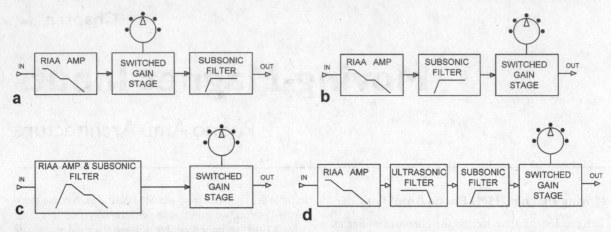

Figure 5.1 Differing architectures for an MM phono amplifier

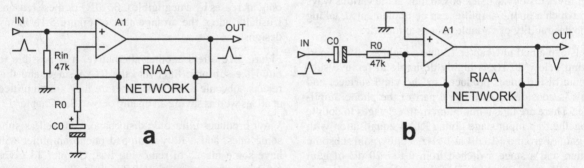

Figure 5.2 Series (a) and shunt-feedback (b) RIAA configurations

The two RIAA networks in Figure 5.2 will have quite different values, and so one configuration cannot easily be converted to the other.

Passive and Semi-Passive RIAA Equalisation

For many years, series-feedback RIAA preamplifiers, as described earlier, were virtually universal, it being accepted by all that they gave the best noise, overload performance, and economy, especially of active components. However, human nature being what it is, some people will always want to do things the hard way, and this is exemplified by the fashion for passive (actually, semi-passive is usually more accurate) RIAA equalisation. The basic notion is to split the RIAA equalisation into separate stages, and I have a dark and abiding suspicion that this approach may be popular simply because it makes the design of accurate RIAA equalisation much easier, as all you have to do is calculate simple time-constants instead of grappling with foot-long equations.

There is a price, and a heavy one; the overload and/or noise performance is inevitably compromised.

a) Clearly a completely passive RIAA stage is a daft idea because a lot of gain is required somewhere to get the 5 mV cartridge signal up to a usable amplitude. The nearest you can get to completely passive is the scheme shown in Figure 5.3a, where the amplification and the equalisation are wholly separate, with no frequency-dependent feedback used at all. R2, R3, and C1 implement T3 and T4, while C2 implements T5. There is no inconvenient T6 because the response carries on falling indefinitely with frequency. This network clearly gives its maximum gain at 20 Hz, and at 1 kHz it attenuates by about 20 dB. Therefore, if we want the modest +30 dB gain at 1 kHz used in the previous example, the A1 stage must have a gain of no less than 50 dB. A 5 mV rms 1 kHz input would therefore result in 1.58 V at the output of A1. This is only 16 dB below clipping, assuming we are using the usual sort of opamps, and an overload margin of 16 dB is much too small to be usable. It is obviously impossible to drive anything like a volume-control or tone-control stage from the passive network, so the buffer stage A2 is shown to emphasise that extra electronics is required with this approach. This is what I call a Passive-Passive configuration, ignoring the opamp stages A1, A2 that do not perform equalisation.

The only way to improve the overload margin is to split the gain so that the A1 stage has perhaps 30 dB, while A2 after the passive RIAA network makes up the loss with 20 dB more gain. Sadly, this second stage of amplification must introduce extra noise, and there is always the point that you now have to put the signal through two amplifiers instead of one, so there is the potential for increased distortion.

This configuration does not even have the advantage of separate time-constants for easy calculation, as all the equalisation is done in one network. It is deprecated.

b) The most popular architecture that separates the high and low RIAA sections is seen in Figure 5.3b. Here there is an active LF RIAA stage using feedback to implement T3 and T4 with R1, C1, R2, followed by R3, C2, which give a passive HF cut for T5. Therefore it is an Active-Passive configuration. The values shown give an RIAA curve correct to within 0.04 dB from 20 Hz to 20 kHz. Note that because of the lack of time-constant interaction, we can choose standard values for both capacitors, but we are still left with awkward resistor values. These can be easily

addressed by using parallel resistor pairs in the 2xE24 format.

As always, amplification followed by attenuation means a headroom bottleneck, and this passive HF roll-off is no exception. Signals direct from disc have their highest amplitudes at high frequencies, so both these configurations give poor HF headroom, overload occurring at A1 output before passive HF cut can reduce the level. Figure 5.4 shows how the level at A1 output (Trace B) is higher at HF than the output signal (Trace A). The difference is Trace C, the headroom loss; from 1 dB at 1 kHz this rises to 14 dB at 10 kHz and continues to increase in the ultrasonic region. The passive circuit was driven from an inverse RIAA network, so a totally accurate disc stage would give a straight line just below the +30 dB mark.

A related problem in this Active-Passive configuration is that the opamp A1 must handle a signal with much more HF content than the opamp in the single-stage series-feedback configuration, worsening any difficulties with slew-limiting and HF distortion. It uses two amplifier stages rather than one, and more precision components, because of the extra resistor. Another difficulty is that A1 is more likely to run out of open-loop gain or slew rate at HF, as the response plateaus above 1 kHz rather than being steadily reduced by increasing negative feedback. Once again a buffer stage A2 is required to isolate the final time-constant from loading.

c) A third method of equalisation is shown in Figure 5.3c, where the T5 roll-off is done by feedback via R4, C2 rather than by passive attenuation, making

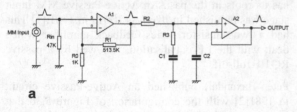

Figure 5.3 Passive and semi-passive RIAA configurations

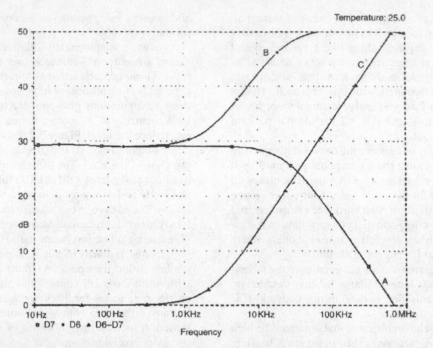

Figure 5.4 Headroom loss with passive RIAA equalisation. The signal level at A1 (Trace B) is greater than at A2 (Trace A) so clipping occurs there first. Trace C shows the headroom loss, which reaches 18 dB at 20 kHz.

it an Active-Active configuration. It is not really passive in any way, as the equalisation is done in two active stages, but it does share the crucial feature of splitting up the time-constants for easier design. As with the previous circuit, A1 is running under unfavourable conditions because it has to handle a larger HF content than in the series-feedback version, and there is now an inconvenient phase reversal. The values shown give the same gain and RIAA accuracy as the previous circuit, though in this case the value of R3 can be scaled to change the gain.

d) There are many other alternative arrangements that can be used for passive or semi-passive equalisation. There could be a flat input stage followed by a passive HF cut and then another stage to give the LF boost, as in Figure 5.3d, which has even more headroom problems and uses yet more components. I call this a Passive-Active configuration.

In contrast, the "all-in-one-go" single-stage series-feedback configuration avoids all these headroom restrictions and uses the minimum number of parts. This list is not exhaustive, as there are many ways to rearrange the amplifiers and time-constants.

Passive RIAA is not an attractive option for general use but comes into its own in the archival transcription of recordings, where there are dozens of different pre-RIAA equalisation schemes, and it must be possible to adjust the turnover frequencies $f3$, $f4$ and $f5$ independently. This is done most straight-forwardly by a fifth Passive-Passive equalisation configuration, which is described in detail in Chapter 8.

As is so often the case, what you think is a recent trend has its roots in the past. An Active-Passive MM input stage was published in *Wireless World* in 1961.[1] This had a two-transistor series-feedback amplifier which dealt with the LF equalisation, followed by a passive RC HF roll-off.

Peter Baxandall published an Active-Passive circuit in 1981[2] with the configuration of Figure 5.3b that gave easy switched-gain control and allowed the use of preferred values, with only two of them in the E24 series. Like all Peter's ideas it is well worth studying and is shown in Figure 5.5. The gains are +20, +30, and +40 dB; the switchable gain largely avoids the headroom problems of passive RIAA equalisation. The RIAA accuracy is within ±0.03 dB between 1 kHz and 20 kHz for each gain setting, falling off to about −0.1 dB at 100 Hz. This is due to the way that R0 and C0 implement the IEC amendment, giving f2 = 21.22 Hz rather than the correct 20.02 Hz; that is as close as you can get

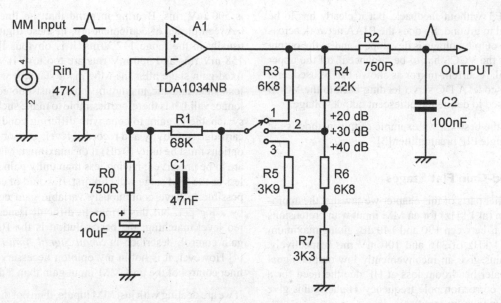

Figure 5.5 Active-passive RIAA stage with switched gain by Peter Baxandall

with a single 750 Ω E24 resistor for R0. It results in a response 0.34 dB too low at 20 Hz. The correct value for R0 is 795 Ω, so f2 could be made much more accurate by using the 2xE24 parallel pair 1 kΩ and 3.9 kΩ, which is only 0.1% too high. However, there is the tolerance of C0 to be considered, and when Peter was writing (1981) that would have been larger than we would expect today, so 750 Ω was no doubt considered close enough.

The passive HF equalisation means that no HF correction pole is necessary at any gain setting. No buffering of this R2–C2 network is shown; in many cases this will be necessary to keep the HF roll-off accurate.

One problem with this circuit suggests itself. When the gain switch is between contacts, A1 has no feedback and will hit the rails. Very likely Peter was thinking of a make-before break switch. Another possible way of solving this is given in Figure 5.7, where feedback is maintained when the switch moves. The TDA1034B was an early version of the 5534 and capable of driving the relatively low impedance of the R2–C2 combination.

There is much more on RIAA gain switching in Chapter 7, for example showing how to get the RIAA absolutely accurate at two different gain settings.

Transconductance RIAA Stages

Transconductance RIAA stages are a way of implementing what might be loosely called "passive RIAA

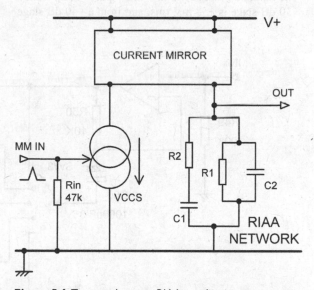

Figure 5.6 Transconductance RIAA equalisation stage

equalisation" while avoiding some of the noise/headroom problems described earlier. In Figure 5.6 the VCCS is a voltage-controlled current source, in other words a transconductance stage (see Chapter 1). This has its output in the form of a current rather than a voltage, and that current is not affected by the voltage produced when it flows through an impedance, such as the Config-C RIAA network shown. The VCCS may be a

BJT or FET without feedback, but it clearly has to be referenced to ground. So does the RIAA network across which the output voltage is developed, and so the output current of the VCCS has to be "bounced" off the upper supply rail. A current-mirror as shown can be used, or a folded-cascode. A DC servo feeding back to the VCCS is often used to define the quiescent output voltage.

Probably the best-known example of this method is the Pink Triangle PIP preamplifier.[3]

Switched-Gain Flat Stages

In the earlier parts of this chapter we saw that the appropriate gain (at 1 kHz) for an MM input with pretensions to quality is between +30 and +40 dB, giving maximum inputs at 1 kHz of 316 and 100 mV rms respectively. Lower gains give an inconveniently low output signal and a greater headroom loss at HF due the need for a lower HF correction pole frequency. Higher gains give too low a maximum input.

The nominal output for 5 mV rms input (1 kHz) from a +30 dB stage is 158 mV rms, and from a +40 dB stage is 500 mV rms. Bearing in mind that the line signals between pieces of equipment are, in these digital days, usually in the range 1–2 Vrms, it is obvious that both 158 mV rms and 500 mV rms are too low. If we put a fixed-gain stage after the MM input stage, it will overload first, and the maximum inputs just quoted are no longer valid. It is therefore desirable to make such a stage switchable in gain, to cope with differing conditions of cartridge sensitivity and recorded level. One of the gain options must be unity (0 dB) if the maximum MM inputs are to be preserved; having less than unity gain is pointless as the MM stage will clip first. It would of course be possible to have continuously variable gain controlled by a log pot, but this brings in difficult issues of stereo level matching; the best solution is the Baxandall gain control, described in *Small Signal Audio Design*. [4] However, it is not in my opinion necessary to have finer control of the post-MM-input gain than 5 dB steps.

If we are dealing with just MM inputs, then not many gain options are required. If we assume a +30 dB (1 kHz) MM stage with its nominal 158 mV rms output, then we need 6.3 times or +16 dB of gain to raise that level to 1 Vrms.

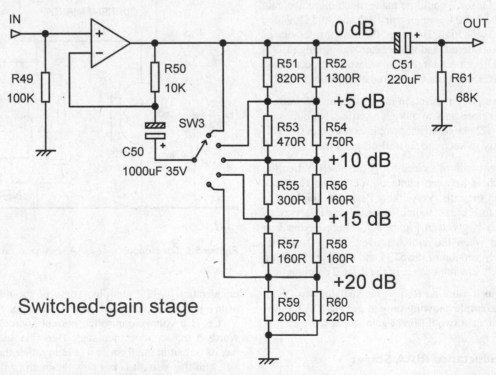

Figure 5.7 A flat gain stage with accurate switched gains of 0, +5, +10, +15, and +20 dB. Resistor pairs are used to get the exact gains wanted and to reduce the effect of tolerances.

This suggests that gain options of 0 dB, +5 dB, +10 dB, and +15 dB are all that are needed, with the lower gains allowing for more sensitive cartridges and elevated recording levels.

However, it will be seen in Chapter 11 that MC cartridges have a much wider spread of sensitivities than the MM variety, and if the MM input stage followed by the flat switched-gain stage are going to be used to perform the RIAA equalisation after a flat +30 dB MC head amp, a further +20 dB gain option in the switched-gain stage is required to ensure that even the most insensitive MC cartridges can produce a full 1 Vrms nominal output.

The switched-gain stage in Figure 5.7 is derived from my Elektor 2012 preamp[5] and gives the same gain options. The AC negative feedback is tapped from the divider R51–R60, which is made up of 2xE24 pairs of resistors to achieve the exact gain required and to reduce the effect of the resistor tolerances. There is always DC feedback for the opamp through R50 to prevent the opamp hitting the rails when switching the gain; I am assuming a break-before-make switch as they are much more common than make-before-break. The blocking capacitor C50 is more than large enough at 1000 uF to prevent any frequency response irregularities in the audio band, and it probably could be reduced in size. Assuming the source impedance is reasonably low, an LM4562 will give better noise and distortion results than a 5532 section.

The correct setting for the gain switch can be worked out by considering cartridge sensitivity specs and recording levels, but the latter are usually unknown, so some form of level indicator is very useful when setting up. A bar-graph meter seems a bit over the top for a facility that will not be used very often, and a single LED indication makes more sense. For this reason the Log-Law Level LED was developed, giving about as much level information as can be had from one LED. It is fully described in Chapter 15 on metering. It is desirable that any level indication can be switched off, as not everyone thinks that flashing lights add to the musical experience.

Line Outputs

The phono amplifier architectures shown in Figure 5.1 assume a simple unbalanced output. This can give trouble with ground loops and consequent buzz and hum currents flowing down the ground of the collector cable. These problems should disappear if the driven equipment has a balanced input, and its cold pin is correctly connected to the phono amplifier ground, i.e. connected at the sending end. A balanced output impedance will improve the common-mode rejection and so the suppression of hum, while spending just a little more (for one opamp section) to get a true balanced output with anti-phase signals will improve the signal/noise ratio of the balanced link by 6 dB. A more subtle option is the ground-cancelling output.

All of these options are comprehensively described in Chapter 14.

References

1. Lewis, T. M. A. "Accurate Record Equaliser" *Wireless World*, Mar 1961, p. 121.

2. Baxandall, P. Letter to Editor. "Comments on 'On RIAA Equalisation Networks'" *JAES*, Volume 29, #1/2, Jan/Feb 1981, pp. 47–53.

3. Miller, P. "Pip! Pip!" *Hifi News*, Dec 2016, pp. 124–129.

4. Self, D. *Small Signal Audio Design*. 2nd edition. Focal Press, 2015, pp. 607–608. ISBN: 978-0-415-70974-3 (hbk) ISBN: 978-0-415-70973-6 (pbk) ISBN: 978-1-315-88537-7 (ebk).

5. Self, D. "Precision Preamplifier 96" *Electronics World*, July/Aug and Sept 1996.

Chapter 6

Signals From Vinyl

Levels and Limitations

Cartridge Types

This chapter deals with the signal levels generated by moving-magnet (MM) cartridge inputs and how this interacts with their special loading requirements and the need for RIAA equalisation. MM cartridges have been for many years less popular than moving-coil (MC) cartridges, but seem to be staging a comeback. However it would be relatively unusual nowadays to design a phono input that accepted MM inputs only. There are several ways to design a combined MM/MC input, but the approach that gives the best results is to design an MM preamp that incorporates the RIAA equalisation and put a flat-response low-noise head amplifier in front of it to get MC inputs up to MM levels. This allows the head amp to operate in the best conditions for low noise. A large part of this chapter is devoted to the tricky business of RIAA equalisation and so is equally relevant to MC input design. Almost all modern cartridges are of these two types, though Grado makes a moving-iron (MI) series; this is essentially a variation on MM.

Ceramic cartridges were still very much around when I first got into the audio business, but they now seem to be a rare example of an obsolete audio technology that no-one wants to revive. Ceramic cartridges have elements of Rochelle salt (in early versions, often called crystal pickups) or PZT (in ceramic versions) which generate electricity when flexed by the stylus. They look like a pure capacitance of around 200 pF to an amplifier input, and given a suitably high-impedance load, greater than 2 MΩ, they respond to stylus displacement rather than velocity. This led people to say that RIAA equalisation was not required, though this seems to overlook the presence of the 2-octave plateau in the middle of the RIAA characteristic. With 2 MΩ loading the output was in the range of 200–600 mV rms, much higher than the output of MM cartridges. This did not necessarily

simplify preamplifier circuitry because of the need to establish a 2 MΩ input impedance; there is no real difficulty in doing this even with low-beta BJTs, as described in Chapter 3, but a few more parts are needed. Alternatively the cartridge can be more heavily loaded and equalisation applied. These two philosophies of ceramic cartridge termination were described in a 1969 article by Linsley-Hood,[1] and there was a thorough discussion of the whole business by Burrows in 1970.[2], [3]

The tracking force for ceramic cartridges is usually higher than that of MM or MC, often being on the order of 3 to 5 grams, and the increased groove wear is one powerful reason for not using them today. Sonotone and Acos were major ceramic cartridge manufacturers. The first cartridge I ever owned was a Sonotone 9TA; I still have it. Like most of its kind, it was a turnover cartridge; in other words there were two styli, one on each side of the cantilever. One was for microgroove LPs and the other for coarse-groove 78s, and they were selected by turning over the cantilever with a little plastic tab. This type of cartridge is still sometimes used for transcribing old 78 rpm records. An excellent account of the history of Sonotone cartridges can be found in and online history of the company.[4]

Strain-gauge cartridges also have a long history and are still with us today. Those made by Sound-Smith[5] appear to have a good reputation, though I have no experience with them myself. They are also sensitive to stylus displacement rather than stylus velocity (as MM and MC cartridges are). Naturally they require a specialised preamplifier, and the whole setup is not cheap.

Capacitance pickups, aka "FM pickups" or "electrostatic pickups," consist of a small capacitor, one plate of which is wiggled by the stylus; the change in capacitance frequency modulates an oscillator, and the signal is decoded by standard FM-receiver technology. These were made by Weathers; the relevant patent appears to be US4,489,278, granted to Paul Weathers in 1984. Stax also made them; their first capacitance cartridge was the mono CP-20 introduced in 1952, the stereo CPS-40 not appearing until 1962. In 1977 Stax introduced a new technology, the CP-Y cartridge, having a permanent electret element and a head-amplifier IC built into the cartridge body. Shortly afterwards, their FM system was discontinued.

Moving-Magnet and Moving-Coil Cartridges

Moving-magnet (MM) cartridges create a signal voltage by creating a moving magnetic field which induces a voltage in fixed coils. The coils do not move and so can be relatively heavy, with a large number of turns to generate a relatively large voltage, around 5 mV rms. In contrast moving-coil cartridges move lightweight coils with few turns in a fixed magnetic field, so generally the output is much lower, with two clusters around 100–300 uV and 500–700 uV. Moving-coil cartridges are dealt with in more detail in Chapter 11.

Both types are generally connected as shown in Figure 6.1. The major variation is whether or not the metal screening of the cartridge (to screen electric fields rather than magnetic ones, which is much less practicable, especially given the weight limitations) is connected to one of the signal grounds or not. If it is, as at A, then use of a metal headshell on a grounded arm will create a ground loop via the fixing screws B; nylon fixing screws solve this effectively. If the metal screening is not grounded and a nonmetal headshell is used, the screening is ungrounded, and electric fields from the turntable electrics will cause severe hum. Connection A was/is

fitted to many Shure cartridges, as a metal tab which could be removed with care. At least one manufacturer has not fitted any electric screening to its cartridges, resulting in endemic hum problems.

The Vinyl Medium

The vinyl disc as a medium for music delivery in its present form dates back to 1948, when Columbia introduced microgroove 33⅓ rpm LP records. These were followed soon after by microgroove 45 rpm records from RCA Victor. Stereo vinyl did not appear until 1958. The introduction of Varigroove technology, which adjusts groove spacing to suit the amplitude of the groove vibrations, using an extra look-ahead tape head to see what the future holds, allowed increases in groove packing density. This density rarely exceeded 100 grooves per inch in the 78 rpm format, but with Varigroove 180–360 grooves/inch could be used at 33⅓ rpm.

While microgroove technology was unquestionably a considerable improvement on 78 rpm records, any technology that is 60 years old is likely to show definite limitations compared with contemporary standards, and indeed it does. Compared with modern digital formats, vinyl has a restricted dynamic range and poor linearity (especially at the end of a side) and is very vulnerable to permanent and irritating damage in the form of scratches. Even with the greatest care, scratches are likely to be inflicted when the record is removed from its sleeve. This action also generates significant static charges which attract dust and lint to the record surface. If not carefully removed, this dirt builds up on the stylus and not only degrades the reproduction of high-frequency information today but may also damage it in the future if it provokes mistracking.

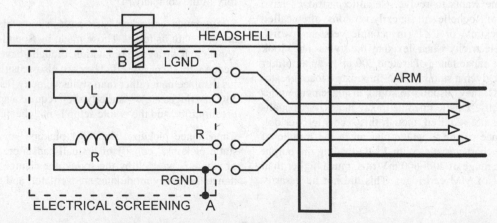

Figure 6.1 The wiring of a typical MM cartridge, with possible screen grounding at A or B; not both, or you get a ground loop

Vinyl discs do not shatter under impact like the 78 shellac discs, but they are subject to warping by heat, improper storage, or poor manufacturing quality control. Possibly the worst feature of vinyl is that the stored material is degraded every time the disc is played, as the delicate high-frequency groove modulations are worn away by the stylus. When a good turntable with a properly balanced tone arm and correctly set up low-mass stylus is used this wear process is relatively slow, but it nevertheless proceeds inexorably. The stylus suffers wear too.

However, for reasons that have very little to do with logic or common-sense, vinyl is still very much alive. Even if it is accepted that as a music-delivery medium it is technically as obsolete as wax cylinders, there remain many sizable album collections that it is impractical to replace with CDs and would take an interminable time to transfer to the digital domain. I have one of them. Disc inputs must therefore remain part of the audio designer's repertoire for the foreseeable future, and the design of the specialised electronics to get the best from the vinyl medium is still very relevant.

Vinyl Problems: Spurious Signals

It is not easy to find dependable statistics on the dynamic range of vinyl, but there seems to be general agreement that it is in the range 50 to 80 dB, the 50 dB coming from the standard-quality discs and the 80 dB representing direct-cut discs produced with quality as the prime aim. My own view is that 80 dB is rather optimistic.

The most audible spurious noise coming from vinyl is that in the midfrequencies, stemming from the inescapable fact that the music is read by a stylus sliding along a groove of finite smoothness. There is nothing that the designer of audio electronics can do about this; the levels involved are examined in Chapter 9 on noise.

Scratches create clicks that have a large high-frequency content, and it has been shown that they can easily exceed the level of the audio.[6] It is important that such clicks do not cause clipping or slew-limiting, as this makes their subjective impact worse. Lowpass filters can remove the ultrasonic disturbances and should be placed as early as possible in the signal path; see Chapter 13.

The signal from a record deck also includes copious amounts of low-frequency noise, which is often called rumble; it is typically below 30 Hz. This can come from several sources:

1) Mechanical noise generated by the motor and turntable bearings and picked up by the stylus/arm combination. These tend to be at the upper end of the low-frequency domain, extending up to 30 Hz or thereabouts. This is a matter for the mechanical designer of the turntable, as it clearly cannot be filtered out without removing the lower part of the audio spectrum.

2) Room vibrations will be picked up if the turntable and arm system is not well isolated from the floor. This is a particular problem in older houses where the wooden floors are not built to modern standards of rigidity and have a perceptible bounce to them. Mounting the turntable shelf to the wall usually gives a major improvement. Subsonic filtering is effective in removing room vibration.

3) Low-frequency noise from disc imperfections. This is the worst cause of disturbances. They can extend as low as 0.55 Hz, the frequency at which a 33⅓ rpm disc rotates on the turntable, due to large-scale disc warps. Warping can also produce ripples in the surface, generating spurious subsonic signals up to a few hertz at surprisingly high levels. These can be further amplified by a poorly controlled resonance of the cartridge compliance and the pickup arm mass. When woofer speaker cones can be seen wobbling and bass reflex designs with no cone loading at very low frequencies are the worst for this—disc warps are usually the cause. Subsonic filtering is again effective in removing this.

(As an aside, I have heard it convincingly argued that bass reflex designs have only achieved their current popularity because of the advent of the CD player, with its greater bass signal extension but lack of subsonic output.)

Some fascinating data on the subsonic output from vinyl was given in an article by Tomlinson Holman [7] which shows that the highest warp signals occur in the 2 to 4 Hz region, being some 8 dB less at 10 Hz. By matching these signals with a wide variety of cartridge-arm combinations, he concluded that to accommodate the very worst cases, a preamplifier should be able to accept not less than 35 mV rms in the 3–4 Hz region. This is a rather demanding requirement, driven by some truly diabolical cartridge-arm setups that accentuated subsonic frequencies by up to 24 dB.

Since the subsonic content generated by room vibrations and disc imperfections tends to cause vertical movements of the stylus, the resulting electrical output will be out of phase in the left and right channels. The use of a central mono subwoofer system that sums the two channels will provide partial cancellation, reducing the amount of rumble that is reproduced. It is however still important to ensure that subsonic signals do not reach the left and right speakers.

For a great deal of information on subsonic filtering, see Chapter 12.

Vinyl Problems: Distortion

The reproduction of vinyl involves other difficulties apart from the spurious signals mentioned earlier: Distortion is a major problem. It is pretty obvious that the electromechanical processes involved are not going to be as linear as we now expect our electronic circuitry to be. Moving-magnet and moving-coil cartridges add their own distortion, which can reach 1–5% at high levels.

Distortion gets worse as the stylus moves from the outside to the inside of the disc. This is called "end of side distortion" because it can be painfully obvious in the final track. It occurs because the modulation of the inner grooves is inevitably more compressed than those of the outer tracks, due to the constant rotational speed of a turntable. I can well recall buying albums and discovering to my chagrin that a favourite track was the last on a side.

It is a notable limitation of the vinyl process that the geometry of the recording machine and that of the replay turntable do not match. The original recordings are cut on a lathe where the cutting head moves in a radial straight line across the disc. In contrast, almost all turntables have a pivoting tone arm about 9 inches in length. The pickup head is angled to reduce the mismatch between the recording and replay situations, but this introduces side forces on the stylus and various other problems, increasing the distortion of the playback signal. A recent article in *Stereophile*[8] shows just how complicated the business of tone arm geometry is. SME produced a 12-inch arm to reduce the angular errors; I have one and it is a thing of great beauty, but I must admit I have never put it to use.

The vinyl process depends on a stylus faithfully tracking a groove. If the groove modulation is excessive, with respect to the capabilities of the cartridge/arm combination, the stylus loses contact with the groove walls and rattles about a bit. This obviously introduces gross distortion and is also very likely to damage the groove.

Vinyl Problems: Wow

A really disabling problem is "wow"; the slow cyclic pitch change resulting from an off-centre hole. Particularly bad examples of this used to be called "swingers" because they were so eccentric that they could be visibly seen to be rotating off-centre. I understand that nowadays the term means something entirely different and relates to an activity which sounds as though it could only be a distraction from critical listening.

Most of the problems that vinyl is heir to are supremely unfixable, but this is an exception; do not underestimate the ingenuity of engineers. In 1983 Nakimichi introduced the extraordinary TX-1000 turntable that measured the disc eccentricity and corrected for it.[9] A secondary arm measured the eccentricity of the run-out groove and used this information to mechanically offset the spindle from the platter-bearing axis. This process took 20 seconds, which I imagine could get a bit tedious once the novelty has worn off. This idea deserved to prosper, but CDs were coming; the timing was bad.

Vinyl Problems: Flutter

Flutter is rapid changes in pitch, rather than the slow ones that constitute wow. You would think that a heavy platter (and some of them are quite ridiculously massive) would be unable to change speed rapidly, and you would be absolutely right. But . . . the other item in the situation is the cartridge/arm combination, which moves up and down but does not follow surface irregularities because of the resonance between cartridge compliance and arm+cartridge mass. The stylus therefore moves back and forward in the groove, frequency-modulating the signal.

This and the other mechanical issues of turntables and arms are described in an excellent article by Hannes Allmaier,[10] who makes the important point that the ear is most sensitive to flutter at around 4 Hz, uncomfortably close to the cart/arm resonance region of 8–12 Hz.

Vinyl Problems: Tracking Force

The tracking force of a stylus in a groove is a highly important parameter. Too much force causes excessive groove wear, whereas too little permits mistracking, which sounds terrible and causes even more excessive groove wear. However, the tracking force varies significantly even with apparently flat vinyl, due to . . . yes, the cart/arm resonance again.

Attempts to reduce the multiple bad effects of the cart/arm resonance include damping the arm with a silicone-filled dashpot, but this is messy business and has never caught on. The Shure M97xE carries a little carbon-fibre brush which can be swung to contact the vinyl surface and reduce the vertical forces on the stylus and hence its movement.

Vinyl Problems: Surface Damage

A vinyl disc is horribly vulnerable to damage, and the resulting clicks and ticks are thoroughly annoying. Scratches usually occur when sliding the disc in or out of the inner sleeve. While you cannot, as far as I know, mend groove damage, various electronic devices have been introduced with the aim of suppressing clicks. This obviously depends on being able to detect clicks and reliably distinguish them from musical transients. Clicks normally have a much higher slew rate than music, and this is the basis of their detection; they also tend to be out-of-phase, and this information is sometimes also used. When detected the click is suppressed by very briefly reducing the gain, ideally so that the resulting waveform "smears over" the site of the click on the waveform. In some versions the audio is delayed (a 5 us delay implemented by four cascaded allpass filters has been used), so the suppression circuitry has time to operate.

The best known of the declicking devices is the SAE 5000A Impulse Noise Reduction system introduced in 1980; many of these are still in use. The general view seems to be that given careful adjustment of the sensitivity control it works reasonably well but by no means perfectly. Declicking can also be done using DSP software; an excellent overview of this is given in a PDF by Godsill et al.[11]

Maximum Signal Levels on Vinyl

There are limits to the signal level possible on a vinyl disc, and they impose maxima on the signal that a cartridge and its associated electronics will be expected to reproduce. The exact values of these limits may not be precisely defined, but the way they work sets the ways in which maximum levels vary with frequency, and this is of great importance.

There are no variable-gain controls on RIAA inputs because implementing an uneven but very precisely controlled frequency response and a suitably good noise performance are quite hard enough without adding variable gain as a feature. No doubt it could be done, but it would not be easy, and the general consensus is that it is not necessary. The overload margin, or headroom, is therefore of considerable importance, and it is very much a case of the more the merrier when it comes to the numbers game of specmanship. The issue can get a bit involved, as a situation with frequency-dependent vinyl limitations, and frequency-dependent gain is often further complicated by a heavy frequency-dependent load in the shape of the feedback network, which can

put its own limit on amplifier output at high frequencies. Let us first look at the limits on the signal levels which stylus-in-vinyl technology can deliver. In the diagrams that follow the response curves have been simplified to the straight-line asymptotes.

Figure 6.2a shows the physical groove amplitudes that can be put onto a disc. From subsonic up to about 1 kHz, groove amplitude is the constraint. If the sideways excursion is too great, the groove spacing will need to be increased to prevent one groove breaking into another, and playing time will be reduced. Well before actual breakthrough occurs, the cutter can distort the groove it has cut on the previous revolution, leading to "pre-echo" in quiet sections, where a faint version of the music you are about to hear is produced. Time travel may be fine in science fiction, but it does not enhance the musical experience. The ultimate limit to groove amplitude is set by mechanical stops in the cutter head.

There is an extra limitation on groove amplitude; out-of-phase signals cause vertical motion of the cutter, and if this becomes excessive it can cause it to cut either too deeply into the disc medium and dig into the aluminium substrate or lose contact with the disc altogether. An excessive vertical component can also upset the playback process, especially when low tracking forces are used; in the worst case the stylus can be thrown out of the groove completely. To control this problem the stereo signal is passed through a matrix that isolates the L-R vertical signal, which is then amplitude limited. This potentially reduces the perceived stereo separation at low frequencies, but there appears to be a general consensus that the effect is not audible. The most important factor in controlling out-of-phase signals is the panning of bass instruments (which create the largest cutter amplitudes) to the centre of the stereo stage. This approach is still advantageous with digital media, as it means that there are two channels of amplification to reproduce the bass information rather than one.

From about 1 kHz up to the ultrasonic regions, the limit is groove velocity rather than amplitude. If the disc cutter head tries to move sideways too quickly compared with its relative forward motion, the back facets of the cutter destroy the groove that has just been cut by its forward edges.

On disc replay, there is a third restriction—that of stylus acceleration, or to put it another way, groove curvature. This sets a limit on how well a stylus of a given size and shape can track the groove. Allowing for this at cutting time places an extra limitation on signal level, shown by the dotted line in Figure 6.2a. The severity of this

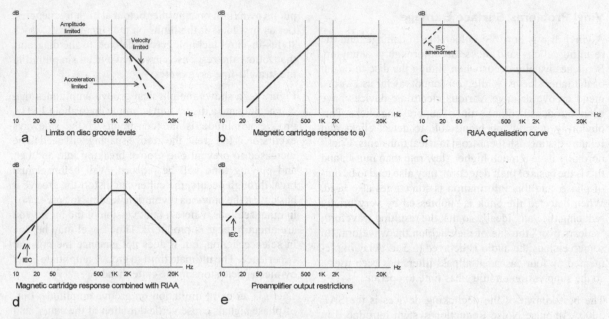

Figure 6.2 a) The levels on a vinyl disc; b) the cartridge response combined with the disc levels; c) the RIAA curve; d) the RIAA combined with curve b; e) possible preamplifier output restrictions

restriction depends on the stylus shape; an old-fashioned spherical type with a tip diameter of 0.7 mil requires a roll-off of maximum levels from 2 kHz, while a (relatively) modern elliptical type with 0.2 mil effective diameter postpones the problem to about 8 kHz. The limit however still remains.

Thus disc-cutting and playback technology put at least three limits on the maximum signal level. This is not as bad a problem as it might be, because the distribution of amplitude with frequency for music is not flat with frequency; there is always more energy at LF than HF. This is especially true of the regrettable phenomenon known as rap music. For some reason there seems to be very little literature on the distribution of musical energy versus frequency, but a very rough rule is that levels can be expected to be fairly constant up to 1 kHz and then fall by something like 10 dB/octave. The end result is that despite the limits on disc levels at HF, it is still possible to apply a considerable amount of HF boost which, when undone at replay, reduces surface noise problems. At the same time the LF levels are cut to keep groove amplitude under control. Both functions are implemented by applying the inverse of the familiar RIAA replay equalisation at cutting time. More on the limitations affecting vinyl levels can be found on Jim Lesurf's website.[12]

A reaction to the limitations of the usual 7-inch single was the 12-inch single, which appeared in the mid-1970s,

before CDs arrived. The much greater playing area allowed greater groove spacing and higher recording levels. I bought several of these, in the 45 rpm format, and I can testify that the greater groove speed gave a much clearer and less distorted high end, definitely superior to 33 rpm LPs.

Having looked at the limitations on the signal levels put onto disc, we need to see what we will get back when we replay it. This obviously depends on the cartridge sensitivity. That issue is dealt with in the next section, but it might as well be said now that in general MM cartridge sensitivity varies over a limited range of about 7 dB, while MC sensitivity variation is much greater.

Since MM input stages do not normally have gain controls, it is important that they can accept the whole range of input levels that occur. A well-known paper by Tomlinson Holman[13] quotes a the worst-case peak voltage from an MM cartridge of 135 mV at 1 kHz given by Huntley in Reference [14]. This is equivalent to 95 Vrms at 1 kHz. He says: "[T]his is a genuinely worst-case combination which is not expected to be approached typically in practice."

Shure are a well-known manufacturer of MM cartridges, and their flagship V15 phonograph cartridge series (the 15 in each model name referred to the cartridges' 15-degree tracking angle), for many years set the standard for low tracking force and high tracking ability. Its development necessitated much research into the maximum levels on vinyl. Many other workers also contributed in this field. The results are usually expressed in velocity (cm/s) as this eliminates the effect of cartridge sensitivity. I have boiled down the Shure velocity data into Table 6.1. I have included the acceleration of the stylus tip required for the various frequency/velocity pairs; this is not of direct use, but given that the maximum sustained acceleration the human body can withstand is around 3 g, it surely makes you think. Since the highest MM cartridge sensitivity for normal use is 1.6 mV per cm/s, (see next section) Table 6.1 tells us that we need to be able to handle an MM input of 1.6 × 38 = 61 mV rms. This is not far out of line with the 95 mV rms quoted by Holman, being only 3.8 dB lower.

The website of Jim Lesurf[15] has many contemporary measurements of maximum groove velocities. The maximum quoted is 39.7 cm/s, which gives 1.6 × 39.7 = 63.5 mV rms. Rooting through the literature, the Pressure Cooker discs by Sheffield Labs were recorded direct to disc and are said to contain velocities up to 40 cm/s, giving us 1.6 × 40 = 64 mV rms. It is reassuring that these maxima do not differ very much. On the other hand, the jazz record *Hey! Heard The Herd* by Woody Herman (Verve V/V6 8558, 1953) is said to have a peak velocity of 104 cm/s at 7.25 kHz,[16] but this seems out of line with all other data. If it is true, the input level from the most sensitive cartridge would be 1.6 × 105 = 166 mV rms.

So we may conclude that the greatest input level we are likely to encounter is 64 mV rms, though that 166 mV rms should perhaps not be entirely forgotten.

The maximum input a stage can accept before output clipping is set by its gain and supply rails. If we are using normal opamps powered from ±17V rails, we can assume an output capability of 10 Vrms. Scaling this down by the gain in each case gives us Table 6.2, which also shows the output level from a nominal 5 Vrms input.

Clearly if we want to accept a 64 mV rms input the gain cannot much exceed +40 dB. In fact a gain of +43.8 dB will just give clipping for 64 mV rms in. If we want to accept the Woody Herman 166 mV rms, then the

Table 6.2 Maximum input, overload margin, and nominal output for various MM preamp gains; all at 1 kHz

Gain dB	Gain times	Max input mV rms	Overload margin dB	5 V rms would be raised to:
50	316	32 mV	16 dB	1580 Vrms
45	178	56 mV	21 dB	890 Vrms
40	100	100 mV	26 dB	500 Vrms
35	56.2	178 mV	31 dB	281 Vrms
30	31.6	316 mV	36 dB	158 Vrms
25	17.8	562 mV	41 dB	89 Vrms
20	10	1000 mV	46 dB	50 Vrms

maximum permissible gain is +35.6 dB. I would suggest that a safety margin of at least 5 dB should be added, so we conclude that 30 dB (1 kHz) is an appropriate gain for an MM input stage; this will accept 316 mV rms from a cartridge before clipping. A recent review of a valve phono stage[17] described an MM input capability of 300 mV rms as "extremely generous", and I reckon that an input capability around this figure will render you immune to overload forever, and be more than adequate for the highest quality equipment. The stage output with a nominal 5 mV rms input is only 158 mV rms, which is not enough to operate your average power amplifier, and so there will have to be another amplifying stage after it. This must have variable gain or be preceded by a passive volume control, for otherwise it will clip before the first stage and reduce the overload margin.

While we must have a relatively low gain in the MM stage to give a good maximum signal capability, we do not want it to be too low, or the signal/noise ratio is likely to be degraded as the signal passes through later stages. It is one of the prime rules of audio that you should minimise the possibility of this by getting the signal up to a decent level as soon as possible, but it is common practice and very sensible for the MM output to go through a unity-gain subsonic filter before it receives any further amplification; this is because the subsonic stuff coming from the disc can be at disturbingly high levels.

Table 6.1 Maximum groove velocities from vinyl (after Shure)

	400 Hz	500 Hz	2 kHz	5 kHz	8 kHz	10 kHz	20 kHz
Velocity cm/s	26	30	38	35	30	26	10
Acceleration	66.5 g	96 g	487 g	1120 g	1535 g	1665 g	1281 g

In the history of preamplifiers MM input overload margin used to be a test of macho—this is less true now, as the changeover to MC cartridges with a wide range of output levels makes a single input overload figure much less meaningful.

Moving-Magnet Cartridge Sensitivities

Having looked at the limitations on the signal levels put onto disc, we need to see what levels we will get back when we replay it. The level reaching the preamplifier is clearly proportional to the cartridge sensitivity. Due to their electromagnetic nature MM cartridges respond to stylus velocity rather than displacement (the same applies to MC cartridges), so output voltage is usually specified at a velocity of 5 cm/sec. That convention is followed throughout this chapter.

A survey of 72 MM cartridges on the market in 2012 showed that they fall into two groups—what might be called normal hi-fi cartridges (57 of them) and specialised cartridges for DJ use (15 of them). The DJ types have a significantly higher output than the normal cartridges—the Ortofon Q-Bert Concorde produces no less than 11 mV, the highest output I could find. It seems unlikely that the manufacturers are trying to optimise the signal-to-noise ratio in a DJ environment, so I imagine there is some sort of macho "my cartridge has more output than yours" thing going on. Presumably DJ cartridges are also designed to be exceptionally mechanically robust. We will focus here on the normal cartridges, but to accommodate DJ types all you really need to do is allow for 6 dB more input level to the preamplifier.

The outputs of the 58 normal cartridges at a velocity of 5 cm/sec are summarised in the histogram of Figure 6.3.

The range is from 3.0 mV to 8.0 mV, with significant clumps around 4–5 mV and 6.5 mV. If we ignore the single 8.0 mV cartridge, the output range is restricted to 3.0 to 6.5 mV, which is only 6.7 dB. This is a very small range compared with the very wide one shown by MC cartridges and makes the design of a purely MM input a simpler matter. There is no need to provide different amplifier sensitivities, as a 6.7 dB range can be easily accommodated by adjustment of a volume control later in the audio path.

Overload Margins and Amplifier Limitations

The safety factor between a nominal 5 mV rms input and the clipping point may be described as either the input headroom in mV rms or the "overload margin", which is the dB ratio between the nominal 5 mV rms input and the maximum input. Table 6.2 shows that an MM stage with a +35 dB gain (1 kHz) gives an output of 280 mV, an input overload level of 178 mV rms and an overload margin of 31 dB, which might be called very good. A +30 dB (1 kHz) stage gives a nominal 158 mV out, an input overload level of 316 mV rms, and an overload margin of 36 dB, which is definitely excellent, giving 5 dB more headroom.

The maximum input capability of an MM stage is not always defined by simple frequency-independent clipping at its output. Things may be complicated by the stage output capability varying with frequency. An RIAA feedback network, particularly one designed with a relatively low impedance to reduce noise, presents a heavier load as frequency rises because the impedance of the capacitors falls. This heavy loading

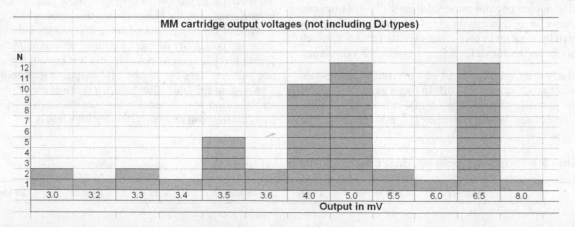

Figure 6.3 The output voltages for 57 MM cartridges, excluding specialised DJ types

at HF was very often a major cause of distortion and headroom-limitation in discrete RIAA stages that had either common-collector or emitter-follower output topologies with highly asymmetrical drive capabilities; for example an NPN emitter-follower is much better at sourcing current than sinking it. With conventional discrete designs the 20 kHz output capability, and thus the overload margin, was often reduced by 6 dB or even more. Replacing the emitter resistor of an emitter-follower with a current source much reduces the problem, and the very slight extra complication of using a push-pull Class-A output structure can bring it down to negligible proportions; for more details see Chapter 10 on discrete MM input design. Earlier opamps such as the TL072 also struggled to drive RIAA networks at HF, as well as giving a very poor noise performance. It was not until the advent of the 5532 opamp, with its excellent load-driving capabilities, that the problem of driving low-impedance RIAA networks was solved; the noise performance was much better, too. However, if a low-impedance HF correction pole (more on this later) is being driven as well, there may still be some slight loss of output capability at 20 kHz.

We saw in the earlier section on spurious signals that Tomlinson Holman concluded that to accommodate the worst of the worst, a preamplifier should be able to accept not less than 35 mV rms in the 3–4 Hz region.[7] If the IEC Amendment is after the preamplifier stage and C0 is made very large so it has no effect (see Chapter 7), the gain in the 2–5 Hz region will have flattened out at +19.9 dB, so the equivalent overload level at 1 kHz will be need to be 346 mV rms, which is rather high. The +30 dB (1 kHz) gain stages examined in Chapter 7 have a 1 kHz overload level of 316 mV rms, which is only 0.8 dB below this rather extreme criterion; we are good to go. Using a +35 dB (1 kHz) gain stage instead would significantly reduce the safety margins.

Further headroom restrictions may occur when not all of the RIAA equalisation is implemented in one feedback loop. Putting the IEC Amendment roll-off after the preamplifier stage (as in Figure 7.3) means that very low frequencies are amplified by 3 dB more at 20 Hz than they otherwise would be, and this is then undone by the later roll-off. This sort of audio impropriety always carries a penalty in headroom as the signal will clip before it is attenuated, and the overload margin at 20 Hz is reduced by 3.0 dB. This effect reduces quickly as frequency increases, being 1.6 dB at 30 Hz and only 1.0 dB at 40 Hz. Whether this loss of overload margin is more important than providing an accurate IEC Amendment response is a judgement call, but in my experience it creates no trace of any problem in an MM stage with

a gain of +30 dB (1 kHz). Passive-equalisation input architectures that put flat amplification before an RIAA stage suffer much more severely from this kind of headroom restriction, and it is quite common to encounter preamplifiers that claim to be high end, with a high-end price-tag but a very low-end overload margin of 20 to 22 dB. Bad show, chaps.

At the other end of the audio spectrum, adding an HF correction pole after the preamplifier to correct the RIAA response with low gains (see Figure 7.3 again) also introduces a compromise in the overload margin, though generally a much smaller one. A 30 dB (1 kHz) stage has a mid-band overload margin of 36 dB, falling to +33 dB at 20 kHz. Only 0.4 dB of this loss is due to the amplify-then-attenuate action of the HF correction pole, the rest being due to the heavy capacitative loading on A1 of both the main RIAA feedback path and the pole-correcting RC network. This slight compromise could be eliminated by using an opamp structure with greater load-driving capabilities, so long as it retains the low noise of a 5534A.

An attempt has been made to show these extra preamp limitations on output level in Figure 6.2e, and comparing Figure 6.2d, it appears that in practice they are almost irrelevant because of the falloff in possible input levels at each end of the audio band.

When the RIAA equalisation of Figure 6.2c is applied to the cartridge output of Figure 6.2b, the result looks like Figure 6.2d, with the maximum amplitudes occurring around 1–2 kHz. This is in agreement with Tomlinson Holman's data.[7]

Figure 6.2e shows some possible output level restrictions that may affect Figure 6.2d. If the IEC Amendment is implemented after the first stage, there is a possibility of overload at low frequencies which does not exist if the amendment is implemented in the feedback loop by restricting C0. At the high end, the output may be limited by problems driving the RIAA feedback network, which falls in impedance as frequency rises.

To put all this into some sort of perspective, here are the 1 kHz overload margins for a few of my published designs. My first preamplifier, the "Advanced Preamplifier",[18] achieved +39 dB in 1976, partly by using all-discrete design and ± 24V supply rails. A later discrete design in 1979[19] gave a tour-de-force +47 dB, accepting over 1.1 Vrms at 1 kHz, but I must confess this was showing off a bit and involved some quite complicated discrete circuitry, including the push-pull Class-A output stages mentioned in Chapter 10. Later designs such as the Precision Preamplifier[20] and its linear descendant the Precision Preamplifier '96[21] accepted the limitations of opamp output voltage in exchange for much greater convenience in most other directions and still have an excellent overload margin of 36 dB.

References

1. Linsley-Hood, J. "Modular Pre-Amplifier Design" *Wireless World*, July 1969, p. 306.

2. Burrows, B. "Ceramic Pickup Equalisation: 1" *Wireless World*, July 1971, p. 321.

3. Burrows, B. "Ceramic Pickup Equalisation: 2" *Wireless World*, Aug 1971, p. 379.

4. Sonotone history. www.roger-russell.com/sonopg/ sonopg.htm Accessed Nov 2016.

5. Sound-Smith. www.sound-smith.com/cartridges/ sg.html Accessed Nov 2016.

6. Jones, M. "Designing Valve Preamps: Part 1" *Electronics World*, Mar 1996, p. 192.

7. Holman, T. "Dynamic Range Requirements of Phonographic Preamplifiers" *Audio*, July 1977, p. 74.

8. Howard, K. www.stereophile.com/reference/arc_ angles_optimizing_tonearm_geometry/index.html Accessed June 2013.

9. Smith, A, and Miller, P. "Nakimichi TX-1000 turntable" *Hi-Fi News*, Aug 2016, pp. 118–123 (self-centering turntable).

10. Allmaier, H. "The Ins & Outs of Turntable Dynamics—and How They Mess Up Your Vinyl Playback" *Linear Audio*, Volume 10, Sept 2015, p. 14.

11. Godsill et al. www2.ece.ohio-state.edu/~schniter/ee 597/handouts/restoration_chapter.pdf Accessed Dec 2016.

12. Lesurf, J. www.audiomisc.co.uk/HFN/LP1/KeepIn Contact.html Accessed June 2013.

13. Holman, T. "New Factors in Phonograph Preamplifier Design".

14. Huntley, C. "Preamp Overload" *Audio Scene Canada*, Nov 1975, pp. 54–56.

15. Lesurf, J. www.audiomisc.co.uk/HFN/LP2/OnThe Record.html Accessed June 2013.

16. Boston Audio Society. www.bostonaudiosociety. org/pdf/bass/BASS-05-03-7612.pdf Accessed June 2013.

17. Miller, P. *Hi-Fi News*, Review of Canor TP306 VR+ phono stage, Aug 2013, p. 25.

18. Self, D. "An Advanced Preamplifier Design" *Wireless World*, Nov 1976.

19. Self, D. "High Performance Preamplifier" *Wireless World*, Feb 1979.

20. Self, D. "A Precision Preamplifier" *Wireless World*, Oct 1983.

21. Self, D. "Precision Preamplifier 96" *Electronics World*, July/Aug and Sept 1996.

Chapter 7

RIAA Equalisation

Equalisation and Its Discontents

Both moving-magnet and moving-coil cartridges operate by the relative motion of conductors and magnetic field, so the voltage produced is proportional to rate of change of flux. The cartridge is therefore sensitive to groove velocity rather than groove amplitude, and so its sensitivity is expressed as X mV per cm/sec. This velocity-sensitivity gives a frequency response rising steadily at 6 dB/octave across the whole audio band for a groove of constant amplitude.

The RIAA replay equalisation curve is shown in Figure 7.1. It has three basic corners in its response curve, with frequencies at 50.05 Hz, 500.5 Hz, and 2.122 kHz, which are set by three time-constants of 3180 μsec, 318 μsec, and 75 μsec. This is shown here diagrammatically, and in reality the response is a smooth curve with a wiggle around 1 kHz. The RIAA curve was of US origin but was adopted internationally with surprising speed, probably because everyone concerned was heartily sick of the ragbag of equalisation curves that existed previously. For example, the Leak Varislope II preamplifier (circa 1954) had equalisation options on the front panel labelled BRIT-78, BRIT-LP, COL-LP, RCA-Ortho, AES, and NARTB. Flat Tuner, Tape, Mic, and Auxiliary inputs were also provided. The RIAA became part of the IEC 98 standard, first published in 1964, and is now enshrined in IEC 60098, "Analogue Audio Disk Records and Reproducing Equipment".

In Figure 7.1, T3, T4, and T5 are the time-constants that define the basic RIAA curve, while $f3$, $f4$, and $f5$ are the equivalent frequencies. This is the naming convention used by Stanley Lipshitz in his landmark paper[1] and is used throughout this book. T2 is the extra time-constant for the IEC Amendment to the RIAA, introduced in 1976; and T1 shows where its effect ceases at very low frequencies when the gain is approaching unity at the low-frequency end due to C0. When the IEC is not implemented the LF roll-off point A is usually much lower in frequency, as shown here, to prevent it intruding on the defined RIAA response band of 20 Hz–20 kHz; T1 is therefore at an even lower frequency.

At the HF end, the final zero is at frequency f6, with associated time-constant T6, and because the gain of Figure 7.1 was chosen to be +35 dB at 1 kHz, it is quite a long way from 20 kHz and has very little effect at this frequency, giving an excess gain of only 0.10 dB. This error quickly dies away to nothing as frequency falls below 20 kHz.

Note the flat shelf between 500 Hz and 2 kHz. It may occur to you that a constant downward slope across the audio band would have been simpler, required fewer precision components to accurately replicate, and would have saved us all a lot of trouble with the calculations. But . . . such a response require 60 dB more gain at 20 Hz than at 20 kHz, equivalent to 1000 times. The minimum open-loop gain at 20 Hz would have to be 70 dB (3000 times) to allow even a feeble 10 dB of negative feedback at that frequency, and implementing that with a simple two-transistor preamplifier stage would have been difficult if not impossible. (Must try it sometime.) The 500 Hz–2 kHz shelf in the RIAA curve reduces the 20 Hz–20 kHz gain difference by 12 dB to only 48 dB, making a one-valve or two-transistor preamplifier stage practical, if not exactly a model of linearity. One has to conclude that the people who established the RIAA curve knew what they were doing there.

The Unloved IEC Amendment

Figure 7.1 shows an extra response corner at 20.02 Hz, corresponding to the T2 time-constant of 7950 μs. This extra LF roll-off is called the "IEC Amendment" and it was added to what was then IEC 98 in 1976. Its apparent intention was to reduce the subsonic output from the preamplifier, but its introduction is something of a mystery. It was certainly not asked for by either equipment manufacturers or their customers, and it was unpopular

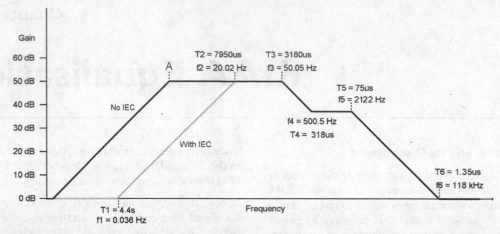

Figure 7.1 The response for series-feedback RIAA equalisation, with and without IEC Amendment T2, which gives an extra roll-off at 20.02 Hz. Frequency not to scale.

with both, with some manufacturers simply refusing to implement it. The likeliest explanation seems to be that several noise reduction systems, for example dbx, were being introduced for use with vinyl at the time, and their operation was badly affected by subsonic disturbances. None of these systems caught on.

The IEC Amendment still attracts negative comments today. On one hand it is pointed out that as an anti-rumble measure it is ineffective, as its slow 1st-order roll-off meant that the extra attenuation at 13 Hz, a typical cartridge-arm resonance frequency, was a feeble −5.3 dB; however at 4 Hz, a typical disc warp frequency, it did give a somewhat more useful −14.2 dB, reducing the unwanted frequencies to a quarter of their original amplitude. On the other hand there were loud complaints that the extra unwanted replay time-constant caused significant frequency response errors at the low end of the audio band, namely −3.0 dB at 20 Hz and −1.0 dB at 40 Hz.

Some of the more sophisticated preamplifiers allow The Amendment to be switched in or out; one example is the Audiolab 8000PPA phono preamplifier. It is hard to be sure because the topic rarely surfaces in reviews, but the general view is that the IEC Amendment has effectively been abandoned.

The "Neumann Pole"

The RIAA curve is only defined to 20 kHz, but by implication carries on down at 6 dB/octave forever. This implies a recording characteristic rising at 6 dB/octave forever, which could clearly endanger the cutting head

if ultrasonic signals were allowed through. From 1995 a belief began to circulate that record lathes incorporated an extra unofficial pole at 3.18 µs (50.0 kHz) to limit HF gain. This would cause a loss of 0.17 dB at 10 kHz and 0.64 dB at 20 kHz and would require compensation if an accurate replay response was to be obtained. The name of Neumann became attached to this concept simply because they are the best-known manufacturers of record lathes.

The main problem with this story is that it is not true. The most popular cutting amplifier is the Neumann SAL 74B, which has no such pole. For protection against ultrasonics and RF it has instead a rather more effective 2nd-order lowpass filter with a corner frequency of 49.9 kHz and a Q of 0.72,[2] giving a Butterworth (maximally flat) response rolling-off at 12 dB/octave. Combined with the RIAA equalisation this gives a 6dB/octave roll-off above 50 kHz. The loss from this filter at 20 kHz is less than −0.1 dB, so there is little point in trying to compensate for it, particularly because other cutting amplifiers are unlikely to have identical filters.

Opamp MM Disc Input Stages

Satisfactory discrete MM preamplifier circuitry is not that straightforward to design, and there is a lot to be said for using a good opamp, which if well chosen will have enough open-loop gain to implement the RIAA bass boost without introducing detectable distortion at normal operating levels. The 5534/5532 opamps have input noise parameters that are well suited to moving-magnet

(MM) cartridges. Figure 7.2 shows a basic phono amplifier using an opamp; be aware that it is by no means optimised, so don't warm up the soldering iron just yet. We will optimise it later; we will lower the gain and reduce the value of R0 to reduce its Johnson noise contribution, and the effect of opamp current noise flowing in it. The stage is designed for a gain of 35.0 dB at 1 kHz, which means a maximum input of 178 mV at the same frequency. With a nominal 5 mV rms input at 1 kHz the output is 280 mV. The RIAA accuracy is within ±0.1 dB from 20 Hz to 20 kHz, and the IEC Amendment is implemented by making C0 a mere 7.96 uF.

The component designations Cin, Rin, R0, R1, R2, R3, C0, C1, C2, and C3 are adhered to throughout this book. R3 and C3 are reserved for the HF correction pole.

Here some requirements for a good RIAA preamplifier:

1) Use a series-feedback RIAA network, as shunt feedback is approximately 14 dB noisier. See Chapter 9.

2) Chose a suitable gain at 1 kHz. To give a first-class overload margin, +30 dB is the upper limit. Chapter 6 discusses this in depth.

3) Make the RIAA equalisation accurate. My 1983 preamplifier was designed for ±0.2 dB accuracy from 20–20 kHz, the limit of the testgear I had at the time. This was tightened to +0.05 dB without using rare parts in my 1996 preamplifier. With software tools to evaluate the Lipshitz equations and intelligent tweaking of values to allow for finite open-loop gains, it is simple to achieve ±0.01 dB in SPICE simulation.

4) Use easily obtainable parts. Resistors are best from the E24 series, using two or three in parallel as required. The E96 series is now readily available but

less convenient; see Chapter 2. Capacitors will probably be from the E3, E6, or E12 series, so awkward values require a different approach, with more paralleling to get the required value.

5) R0 in Figure 7.2 should be kept low as its Johnson noise is effectively in series with the input signal. However its noise contribution is usually small compared with other noise sources. The value of R0 is more important when the MM preamplifier is fed from a low impedance, which typically only occurs when it is providing RIAA equalisation for the output of a flat MC preamplifier, rather accepting input direct from an MM cartridge with its high inductance.

6) The feedback RIAA network impedance to be driven must not be so low as to increase the distortion or limit the output swing of the amplifier. This impedance can fall to low values at HF.

7) The resistive path through the feedback arm (R1, R2 in Figure 7.1) should ideally have the same DC resistance as input bias resistor Rin, to minimise offsets at A1 output. This is clearly not the case in Figure 7.1. Rin can be split into two so the main part of the cartridge loading is inside the DC-blocking capacitor Cin, giving more freedom of choice in the bias resistor; this is described in the next section on input networks. It is something of a minor point as the voltage offset would have to be quite large to significantly affect the output voltage swing. It may not be possible to meet this constraint as well as other more important requirements like the stage gain. There is little point in trying to be super-precise in this because the bias currents for the two opamp inputs will not be exactly equal.

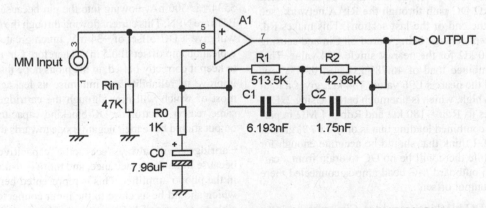

Figure 7.2 Series-feedback RIAA equalisation Configuration-A, IEC Amendment implemented by C0. Component values for 35.0 dB gain (1 kHz). Maximum input 178 mV rms (1 kHz). RIAA accuracy is within ± 0.1 dB from 20 Hz to 20 kHz, without an HF correction pole.

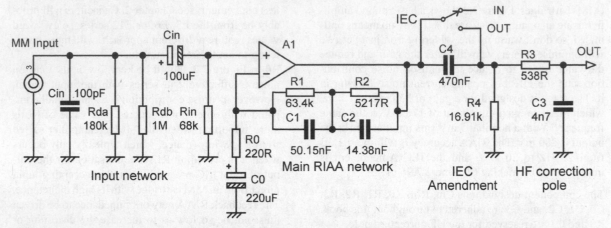

Figure 7.3 Series-feedback RIAA equalisation Configuration-A, redesigned for +30.0 dB gain (1 kHz) allowing a maximum input of 316 mV rms (1 kHz). The switchable IEC Amendment is implemented by C3, R3. HF correction pole R4, C4 is added. RIAA accuracy within ± 0.1 dB 20 Hz to 20 kHz.

Paying attention to these points, we can show an improved and more complete circuit in Figure 7.3.

The Input Network

Figure 7.2 shows the 47 kΩ input resistor that gives the cartridge its correct loading. However, details that are essential for practical use, like input DC-blocking capacitors, DC drain resistors, and EMC/cartridge-loading capacitors, have been omitted to keep things simple. These components are added in Figure 7.3. Cin prevents the bias current of the opamp flowing through the cartridge. The total 47 kΩ load is split between Rin and DC drain resistor Rd, so that Rin is 68 kΩ to match the 68.62 kΩ DC path through the RIAA network (see Point 7 at the end of the last section). This makes Rd, which prevents charge being stored on Cin and causing thumps, 150 kΩ for the nearest single E24 value. That gives a combined load of 46.789 kΩ, just 0.44% low. Making Rd the nearest E96 value, 154 kΩ, gives a total load 0.36% high, which is not much better. Using 2xE24 for Rd gives us Rda = 180 kΩ and Rdb = 1 MΩ in parallel, and a combined loading that is only 0.07 % greater than 47 kΩ. I think that should be accurate enough for anyone. While there will be no DC to drain from a cartridge, if an outboard MC head amp is connected there may be an output offset.

The input DC-blocking capacitor Cin must be large enough to have negligible impedance at 20 Hz compared with the cartridge. This will prevent a rise in low-frequency voltage noise due to the current noise passing through an increased impedance. The reactance of a 47 uF capacitor at 20 Hz is 169 Ω, compared with a typical MM cartridge resistance of 500 Ω, implying an increase in the effect current of current noise of 0.5 dB relative to 100 Hz, where the effect is negligible (remember that reactance is at right angles to resistance, so to speak). At 40 Hz the increase is less than 0.1 dB. Increasing Cin to 100 uF gives an increase of only 0.1 dB at 20 Hz, which seems to me to be quite good enough. Avoid making Cin too big physically, as it is susceptible to picking up hum capacitively.

Cin in Figure 7.3 has its negative side facing the opamp, and this no accident. The typical input bias current for a 5534A is 500 nA, flowing into the pin because the input BJTs are NPN. This current flowing through the 68 kΩ Rin will give a DC offset of −34 mV, much greater than the typical opamp offset of 0.5 mV. Therefore Cin is oriented to keep it correctly biased, in the pious hope that this will improve its reliability and minimise its leakage current, most of which will flow through the cartridge. For the same reason electrolytic DC-blocking capacitors on the output should have their negative side towards the 5534A.

Cartridges will always see some capacitive loading because of cable capacitance, and more is usually added in the phono amplifier. This is represented here by Cin, which should be as close to the input connector as possible to optimise EMC immunity. There is often provision for switching in extra capacitance. There is more on capacitance loading later in this chapter.

The input network components are omitted in most of the schematics that follow, for clarity.

Calculating the RIAA Equalisation Components

Calculating the values required for series-feedback configuration is not straightforward. You absolutely cannot take Figure 7.1 or 7.3 and calculate the time-constants of R2, C2 and R3, C3 as if they were independent of each other; the answers will be wrong. Empirical approaches (cut-and-try) are possible if no great accuracy is required, but attempting to reach even ±0.2 dB by this route is tedious, frustrating, and generally bad for your mental health.

The definitive paper on this subject is by Stanley Lipshitz.[2] This heroic work covers both series and shunt-feedback configurations and much more besides, including the effects of low open-loop gain. It is relatively straightforward to build a spreadsheet using the Lipshitz equations that allows extremely accurate RIAA networks to be designed in a second or two; the greatest difficulty is that some of the equations are long and complicated—we're talking real turn-the-paper-sideways algebra here—and some very careful typing is required.

My spreadsheet model takes the desired gain at 1 kHz and the value of R0, which sets the overall impedance level of the RIAA network. In my preamplifier designs the IEC Amendment is definitely *not* implemented by restricting the value of C0; this component is made large enough to have no significant effect in the audio band, and The Amendment roll-off is realised in the next stage.

Exact RIAA equalisation cannot be achieved with preferred component values, and that extends to E24 resistors and E12 capacitors. If you see any single-stage RIAA preamp where the equalisation is achieved by two E24 resistors and two E6 capacitors in the same feedback loop, you can be sure it is not very accurate. One wonders why the RIAA specifiers did not give us time-constants that could be implemented with preferred values.

Implementing RIAA Equalisation

It can be firmly stated from the start that the best way to implement RIAA equalisation is the traditional series-feedback method. So-called passive (usually only semi-passive) RIAA configurations suffer from serious compromises on noise and headroom. For completeness they are dealt with in Chapter 5.

There are several different ways to arrange the resistors and capacitors in an RIAA network, all of which give identically exact equalisation when the correct component values are used. Figures 7.1 and 7.3 show a series-feedback MM preamp built with what I call RIAA Configuration-A, which has the advantage that it makes the RIAA calculations somewhat easier, but otherwise is not the best; there will be much more on this topic later in this chapter. You will note with apprehension that in both circuits only one of the RIAA components, R0, is a standard value, and that is because it was used as the input to the RIAA design calculations that defined the overall RIAA network impedance and so could be chosen arbitrarily. This is always the case for accurate RIAA networks. Here, even if we assume that capacitors of the exact value could be obtained and we use the nearest E96 resistor values, systematic errors of up to ±0.06 dB will be introduced. Not a long way adrift, it's true, but if we are aiming for an accuracy of ± 0.1 dB it's not exactly a good start. If E24 resistors are the best available the errors grow to a maximum of ±0.12 dB, and we have not considered component tolerances—we are assuming the components are exact. If we resort to the nearest E12 value (which really shouldn't be necessary these days), then the errors exceed ±0.7 dB at the HF end. And what about those capacitors?

The answer to this is that by using multiple components in parallel or series we can get pretty much what value we like, and it is perhaps surprising that this approach is not adopted more often. The reason is probably cost—a couple of extra resistors are no big deal but extra capacitors make more of an impact on the costing sheet. The use of multiple components also improves the effective tolerance of the total value, as described in Chapter 2. There is also more on this important topic later in this chapter.

The only drawback to the series-feedback RIAA configuration is what might be called the unity-gain problem. While the RIAA equalisation curve is not specified above 20 kHz, the implication is clear that it will go on falling indefinitely at 6 dB/octave. A series-feedback stage cannot have a gain of less than unity, so at some point the curve will begin to level out and eventually become flat at unity gain.

The HF Correction Pole

However . . . if the gain of the stage is set lower than about +40 dB (1 kHz) to maximise the input overload margin, the 6 dB/octave fall tends to level out at unity early enough to cause significant errors in the audio band. Adding a HF correction pole (i.e. lowpass time-constant) just after the input stage makes the simulated and measured frequency response exactly correct. It is not a question of bodging the response to make it roughly right. If the correction pole frequency is correctly chosen then its roll-off in amplitude and phase cancels *exactly* with the "roll-up" of the final zero at f6.

An HF correction pole R4, C4 is demonstrated in Figure 7.3, where several other important changes have been made compared with Figure 7.1.

1) Most importantly, the gain has been reduced to +30 dB (1 kHz) to get more overload margin. With a nominal 5 mV rms input at 1 kHz the output will now be 158 mV.
2) This reduction in gain means that the final zero f6 in Figure 7.3 is now at 66.4 kHz, much closer in, and it introduces an excess gain at 20 kHz of 0.38 dB, which is too much to ignore if you are aiming to make high-class gear. The HF correction pole R4, C4 is therefore added, which solves the problem completely. Since there are only two components and no interaction with other parts of the circuit, we have complete freedom in choosing C4, so we use a standard E3 value and then get the pole frequency exactly right by using a 2xE24 pair of resistors for R3. Since these components are only doing a little fine tuning at the top of the frequency range, the tolerance requirements are somewhat relaxed compared with the main RIAA network. The design considerations are a) that the resistive section R4 should be as low as possible in value to minimise Johnson noise and, on the other hand, b) that the shunt capacitor C4 should not be large enough to load the opamp output excessively at 20 kHz.
3) The IEC Amendment is no longer implemented by C0; if it was then the correct value of C0 would be 36.18 uF, and instead it has been made 220 uF so that its associated −3 dB roll-off does not occur until 3.29 Hz. Even this wide spacing introduces an unwanted 0.1 dB loss at 20 Hz, and perfectionists will want to use 470 uF here, which reduces the error to 0.06 dB.
4) The overall impedance of the RIAA network has been reduced by making R0 220 Ω to reduce Johnson noise from the resistors and the effect of opamp current noise; the component values are no less awkward.

Implementing the IEC Amendment

The unloved IEC Amendment was almost certainly intended to be implemented by restricting the value of

the capacitor at the bottom of a series-feedback arm, i.e. C0 in Figures 7.1 and 7.3. While electrolytic capacitors nowadays (2013) have relatively tight tolerances of ±20%, in the 1970s you would be more likely to encounter −20% +50%, the asymmetry reflecting the assumption that electrolytics would be used for noncritical coupling or decoupling purposes where too little capacitance might cause a problem, but more than expected would be fine. This wide tolerance meant that there could be significant errors in the LF response due to C0. Figure 7.4 shows the effect of a ±20% C0 tolerance on the RIAA response of a preamplifier similar to Figure 7.3, with a gain of +30 dB (1 kHz) and C0 = 36.13uF. The gain will be +0.7 dB up at 20 Hz for a +20% C0, and −1.1 dB down at 20 Hz for a −20% C0. The effect of C0 is negligible above 100 Hz, but this is clearly not a good way to make accurate RIAA networks.

To get RIAA precision it is necessary to implement the IEC amendment separately with a nonelectrolytic capacitor, which can have a tolerance of ±1 % if necessary. In several of my designs the IEC amendment has been integrated into the response of the subsonic filter that immediately follows the RIAA preamplifier; this gives economy of components but means that it is not practicable to make it switchable in and out. Unless buffering is provided, R3 of the HF correction pole will be loaded by the subsonic filter, causing an early roll-off that degrades RIAA accuracy in the 20–100 Hz region.

The best solution is a passive CR highpass network after the preamplifier stage. We make C0 large to minimise its effect and add a separate 7950 μs time-constant after the preamplifier, as shown in Figure 7.3, where R4 and C4 give the required −3 dB roll-off at 20.02 Hz. Once again we can use a standard E3 capacitor value, and 470 nF has been chosen here, and once again an unhelpful resistor value results; in this case 16.91 kΩ. With E24 values, this can be implemented exactly as 16 kΩ + 910 Ω, but using near-equal series or parallel resistor pairs will reduce the effective tolerance, so a better solution would be 30 kΩ in parallel with 39 kΩ, which is only +0.27% high and gives almost all the improvement in effective tolerance that is possible. The IEC switch as shown will not be entirely click-free because of the offset voltage at A1 output, but that is relatively unimportant as it will probably only be operated a few times in the life of the equipment. If ever.

When C0 is made large and the IEC amendment is done later, we find C0 still has some effect. Since it is not infinite in value it will cause a roll-off of gain at some frequency. If we make C0 220 uF, which will

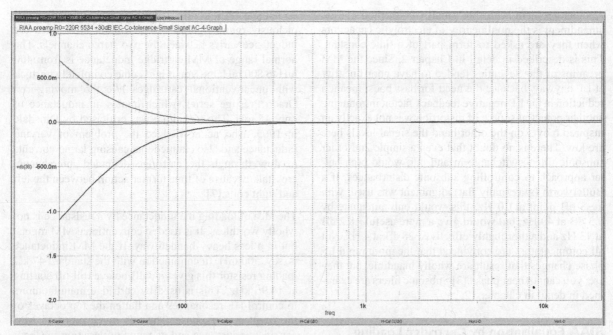

Figure 7.4 The effect of a ±20% tolerance for C0 when it is used to implement the IEC amendment

be a handily compact component, there is an error of −0.128 dB at 20 Hz (assuming no IEC amendment is used). If C0 is +20% high in value the error is reduced to −0.094 dB, and if it is −20% low the error is increased to −0.192 dB. Making C0 larger, such as 470 uF, reduces the basic error to a rather small −0.040 dB, and the variability due to its tolerance becomes negligible. A C0 of 470 uF is a reasonable size at 6V3 rating, which is quite adequate for a component that is only exposed to opamp offset voltages. Going to 1000 uF or even 2200 uF starts to make significant demands on PCB area and gains very little extra precision. This is summarised in Table 7.1. Most of the design examples in this chapter use C0 = 220 uF, but feel free to use 470 uF if you prefer; no other changes are required.

It is possible to compensate for the effect of C0 by tweaking the IEC amendment. In Figure 7.3 the 220 uF value for C0 gives an error of −0.128 dB at 20 Hz. If R3 is changed from 16.91 kΩ to 17.4 kΩ the overall response is made accurate to ±0.005 dB. The compensation is not mathematically exact—there is a +0.005 dB hump around 20 Hz—but I suggest it is good enough for most of us. This process does not of course do anything to reduce the effects of the tolerance of C0 and is not usable if it desired to make the IEC amendment switchable in/out. If a subsonic filter is used, it is probably starting to

Table 7.1 Effect of C0 with ±20% tolerance on RIAA accuracy at 20 Hz. Preamp gain +30 dB, (1 kHz) R0 = 220R

Nominal C0 value	C0 nominal	C0 +20%	C0 −20%
100 uF	−0.542 dB	−0.385 dB	−0.806 dB
220 uF	−0.128 dB	−0.094 dB	−0.192 dB
470 uF	−0.040 dB	−0.032 dB	−0.054 dB
1000 uF	−0.020 dB	−0.019 dB	−0.024 dB
2200 uF	−0.016 dB	−0.015 dB	−0.017 dB

take action at 20 Hz, and so small RIAA errors at this frequency are likely to be irrelevant.

The IEC network should come before the HF correction pole, as in Figure 7.3, so that R4 is not loaded by R3, which would cause a 0.3 dB loss; a small amount, perhaps, but you would have to recover it somewhere. Instead C3 is loaded by C4, but this has much less effect. The −0.3 dB figure assumes there is no significant external loading on the output at C4. Often the stage will be feeding the high-impedance input of a noninverting gain stage, but if not some sort of buffering may be required so the two output networks behave as designed.

Another problem with the "small C0" method of IEC amendment is the nonlinearity of electrolytic capacitors when they are asked to form part of a time-constant. This is described in detail in Chapter 2. Since the MM preamps of the Seventies tended to have poor linearity at LF anyway, because the need for bass boost meant a reduction in the LF negative feedback factor, introducing another potential source of distortion was not exactly an inspired move; on the other hand the signal levels here are low. There is no doubt that even a simple 2nd-order subsonic filter, switchable in and out, would be a better approach to controlling subsonic disturbances. If a Butterworth (maximally flat) alignment was used, with a −3 dB point at 20 Hz, this would only attenuate by 0.3 dB at 40 Hz, but would give a more useful −8.2 dB at 13 Hz and a thoroughly effective −28 dB at 4 Hz. Not all commentators are convinced that the more rapid LF phase changes that result are wholly inaudible, but they are; you cannot hear phase.[3] Subsonic filters are examined in detail in Chapter 12.

RIAA Equalisation by Cartridge Loading

It is possible to implement the LF boost part of the RIAA characteristic by loading the cartridge inductance with a relatively low amplifier input resistance, giving a 6 dB/octave slope. As frequency increases, the impedance of the inductance increases and the current into the input decreases.

This idea goes back a long way. The first use of it with a transistor amplifier I am aware of is in a 1961 preamplifier design by Tobey and Dinsdale,[4] where the input stage was a single transistor with shunt feedback around it to implement the HF part of the RIAA curve; the cartridge was loaded with a 3.9 kΩ series input resistor rather than the standard 47 kΩ. There was another example in 1963 where a 6.8 kΩ input resistor was used with a two-transistor shunt-feedback amplifier.[5] The idea may well have been used in valve circuitry long before that. The notion has resurfaced many times since, most recently due to Bob Cordell in Jan Didden's *Linear Audio*.[6]

Unfortunately there is a crippling snag; the LF equalisation now depends crucially on the cartridge inductance, so if you change your cartridge you have to redesign your preamplifier. You can of course make the loading variable, with a knob labelled "cartridge inductance", but this assumes you actually know the cartridge inductance, and know it accurately. Any inaccuracy in dialling in the inductance is directly reflected in errors in the RIAA response. Most people would have to rely on the manufacturer's specification for inductance (a few

do not specify it at all), and this is often quoted in suspiciously round figures. I also wonder how much the inductance varies between the two stereo channels. The normal range of MM cartridge inductance is from 400 mH to 800 mH, but you might come completely unstuck with unconventional cartridges like the moving-iron Grado Prestige series, which quotes an inductance of only 45 mH. The technique was criticised by Dinsdale in 1965, who acknowledged the problem of varying inductance and also claimed that causing larger currents to flow through the cartridge degraded interchannel crosstalk because of transformer action between the left and right coils.[7]

The idea of loading the inductance by a resistance is not wholly worthless. It is used in conventional MM inputs, but in a less heavy-handed way. If the MM inductance is, say, 500 mH, in conjunction with the standard 47 kΩ loading resistor this gives a 6 dB/octave roll-off starting at 14.96 kHz. This is used by cartridge manufacturers to control HF resonances and flatten the top octaves of the frequency response. There is much more on MM cartridge inductance and its range of variation in Chapter 11 on MM input noise, because it has a major effect on this area of performance.

It is worth noting that guitar pickups, which have substantial series inductance, are always operated into a high impedance on the order of 500 kΩ to 1 MΩ to avoid the loss of high frequencies. The capacitance of those long curly leads was often a problem, but now many guitarists use radio links.

RIAA Series-Feedback Network Configurations

There are four possible configurations described by Lipshitz in his classic paper.[2] These, with his component values, are shown in Figure 7.5; the same identifying letters have been used. They are all accurate to within ±0.1 dB when implemented with a 5534 opamp, but in the case of Figure 7.5A the error is getting close to −0.1 dB at 20 Hz due to the relatively high closed-loop gain (+46.4 dB at 1 kHz) and the finite open-loop gain of the 5534. All have RIAA networks at a relatively high impedance. They all have relatively high gain and therefore a low maximum input. The notation R0, C0, R1, C1, R2, C2 is as used by Lipshitz; C1 is always the larger of the two. In each case the IEC amendment is implemented by the value of C0.

In recent years I have always used Configuration-A, mainly because long ago I wrote a design tool to

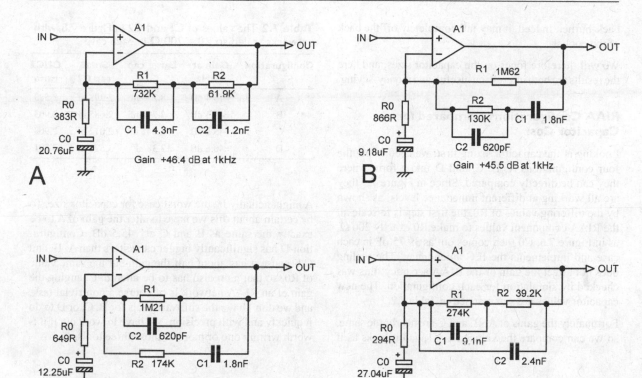

Figure 7.5 The four RIAA feedback configurations in the Lipshitz paper, identified by letter

implement the Lipshitz equations for it. I choose A simply because it was the easiest mathematical case. To repeat that for the other three configurations would be a significant amount of work, so the question arises, do any of the other three configurations have advantages that might make that work worthwhile?

Two things to examine come to mind:

First, each configuration in Figure 7.5 contains two capacitors, a large C1 and a small C2, that set the RIAA response. If they are close tolerance (to get accurate RIAA) and nonpolyester (to prevent capacitor distortion) then they will be expensive, so if there is a configuration that makes the large capacitor smaller, even if it is at the expense of making the small capacitor bigger, it is well worth pursuing. The large capacitor C1 is probably the most expensive component in the RIAA MM amplifier by a large margin.

Second, the signal voltages across each capacitor are going to be different. If polyester capacitors must be used for cost reasons, then if there is a configuration that

puts less voltage across a capacitor, that capacitor will generate less distortion. Capacitor distortion at least triples, and may quadruple, as the voltage across it doubles, so choosing the configuration that minimises the voltage is worthwhile. However, this issue can be dropped right away, as a good deal of simulation tells us that there is actually very little to choose between the four configurations as regards the signal voltage across the capacitors. This is not a helpful result or an interesting result, but sometimes things just are that way.

A related but different question is, if we assume a certain amount of nonlinearity in one or both of the RIAA capacitors, are the configurations different in their sensitivity to that nonlinearity? In other words, how much distortion will appear at the output? This question could be resolved in simulation, by using nonlinear capacitor models constructed with analogue behavioural modelling, but it would be a lot of work, and since the emphasis of this book is on high quality, where we can presumably afford a polypropylene capacitor or two, or a few polystyrene caps, I have put that one on the back-burner. Indeed, it may fall completely off the back of the cooker.

We will therefore focus on the capacitor sizes, and here the results are both interesting, useful, and money-saving.

RIAA Configurations Compared for Capacitor Cost

Looking at the capacitor sizing first, we need to put the four configurations A, B, C, and D into a form where they can be directly compared. Since in Figure 7.5 they are all working at different impedance levels, as shown by the differing values of R0, the first step is to scale all the RIAA component values to make R0 exactly 200 Ω, as in Figure 7.6. C0 then comes out as 39.75 uF in each case and implements the IEC amendment. The scaling does not affect the gain or the RIAA accuracy; this was checked by simulation for each configuration. The new capacitor values are summarised in Table 7.2.

Fortunately the gains of A, B, and C are nearly the same, so we can compare the values for C1, and it looks as if

Table 7.2 The values of C1 and C2 in Figure 7.8, with networks scaled so R0 = 200 Ω in each case

Configuration	Gain at 1 kHz	Large cap C1	Small cap C2	C1/C2 ratio
A	+46.4 dB	8.235 nF	2.298 nF	3.583
B	+45.5 dB	7.794 nF	2.685 nF	2.903
C	+45.5 dB	5.841 nF	2.012 nF	2.903
D	+40.6 dB	13.38 nF	3.528 nF	3.791

A might actually be the worst case for capacitor size. To be certain about this we have to alter the gain of A to be exactly the same as B and C at +45.5 dB. Configuration-D has significantly bigger capacitors than A, B, and C because it has about half the gain but the same value of R0, so that gain also has to be altered. Changing the gain of an RIAA network is of course a nontrivial task, and we don't have the software tools for B, C, or D to do it quickly and with precision. We have to work out if it is worth writing one or more of those three tools.

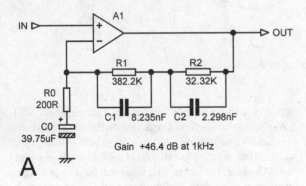

A

Gain +46.4 dB at 1kHz

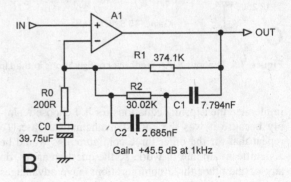

B

Gain +45.5 dB at 1kHz

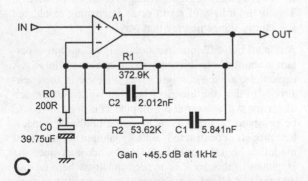

C

Gain +45.5 dB at 1kHz

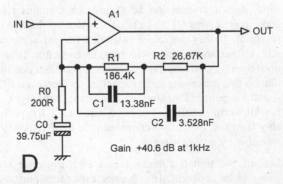

D

Gain +40.6 dB at 1kHz

Figure 7.6 The four RIAA feedback configurations, with component values scaled so that R0 = 200 Ω in each case

Not having the tools, we can change the gain simply by scaling the RIAA network values of A and D, with R0 kept constant. We must accept that the results may not be very accurate but should be good enough for us to judge which configuration is superior. We need to reduce the gain of A by 0.899 dB, or a factor of 1.109 times; we therefore multiply the capacitors C1 and C2 by this factor and divide the resistors R1 and R2 by it. For D, we want to increase the gain by 4.948 dB, or 1.767 times, so now we divide the capacitors C1 and C2 by this factor and multiply the resistors R1 and R2 by it. This gives the values shown in Figure 7.7 and Table 7.3. The C1/C2 ratios are unchanged.

After this process Configuration-A, though less accurate than before, is still within a ±0.1 dB error band; a completely accurate version with the same gain was calculated directly from the Lipshitz equations and is shown in Figure 7.8A; note that the values of C1, C2, and R1. R2 are all slightly different, as you cannot change RIAA gain simply by scaling against R0 and get

Table 7.3 The values of C1 and C2 as in Figure 7.7, after scaling so R0 =200 Ω and gain = +45.5 dB (1 kHz) for all configurations

Configuration	Gain at 1 kHz	Large cap C1	Small cap C2	C1/C2 ratio
A	+45.5 dB	9.132 nF	2.549 nF	3.583
B	+45.5 dB	7.794 nF	2.685 nF	2.903
C	+45.5 dB	5.841 nF	2.012 nF	2.903
D	+45.5 dB	7.567 nF	1.996 nF	3.791

the exactly correct result. For Configuration-D, which has undergone a greater gain change, the RIAA errors now exceed 0.1 dB (though not by much) at several frequencies. This is accurate enough to allow assessment of the configurations.

It is immediately obvious from Figures 7.7 and 7.8, and Table 7.3, that C1 in Configuration-C is only 64% the

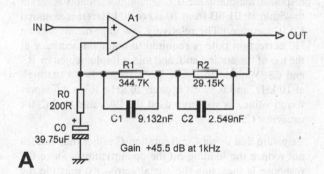

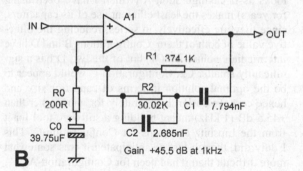

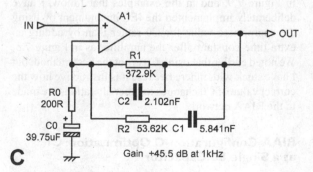

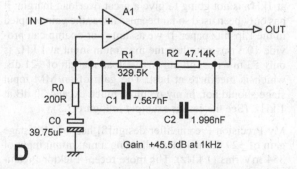

Figure 7.7 The RIAA feedback configurations, with component values scaled so that R0 = 200 Ω and the gain is +45.5 dB at 1 kHz in each case. Note that the RIAA response of A and D here is not wholly accurate. Maximum input in each case is only 53 mV rms (1 kHz), which is not generally adequate.

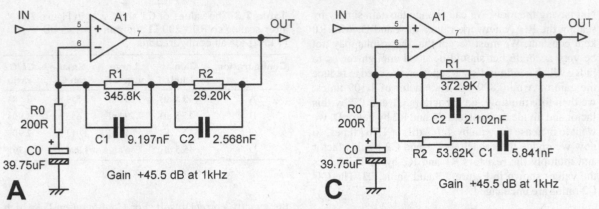

Figure 7.8 Configuration-A with values calculated from the Lipshitz equations to give accurate RIAA response. Configuration-C from Figure 7.7 shown for comparison; C1 in Configuration-A is much larger than C1 in Configuration-C, so the latter is more economical. Gain +45.5 dB at 1 kHz for both.

size of C1 in Configuration-A. I was afraid that this might be accompanied by an increase in C2 in Configuration-C, but this is also smaller at 81%. Unhappily it looks as if Configuration-A (which I have been using for years) makes the least efficient use of its capacitors, since they are effectively in series, reducing the effective value of both of them. Configurations B and D have intermediate values for C1, but of the two D has a significantly smaller C2. Configuration-C would appear to be the optimal solution in terms of capacitor size and hence cost. To design it accurately for gains other than +45.5 dB (1 kHz) meant building a software tool for it from the Lipshitz equations for Configuration-C. This I duly did, though just as anticipated it was somewhat more difficult than it had been for Configuration-A.

While Configuration-C in Figures 7.7 and 7.8 has come out as the most economical, our work here is not done. It will not have escaped you that a gain as high as +45.5 dB at 1kHz is not going to give a great overload margin; it has only been used so far because it was the gain adopted in the Lipshitz paper. If we assume our opamp can provide 10 Vrms out, then the maximum input at 1 kHz is only 53 mV rms, giving an overload margin of +21 dB, which is mediocre at best. The gain of an MM input stage should not, in my opinion, much exceed 30 dB at 1 kHz. (See the earlier example in Figure 7.3).

My Precision Preamplifier design[8] has an MM stage gain of +29 dB at 1 kHz, allowing a maximum input of 354 mV rms (1 kHz). The more recent Elektor Preamplifier 2012[9] has an MM stage gain of +30 dB (1 kHz), allowing a maximum input of 316 mV rms; it is followed by a flat switched-gain stage which allows for the large range in MC cartridge sensitivity.

I used the new software tool for Configuration-C to design the MM input stage in Figure 7.9, which has a gain of +30 dB (1 kHz). This design has an RIAA response, including the IEC amendment, that is accurate to within ±0.01 dB from 20 Hz to 20 kHz. (It is assumed C0 is accurate.) The relatively low gain means that an HF correction pole is required to maintain accuracy at the top of the audio band, and this is implemented by R3 and C3. Without this pole the response is 0.1 dB high at 10 kHz, and 0.37 dB high at 20 kHz. R3 is a nonpreferred value, as we have used the E6 value of 2n2 for capacitor C3.

Be aware that using Configuration-C rather than A does not reduce the loading on the opamp output. Since the response is the same the signal across R0 must be the same, and so the current going through the RIAA network is the same.

In Figure 7.9, and in the examples that follow, I have deliberately implemented the IEC amendment by using the appropriate value for C0 rather than by adding an extra time-constant after the amplifier as in Figure 7.3. We noted earlier that using C0 is not the best method, but I have stuck with it here because it is instructive how the correct value of C0 changes as other alterations are made to the RIAA network.

RIAA Configuration-C Optimisation: C1 as a Single E6 Capacitor

Looking at Figure 7.9, a further stage of optimisation is possible after choosing the best RIAA configuration. There is nothing magical about the value of R0 at 200 Ω (apart from the bare fact that it's an E24 value);

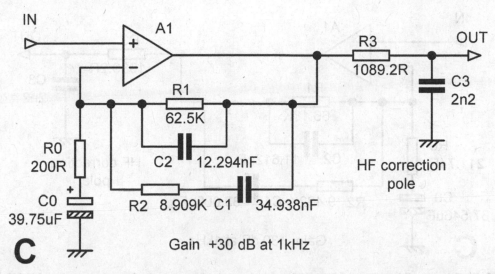

Figure 7.9 Configuration-C with values calculated from the Lipshitz equations to give +30.0 dB gain at 1 kHz and an accurate RIAA response within ±0.01 dB; the lower gain now requires HF correction pole R3, C3 to maintain accuracy at the top of the audio band

it just needs to be suitably low for a good noise performance, so it can be manipulated to make at least one of the capacitor values more convenient, the larger one being the obvious candidate. Compared with the potential savings on expensive capacitors here, the cost of a nonpreferred value for R0 is negligible. It is immediately clear that C1, at 34.9 nF, is close to 33 nF. If we twiddle the new software tool for Configuration-C so that C1 is exactly 33 nF, we get the arrangement in Figure 7.10. R0 has only increased by 6%, and so the effect on the noise performance will be quite negligible. All the values in the RIAA feedback network have likewise altered by about 6%, including C0, but the HF correction pole is unchanged; we would only need to alter it if we altered the gain. The RIAA accuracy of this version is well within ±0.01 dB from 20 Hz–20 kHz when implemented with a 5534.

The circuit of Figure 7.10 now has two preferred-value capacitors, C1 and C3, but that is the most we can manage. All the other component values are, as expected, thoroughly awkward. The best ways of combining components to get any value are described in detail in Chapter 2 and are summarised here:

1) The traditional approach was to use E12 resistor values and keep your fingers crossed. This can lead to quite serious RIAA errors. The tolerance was often 5% at best, which increased the errors. When E24 resistors became freely available this policy was updated to using the nearest E24 values, which was

better but not good enough for quality work. I would call these formats 1xE12 and 1xE24.

2) The most common approach to this problem today is to use the nearest E96 value; this is simple, but the way that requires the least effort is rarely the most effective. The effective tolerance is just that of the resistor series chosen. Despite the close spacing of the nominal values, around 2%, E96 resistors are often available at 1% tolerance. I call this format 1xE96. See Table 7.4; there are errors of almost 1% in the nominal value. Adding in a 1% tolerance makes matters worse.

3) Use two E24 1% resistors in parallel, making them as equal as possible to get the best reduction in effective tolerance. It is often necessary to balance accuracy of nominal value against reduction of effective tolerance.

A criterion that the nominal value should be accurate to better than half of the resistor tolerance was used here; once that was achieved, reduction in effective tolerance was pursued. I call this format 2xE24.

4) Using three E24 1% resistors in parallel not only allows us to get much closer to a desired nominal value but also gives a better chance of getting near-equal resistors that give most of the potential 1/√3 (= 0.577) improvement in effective tolerance, because there are more combinations. I call this format 3xE24. The design process is not obvious; I used a Willmann table, which lists, in order of combined value, all combinations of three E24 resistors that

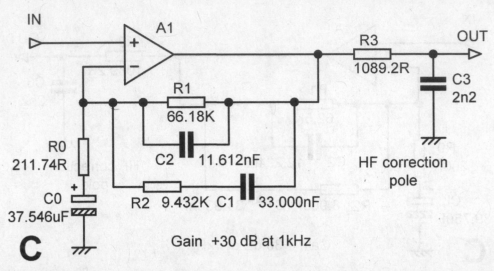

IN

A1

R3
1089.2R

OUT

C3
2n2

R1
66.18K

R0
211.74R

C2   11.612nF

C0
37.546uF

R2 9.432K C1   33.000nF

HF correction
pole

C

Gain +30 dB at 1kHz

Figure 7.10 Configuration-C from Figure 7.9, with R0 tweaked to make C1 exactly the E6 preferred value of 33.000 nF. Gain is still 30.0 dB at 1 kHz, and RIAA accuracy within ±0.01 dB. The HF correction pole R3, C3 is unchanged.

Table 7.4 Approximation to the exact values in Figure 7.10 by using 1xE96 resistor

Component	Desired value	E96 value	Nominal error
R0	211.74 Ω	210 Ω	−0.82 %
R1	66.18 kΩ	66.5 kΩ	−0.48 %
R2	9.432 kΩ	9.53 kΩ	+1.04 %
R3	1089.2 Ω	1100 Ω	+0.99 %

give a combined value within a specified decade. Effectively the very large number of combinations available have already been evaluated, and it is only necessary to pick the best one. This book only makes use of the 3xE24 Willmann table; there are however many more that list E12, E48, and E96 combinations, etc. Gert Willmann intends to make the tables available as free software under the terms of the GNU Lesser General Public License (LGPL); for more details, see www.gnu.org/licenses/. By the time this book is published the tables will be available free of charge either on my website or an alternative site.

In Chapter 2 I describe how to make up arbitrary resistor values by paralleling two or more resistors, and how the optimal way to do this is with resistors of as nearly equal values as you can manage. The individual tolerance errors partly cancel, so two equal resistors, whether in series or in parallel, have the effective tolerance reduced

by a factor of $\sqrt{2}$, while three have it reduced by $\sqrt{3}$, and so on. The resistors are assumed to be E24, and the parallel pairs were selected using a specially written software tool.

In selecting the resistor pairs in this chapter, I used these rules:

1) The nominal value of the combination shall not differ from the desired value by more than half the component tolerance. For ±1% parts this means within ±0.5%.
2) Having satisfied Rule 1, the resistors are to be as near equal as possible to get the maximum improvement possible over the tolerance of a single component.
3) The E24 series of preferred values will be used. The tolerance is assumed to be ±1% unless otherwise stated.

In the 2xE24 Table 7.5, R2 only just squeaks in past Rule 1, but on the other hand its near-equal values of 18 kΩ and 20 kΩ give almost all of the possible $\sqrt{2}$ improvement in effective tolerance. R3 also gets a near-optimal reduction in effective tolerance. Remember we are dealing here with nominal values, and the % error in the nominal value shown in the rightmost column has nothing to do with the resistor tolerances.

Just looking at these four examples you can see that the 2xE24 format gives noticeably more accurate nominal values than the 1xE96 method. This is generally the case; see Chapter 12 on subsonic filters, where the same

Table 7.5 Approximation to the exact values in Figure 7.10 by using 2xE24 resistors, giving Figure 7.11

Component	Desired value	Actual value	Parallel part A	Parallel part B	Nominal error	Effective tolerance
R0	211.74 Ω	211.03 Ω	360 Ω	510 Ω	−0.33 %	0.72%
R1	66.18 kΩ	65.982 kΩ	91 kΩ	240 kΩ	−0.30 %	0.78%
R2	9.432 kΩ	9.474 kΩ	18 kΩ	20 kΩ	+0.44 %	0.71%
R3	1089.2 Ω	1090.9Ω	2 kΩ	2.4 kΩ	+0.16 %	0.71%

issue of awkward resistor values has to be addressed. There 36 nominal values were dealt with, and for these the average absolute error for 1xE96 was 0.805%, for 2xE24 was 0.285%, and for 3xE24 was only 0.025%. So 2xE24 was three times better, and 3xE24 ten times better again. This nicely matches our four results here.

It is also assumed that the capacitors C1, C2, and C3 are of 1% tolerance; if this not the case then the effective tolerance column is simply scaled proportionally. The three-part combination for C2, which I have assumed to be restricted to E6 values, was done by manual bodging, though as you can see from Table 7.6, we have been rather lucky with how the values work out, with only three components getting us very close to the exact value we want for C2. Figure 7.11 shows the circuit that results.

No attempt has been made here to deal with the non-standard value for C0. In practice C0 will be a large value such as 220 uF, so its wide tolerance will have no significant effect on RIAA accuracy. The IEC amendment will be implemented (if at all) by a later time-constant using a nonelectrolytic, as shown earlier in Figure 7.3.

Obviously the slight errors in nominal value seen in Table 7.5 have some effect. Figure 7.12 shows that the gain at 1 kHz now peaks by +0.048 dB at 1 kHz, which is not the end of the world. By pure coincidence 1 kHz is actually where the RIAA accuracy is worst. At higher frequencies the error slowly declines to +0.031 dB at 20 kHz. Most of this deviation is caused by the +0.44%

error in the nominal value of R2; greater accuracy could be got by using a three-resistor combination such as 3xE24.

RIAA Configuration-C Optimisation: C1 as Three 10 nF Capacitors

We have just modified the RIAA network so that the major capacitor C1 is a single preferred value. The optimisation of the RIAA component values can be tackled in another way however; much depends on the initial assumptions about component availability. In many polystyrene capacitor ranges, 10 nF is the highest value that can be obtained with a tolerance of 1%; in other cases the price goes up rather faster than proportionally above 10 nF. Paralleling several 10 nF polystyrene capacitors is usually *much* more cost-effective than using a single precision polypropylene part.

To use this method we need to redesign the circuit of Figure 7.10 so that C1 is either exactly 30 nF or exactly 40 nF. (There is a practical design using Configuration-A with 5 × 10 nF = 50 nF at the end of this chapter, underlining the fact that Configuration-A makes less efficient use of its capacitance.) The 40 nF version costs more than the 30 nF version but gives a total capacitance that is twice as accurate as one capacitor, (because $\sqrt{4} = 2$) while the 30 nF version only improves the effective tolerance by $\sqrt{3}$ (= 1.73) times. Using 40 nF gives somewhat lower general impedance for the RIAA network, which may reduce noise very slightly. Figure 7.13

Table 7.6 Capacitors for Figure 7.11 (1% tolerance)

Component	Desired value	Actual value	Parallel part A	Parallel part B	Parallel part C	Error	Effective tolerance
C1	33 nF	33 nF	33nF	–	–	0.0%	1%
C2	11.612 nF	11.60 nF	4n7	4n7	2n2	−0.10%	0.60%
C3	2n2	2n2	2n2	–	–	0.00%	1%

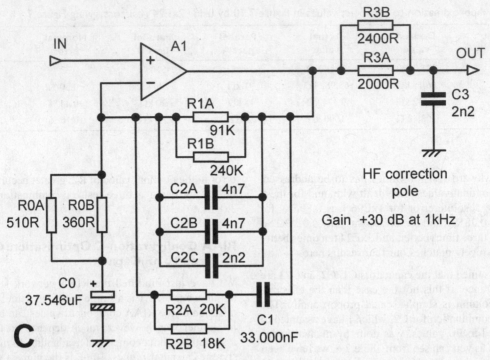

Figure 7.11 Configuration-C from Figure 7.9, with the resistors made up of optimal parallel pairs to achieve the correct value. C2 is now made up of three parts. Gain +30.05 dB at 1 kHz; RIAA accuracy is worsened but still within ±0.048 dB.

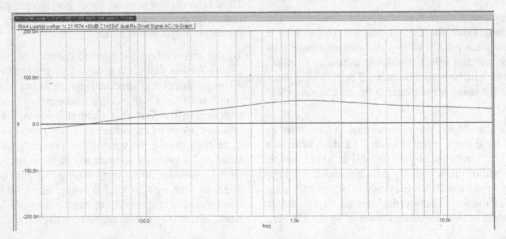

Figure 7.12 The RIAA accuracy of Figure 7.11. Gain is 30.05 dB at 1 kHz, and RIAA error reaches a maximum of +0.048 dB mid-band.

shows the exact resistor values for C1 = 30 nF, and Figure 7.15 shows the exact resistor values for C1 = 40 nF.

Since the gain is unchanged the values for the HF correction pole R3, C3 are also unchanged in each case.

To turn Figure 7.13 into a practical circuit, the awkward resistor values are made up with 1xE96 in Table 7.7,

and with 2xE24 pairs in Table 7.8. The latter gives Figure 7.14.

Once again 2xE24 beats 1xE96 handsomely on nominal accuracy. Note an excellent bit of luck with the value of R1, where the error in the nominal value is only −0.002%.

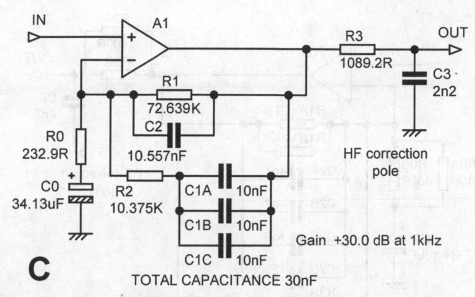

Figure 7.13 Configuration-C from Figure 7.9 redesigned so that C1 is 30 nF, made up with three paralleled 10 nF capacitors. Exact resistor values. Gain +30.0 dB at 1 kHz, RIAA accuracy is worsened but still within ±0.048 dB.

The capacitors for C1 = 30nF are shown in Table 7.9. In this case we have been unlucky with the value of C2, which needs to be trimmed with a 120 pF capacitor to meet the criterion that the error in the nominal value will not exceed half the component tolerance.

This configuration has been built with 1% capacitors and measured, and it works exactly as it should. It gave a parts cost saving of about £2 on the product it was used in; that's real money.

RIAA Configuration-C optimisation: C1 As Four 10nF Capacitors

Alternatively the RIAA components can be calculated to use four 10nF capacitors in parallel for C1; see Figure 7.15. This gives a general reduction in the impedance of the RIAA network. The same component selection processes when applied give the results in Tables 7.10, 7.11, 7.12, and using 2xE24 we get Figure 7.16.

And again 2xE24 shows much better nominal accuracy than 1xE96. Our luck was, however, right out with R0; the best combination to meet Rule 1 is so unequal that the reduction in effective tolerance is negligible.

This time we are much luckier with the value of C2; three 4n7 capacitors in parallel give almost exactly the

Table 7.7 Approximation to the exact values in Figure 7.15 by using 1xE96 resistor

Component	Desired value	E96 value	Nominal error
R0	232.9 Ω	232 Ω	−0.39 %
R1	72.64 kΩ	73.2 kΩ	+0.77 %
R2	10.375 kΩ	10.5 kΩ	+1.20 %
R3	1089.2 Ω	1100 Ω	+0.99 %

Table 7.8 Approximation to the exact values in Figure 7.13 by using 2xE24 resistors, giving Figure 7.14

Component	Desired value	Actual value	Parallel part A	Parallel part B	Nominal error	Effective tolerance
R0	232.9 Ω	233.3 Ω	430 Ω	510 Ω	−0.17 %	0.71%
R1	72.64 kΩ	72.64 kΩ	91 kΩ	360 kΩ	−0.002 %	0.82%
R2	10.375 kΩ	10.359 kΩ	13 kΩ	51 kΩ	−0.15 %	0.82%
R3	1089.2 Ω	1090.9 Ω	2 kΩ	2.4 kΩ	+0.16 %	0.71%

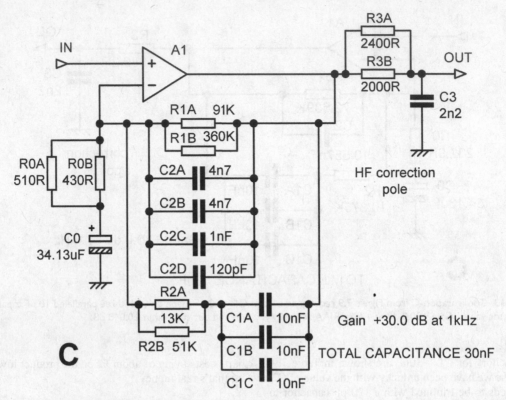

Figure 7.14 Configuration-C from Figure 7.13 with resistors made up of 2xE24 optimal parallel pairs. C2 is made up of four parts. Gain +30.0 dB at 1 kHz, RIAA accuracy is within ±0.01 dB.

Table 7.9 Capacitors for Figure 7.14

Component	Desired value	Actual value	Parallel part A	Parallel part B	Parallel part C	Parallel part D	Nominal error	Effective tolerance
C1	30 nF	30 nF	10 nF	10 nF	10 nF	–	0%	0.58%
C2	10.557 nF	10.52 nF	4n7	4n7	1nF	120 pF	−0.35%	0.64%

required value. On the other hand we are very unlucky with R0, where 180 Ω in parallel with 6.2 kΩ is the most "equal-value" solution that falls within our error criterion.

Both my Precision Preamplifier[8] and the more recent Elektor Preamplifier 2012[9] have MM stage gains close to +30 dB (1 kHz), like these examples, but both use Configuration-A, and five paralleled 10 nF capacitors are required.

Our investigations have shown that there are very real differences in how efficiently the various RIAA networks use their capacitors, and it looks clear that using Configuration-C rather than Configuration-A will cut the cost of the expensive capacitors C1 and C2 in an MM stage by 36% and 19% respectively, which I suggest is both a new result and well worth having. From there we went on to find that different constraints on capacitor availability lead to different optimal solutions for Configuration-C.

Further optimisation of the RIAA networks shown here is possible. For example, we noticed that changing R0 from 200 Ω to 211.74 Ω had a negligible effect on the noise performance; worse by only 0.02 dB. That is well below the limits of hearing, and we could ask

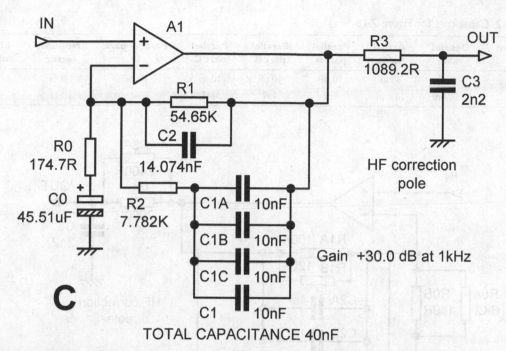

Figure 7.15 Configuration-C redesigned so that C1 is 40 nF, made up with four paralleled 10 nF capacitors. Exact resistor values. Gain +30.0 dB at 1 kHz.

Table 7.10 Approximation to the exact values in Figure 7.17 by using 1xE96

Component	Desired value	Actual value	Nominal error
R0	174.7 Ω	174 Ω	−0.40 %
R1	54.65 kΩ	54.54 kΩ	+0.46 %
R2	7.782 kΩ	7.87 kΩ	+1.31 %
R3	1089.2 Ω	1100 Ω	+0.99 %

what happens if we grit our teeth and accept a 0.1 dB noise deterioration? That is still inaudible. It implies that R0 can be increased to 270 Ω, and the RIAA network impedance is therefore increased by 35%, so we could

for example omit one of the three 10 nF capacitors in Figure 7.13, with of course suitable adjustments to all the other circuit values, and save some more of our hard-earned money. The modest effect on noise of the value of R0 is described in detail in Chapter 9.

I hope you will forgive me for not making public the software tools mentioned in this chapter. They are part of my stock-in-trade as a consultant engineer, and I have invested significant time in their development.

RIAA Configuration-C Optimisation Using Three Resistors in Combination

The 2xE24 examples given in the previous section use two resistors in parallel, and the relatively small number

Table 7.11 Approximation to the exact values in Figure 7.15 by using 2xE24, giving Figure 7.16

Component	Desired value	Actual value	Parallel part A	Parallel part B	Nominal error	Effective tolerance
R0	174.7 Ω	174.9 Ω	180 Ω	6.2 kΩ	+0.13 %	0.97%
R1	54.44 kΩ	54.54 kΩ	100 kΩ	120 kΩ	+0.19 %	0.71%
R2	7.782 kΩ	7.765 kΩ	12 kΩ	22 kΩ	−0.22 %	0.74%
R3	1089.2 Ω	1090.9Ω	2 kΩ	2.4 kΩ	+0.16 %	0.71%

Table 7.12 Capacitors for Figure 7.16

Component	Desired value	Actual value	Parallel part A	Parallel part B	Parallel part C	Parallel part D	Nominal error	Effective tolerance
C1	40 nF	40 nF	10 nF	10 nF	10 nF	10 nF	0 %	0.50%
C2	14.074 nF	14.1 nF	4n7	4n7	4n7	–	+0.18 %	0.58%

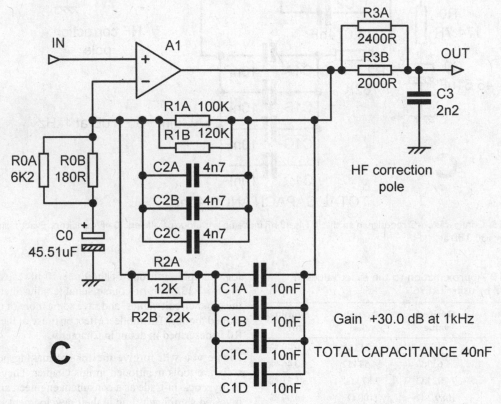

Figure 7.16 Configuration-C from Figure 7.15 with resistors made up of parallel pairs. C2 is made up of three parts. Gain +30.0 dB at 1 kHz; RIAA accuracy is within ±0.01 dB.

of combinations available means that the nominal value is not always as accurate as we would like; for example the 0.44% error in Table 7.5, which only just meets Rule 1, "The nominal value of the combination shall not differ from the desired value by more than half the component tolerance." For the usual 1% parts this means within ±0.5%, and once that is achieved we can pursue the goal of keeping the values as near-equal as possible. Keep in mind that ±0.5% is the error in the nominal value, and the component tolerance, or the effective component tolerance when two or more resistors are combined, is another thing entirely and a source of additional error. It

is usually best to use parallel rather than series combinations of resistors because it makes the connections on a PCB simpler and more compact.

The relatively small number of combinations of E24 resistor values also means that it is difficult to pursue good nominal accuracy and effective tolerance reduction at the same time. This can be addressed by instead using three resistors in parallel. Given the cheapness of resistors, the economic penalties of using three rather than two to approach the desired value very closely are small, and the extra PCB area required is modest. However the design process is significantly harder.

To solve this problem I made use of one of the resistor tables created by Gert Willmann, which he very kindly supplied to me. There are many versions, but the one I used lists in text format all the three-resistor E24 parallel combinations and their combined value. It covers only one decade but is naturally still a very long list, running to 30,600 entries. I applied it first to Figure 7.10, which has +30dB gain at 1 kHz and C1 set to exactly 33nF.

I started with R0, which has a desired value of 211.74 Ω. The Willmann table was read into a text editor, and using the search function to find "211.74" takes us straight to an entry at line 9763 for 211.74396741 Ω, made up of 270 Ω, 1100 Ω, and 9100 Ω in parallel. This nominal value is more than accurate enough, but since the resistor values are a long way from equal, there will be little improvement in effective tolerance; it calculates as 0.808%, which is not much of an improvement over 1%. This area of the Willmann table is shown in Figure 7.17.

Looking up and down the Willmann table by hand, so to speak, better combinations that are more equal than others are easily found. For example, 390 Ω 560 Ω 2700 Ω at line 9774 has a nominal value only 0.012% in error, while the tolerance is improved to 0.667%, and this is clearly a better answer. Scanning the table is a tedious business, but it can be speeded up by a bit of thought. The best result for improving the tolerance would be three equal resistors, but 620 Ω 620 Ω 620 Ω has a nominal value 2.4% too low, and 680 Ω 680 Ω 680 Ω has a nominal value more than 7% too high, so obviously they are no good. But this does suggest that the first resistor should be either 560 Ω or 620 Ω, and armed with this clue searching by eye is faster. I found the best result for R0 is 560 Ω 680 Ω 680 Ω at line 9754, which has a nominal value only −0.09% in error and an effective tolerance of 0.580%, very close to the best possible 0.577% (1/√3). This process could easily be automated in Python (not in JavaScript, as it has no file-handling facilities). The Willmann table and its availability are discussed further in Chapter 2.

In contrast the original two-resistor solution for R0 has a nominal value −0.33% in error and an effective tolerance of 0.718%.

Repeating the process for R1 (66.18 kΩ desired) and searching for "661.8" takes us to 1000 Ω 2000 Ω 91000 Ω. Obviously with resistors outside the range 100 Ω–999 Ω we need to scale by factors of 10; in this

Line	R	r1	r2	r3	Series
1	100	100	-	-	E3
2	100	110	1.1k	-	E24
3	100	110	2.2k	2.2k	E24
:	:	:	:	:	:
9750	211.47068247	220	10k	12k	E12
9751	211.50208097	430	430	13k	E24
9752	211.52542373	300	1.3k	1.6k	E24
9753	211.53846154	220	11k	11k	E24
9754	**211.55555556**	**560**	**680**	**680**	**E12** ◄
9755	211.57127346	330	620	12k	E24
9756	211.59062885	240	3.3k	3.9k	E24
9757	211.63593539	390	470	30k	E24
9758	211.64995936	360	560	6.2k	E24
9759	211.6741501	220	7.5k	22k	E24
9760	211.6838488	220	5.6k	-	E12
9761	211.68437026	360	620	3k	E24
9762	211.73156673	220	8.2k	18k	E12
9763	**211.74396741**	**270**	**1.1k**	**9.1k**	**E24** ◄
9764	211.7577373	220	10k	13k	E24
9765	211.76470588	240	1.8k	-	E24
9766	211.76470588	240	2k	18k	E24
9767	211.76470588	240	3.6k	3.6k	E24
9768	211.76470588	300	750	18k	E24
9769	211.76470588	300	1.2k	1.8k	E24
9770	211.77174906	390	470	33k	E12
9771	211.77416398	220	9.1k	15k	E24
9772	211.81001284	240	2.2k	10k	E24
9773	211.82108626	390	510	5.1k	E24
9774	**211.85600345**	**390**	**560**	**2.7k**	**E12** ◄
9775	211.85860163	330	620	13k	E24
:	:	:	:	:	:
30600	999.3556701	2.2k	3k	4.7k	E24

Figure 7.17 Part of the three-resistor Willmann table. This is the area around 211.7Ω.

case by 100 times. This result has a very accurate nominal value, but once more the tolerance is not brilliant at 0.74%. We need to cast our net wider. 660 Ω × 3 is 1980 Ω, so we look for 1800 Ω as the first resistor, and we get 1800 Ω 2000 Ω 2200 Ω. Scaling that up gives us 180 kΩ 200 kΩ 220 kΩ. The nominal value is only 0.061% low, and the tolerance is 0.579%. It is hard to see how the latter value in particular could be much bettered.

Repeating again for R2 (9432 Ω desired) I got 22 kΩ 33 kΩ 33 kΩ. The nominal value is only 0.036% low, and the effective tolerance is 0.589%. You can see this is working well.

The resistor R3 in the HF correction pole is less critical than the other components, as it only gives a minor tweak at the top of the audio band, but I applied the three-resistor process to it anyway. The desired value is 1089.2 Ω. There are no suitable three-resistor combinations starting with 3000 Ω, and the best I found was 2700 Ω 2700 Ω 5600 Ω, which has a nominal value error of −0.13% and an effective tolerance of 0.602%.

The final result is shown in Table 7.13; the errors in the nominal value column are now much smaller by factors between 12 and 1.2. It is an interesting question as to what the average improvement factor over a large number of two-resistor to three-resistor changes would be. The effective tolerances are shown in the rightmost column,

and you can see that all of them are quite close to the best possible value of 0.577% (1/√3). Figure 7.18 shows the resulting schematic. There is only one E24 resistor out of twelve (200 kΩ), and all the others are E12. This is purely happenstance; no effort whatever was made to avoid E24 values. There is very little point in doing so unless you feel you must use exotic parts that only exist as E12.

The capacitor values are unchanged and are for convenience repeated from Table 7.6 in Table 7.14.

Table 7.13 Table 7.11 redone using paralleled resistor triples (3xE24); see Figure 7.18

Component	Desired value	Actual value	Parallel part A	Parallel part B	Parallel part C	Nominal error	Effective tolerance
R0	211.74 Ω	211.03 Ω	560 Ω	680 Ω	680 Ω	−0.087 %	0.58%
R1	66.18 kΩ	65.982 kΩ	180 kΩ	200 kΩ	220 kΩ	+0.062 %	0.58%
R2	9.432 kΩ	9.474 kΩ	22 kΩ	33 kΩ	33 kΩ	−0.036 %	0.59%
R3	1089.2 Ω	1090.9 Ω	2.7 kΩ	2.7 kΩ	5.6 kΩ	−0.13 %	0.60%

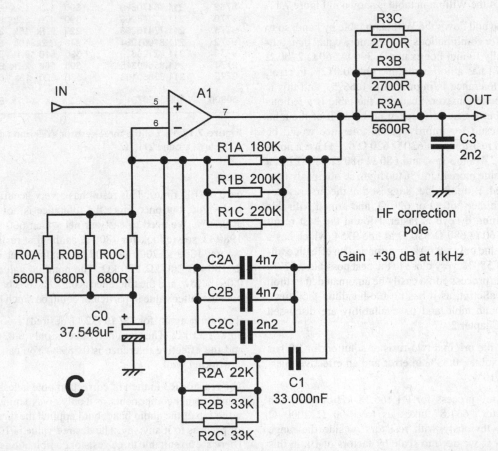

Figure 7.18 The RIAA preamplifier of Figure 7.11, with C1 = 33 nF, redesigned for 3xE24 to approach each non-standard value more closely and have a smaller tolerance

I think it's pretty clear that using three resistors instead of two gives much more accurate nominal values and, at the same time, a usefully smaller tolerance that almost halves the tolerance errors compared with a single resistor. Frequently there are several suitable combinations, and you can choose between a more accurate nominal value or a smaller tolerance percentage.

The obvious question (to me, anyway) is: would four resistors be better? Not really. There is no point in having super-accurate nominal values if you are starting off with 1% parts. The tolerance is now halved, at best, but the improvement depends on the square root of the number of resistors, so we are heading into diminishing returns.

If you want a better tolerance than three 1% resistors can give, the obvious step is to go to 0.1% resistors, which are freely available, though they cost at least ten times as much as the 1% parts. Anything more accurate than that would be a specialised and very expensive item and not obtainable from the usual component distributors.

I applied the 3xE24 process to the 3 × 10 nF design in Figure 7.13, and the result is shown in Figure 7.19. There are now four E24 values out of twelve. The total value of C2 has been made more accurate (now −0.066%) by adjusting the value of C2D; this has nothing to do with the three-resistor process.

I also applied 3xE24 to the 4 × 10 nF design in Figure 7.15, and the result is shown in Figure 7.20. There are now three E24 values out of twelve; I am starting to wonder if there is some mathematical property that means that E24 values are always in the minority. It seems most unlikely, but if anyone with mathematical skills would like to tackle the question, the answer might be enlightening.

Table 7.14 Capacitor values for Figure 7.18 (1% tolerance parts)

C1	33 nF	33 nF	33nF	–	–	0.00 %	1%
C2	11.612 nF	11.60 nF	4n7	4n7	2n2	−0.10 %	0.60%
C3	2n2	2n2	2n2	–	–	0.00 %	1%

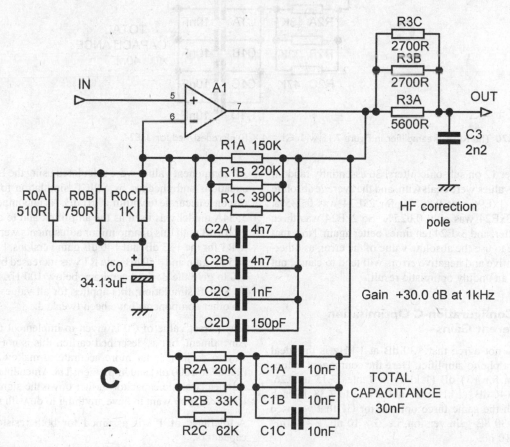

Figure 7.19 The RIAA preamplifier of Figure 7.13, with C1 = 3 × 10 nF, redesigned for 3xE24

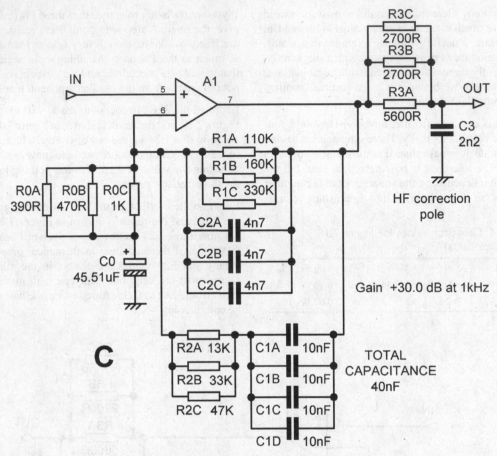

Figure 7.20 The RIAA preamplifier in Figure 7.15, with C1 = 4 × 10 nF, redesigned for 3xE24

In Chapter 12 on subsonic filters, 36 essentially random nominal values were dealt with, and the average absolute error for 1xE96 was 0.805%, for 2xE24 was 0.285%, and for 3xE24 was only 0.025%. So 2xE24 was three times better, and 3xE24 ten times better again. Note that you have to use the absolute value of the error, as otherwise positive and negative errors will tend to cancel out and give an unduly optimistic result.

RIAA Configuration-C Optimisation for Different Gains

You may not agree that +30 dB at 1 kHz is the ideal gain for a phono amplifier. Here the component values are given for +35 dB (1 kHz) in Tables 7.15 to 7.26, and for +40 dB, (1 kHz) in Tables 7.27 to 7.38, in each case with the same three options for C1 that we used for the +30 dB gain version, i.e. 3 × 10 nF, 1 × 33 nF, and 4 × 10 nF.

All component values were calculated using the Lipshitz equations and checked by SPICE simulation to give a response accurate to within ±0.01 dB. An uncompensated 5534A model was used as the amplifier, and to achieve ±0.01 dB with this opamp minor adjustments were made to R1 for the +35 dB and +40 dB gain versions. For both +35 dB gain and +40 dB gain R1 was increased by about 1% to give the desired accuracy below 100 Hz, guided by SPICE simulation; this applies for all values of C1. No other components have been tweaked.

The correct value of C0 is given to implement the IEC Amendment, but as described earlier, this is not a good way to do it; it is far more accurate to make C0 large (220uF to 1000 uF) and implement The Amendment with a nonelectrolytic capacitor farther down the signal path. If, indeed, you want to have anything to do with it at all.

A tolerance of 1% is assumed for both resistors and capacitors.

For Tables 7.15 to 7.18, Gain = +35 dB,
C1 = 30 nF

Table 7.15 +35 dB (1 kHz) C1 =30 nF 1xE96
(1% tolerance)

Component	Desired value	Actual value	Nominal error
R0	130.72 Ω	130 Ω	−0.55%
R1	73.050 kΩ	73.2 kΩ	+0.21%
R2	10.474 kΩ	10.50 kΩ	+0.25%
R3	1344.92 Ω	1330 Ω	−1.12%

Table 7.16 +35 dB (1 kHz) C1 =30 nF 2xE24 (1% tolerance)

Component	Desired value	Actual value	Parallel part A	Parallel part B	Nominal error	Effective tolerance
R0	130.72 Ω	130.154 Ω	180 Ω	470 Ω	−0.43%	0.77%
R1	73.050 kΩ	73.333 kΩ	110 kΩ	220 kΩ	+0.39%	0.74%
R2	10.474 kΩ	10.476 kΩ	20 kΩ	22 kΩ	+0.02%	0.71%
R3	1344.92 Ω	1350.0 Ω	2.7 kΩ	2.7 kΩ	+0.38%	0.71%

Table 7.17 +35 dB (1 kHz) C1 = 30 nF 3xE24 (1% tolerance)

Component	Desired value	Actual value	Parallel part A	Parallel part B	Parallel part C	Nominal error	Effective tolerance
R0	130.72 Ω	130.89 Ω	270 Ω	270 Ω	4.3 k	+0.13%	0.69%
R1	73.050 kΩ	72.973 kΩ	150 kΩ	270 kΩ	300 kΩ	−0.10%	0.61%
R2	10.474 kΩ	10.468 kΩ	24 kΩ	24 kΩ	820 kΩ	−0.06%	0.63%
R3	1344.92 Ω	1346.1 Ω	3600 Ω	4300 Ω	4300 Ω	+0.09	0.58%

Table 7.18 +35 dB (1 kHz) C1 = 30 nF Capacitors (1% tolerance)

Component	Desired value	Actual value	Parallel part A	Parallel part B	Parallel part C	Error	Effective tolerance
C1	30 nF	30 nF	10 nF	10 nF	10 nF	0 %	0.58 %
C2	10.437 nF	10.400 nF	4.7 nF	4.7 nF	1 nF	−0.36 %	0.65 %
C3	1 nF	1 nF	1 nF	–	–	0.0 %	1 %

For all versions of the +35 dB phono amplifier the HF
correction pole R3-C3 is the same, and so the information is not repeated in later tables.

For Tables 7.19 to 7.22, Gain = +35 dB,
C1 = 33 nF

Table 7.19 +35 dB (1 kHz) C1 = 33 nF 1xE96
(1% tolerance)

Component	Desired value	Actual value	Nominal error
R0	118.84 Ω	118 Ω	−0.34%
R1	66.450 kΩ	66.5 kΩ	+0.08%
R2	9.5221 kΩ	9.53 kΩ	+0.08%

Table 7.20 +35 dB (1 kHz) C1 = 33 nF 2xE24 (1% tolerance)

Component	Desired value	Actual value	Parallel part A	Parallel part B	Nominal error	Effective tolerance
R0	118.84 Ω	119.36	160 Ω	470 Ω	+0.44%	0.79%
R1	66.450 kΩ	66.667 kΩ	120 kΩ	150 kΩ	+0.33%	0.71%
R2	9.5221 kΩ	9.5510 kΩ	13 kΩ	36 kΩ	+0.30%	0.78%

Table 7.21 +35 dB (1 kHz) C1 = 33 nF 3xE24 (1% tolerance)

Component	Desired value	Actual value	Parallel part A	Parallel part B	Parallel part C	Nominal error	Effective tolerance
R0	118.84 Ω	118.74 Ω	240 Ω	240 Ω	470 Ω	−0.09%	0.61%
R1	66.450 kΩ	66.383 kΩ	160 kΩ	160 kΩ	390 kΩ	−0.10%	0.61%
R2	9.5221 kΩ	9.4964 kΩ	22 kΩ	24 kΩ	56 kΩ	+0.04%	0.61%

Table 7.22 +35 dB (1 kHz) C1 = 33 nF Capacitors (1% tolerance)

Component	Desired value	Actual value	Parallel part A	Parallel part B	Parallel part C	Parallel part D	Nominal error	Effective tolerance
C1	33 nF	33 nF	33 nF	–	–	–	0 %	1%
C2	11.480 nF	11.530 nF	4.7 nF	4.7 nF	1.8 nF	330 pF	+0.44 %	0.60 %

For Tables 7.23 to 7.25, Gain = +35 dB,
C1 = 40 nF

Table 7.23 +35 dB (1 kHz) C1 = 40 nF 1xE96
(1% tolerance)

Component	Desired value	Actual value	Nominal error
R0	98.045 Ω	97.6 Ω	−0.45%
R1	54.800 kΩ	54.9 kΩ	+0.18%
R2	7.8558 kΩ	7.87 kΩ	+0.18%

Table 7.24 +35 dB (1 kHz) C1 = 40 nF 2xE24 (1% tolerance)

Component	Desired value	Actual value	Parallel part A	Parallel part B	Nominal error	Effective tolerance
R0	98.045 Ω	98.137 kΩ	110 Ω	910 Ω	+0.094%	0.90%
R1	54.800 kΩ	55.000 kΩ	110 kΩ	110 kΩ	+0.36%	0.71%
R2	7.8558 kΩ	7.8787 kΩ	13 kΩ	20 kΩ	+0.29%	0.72%

Table 7.25 +35 dB (1 kHz) C1 = 40 nF 3xE24 (1% tolerance)

Component	Desired value	Actual value	Parallel part A	Parallel part B	Parallel part C	Nominal error	Effective tolerance
R0	98.045 Ω	97.997 Ω	220 Ω	300 Ω	430 Ω	−0.05%	0.60%
R1	54.800 kΩ	54.800 kΩ	130 kΩ	180 kΩ	200 kΩ	0.00%	0.59%
R2	7.8558 kΩ	7.8545 kΩ	16 kΩ	27 kΩ	36 kΩ	−0.02%	0.61%

And again 2xE24 shows much better nominal accuracy than 1xE96. Our luck was, however, right out with R0; the best combination to meet Rule 1 is so unequal that the reduction in effective tolerance is negligible.

Table 7.26 +35 dB (1 kHz) C1 = 40 nF Capacitors (1% tolerance)

Component	Desired value	Actual value	Parallel part A	Parallel part B	Parallel part C	Parallel part D	Nominal error	Effective tolerance
C1	40 nF	40 nF	10 nF	10 nF	10 nF	10 nF	0 %	0.50%
C2	13.915 nF	13.90 nF	4.7 nF	4.7 nF	3.3 nF	1.2 nF	−0.11 %	0.54%

For Tables 7.27 to 7.30, Gain = +40 dB, C1 = 30 nF

Table 7.27 +40 dB (1 kHz) C1 = 30 nF 1xE96 (1% tolerance)

Component	Desired value	Actual value	Nominal error
R0	73.365 Ω	73.2 kΩ	−0.22%
R1	73.55 kΩ	73.2 kΩ	−0.48%
R2	10.529 kΩ	10.5 kΩ	−0.28%
R3	1605.9 Ω	1620 Ω	+0.88%

Table 7.28 +40 dB (1 kHz) C1 = 30 nF 2xE24 (1% tolerance)

Component	Desired value	Actual value	Parallel part A	Parallel part B	Nominal error	Effective tolerance
R0	73.365 Ω	73.333 Ω	110 Ω	220 Ω	−0.04%	0.74%
R1	73.55 kΩ	73.333 kΩ	110 kΩ	220 kΩ	−0.29%	0.74%
R2	10.529 kΩ	10.550 kΩ	13 kΩ	56 kΩ	+0.21%	0.83%
R3	1605.9 Ω	1607.8 Ω	2000 Ω	8.2 kΩ	+0.08%	0.83%

Table 7.29 +40 dB (1 kHz) C1 = 30 nF 3xE24
(1% tolerance)

Component	Desired value	Actual value	Parallel part A	Parallel part B	Parallel part C	Nominal error	Effective tolerance
R0	73.365 Ω	73.333 Ω	220 Ω	220 Ω	220 Ω	−0.04%	0.58%
R1	73.550 kΩ	73.577 kΩ	160 kΩ	180 kΩ	560 kΩ	+0.04%	0.63%
R2	10.529 kΩ	10.516 kΩ	24 kΩ	36 kΩ	39 kΩ	−0.11%	0.59%
R3	1605.9 Ω	1602.8 Ω	4300 Ω	4700 Ω	5600 Ω	−0.19%	0.58%

Table 7.30 +40 dB (1 kHz) C1 = 30 nF Capacitors (1% tolerance)

Component	Desired value	Actual value	Parallel part A	Parallel part B	Parallel part C	Parallel part D	Nominal error	Effective tolerance
C1	30 nF	30 nF	10 nF	10 nF	10 nF	–	0 %	0.58%
C2	10.371 nF	10.37 nF	3.3 nF	3.3 nF	3.3 nF	0.47	−0.01%	0.55%
C3	470 pF	470 pF	470 pF	–	–	–	0 %	1%

For all versions of the +40dB phono amplifier the HF correction pole R3-C3 is the same, and so the information is not repeated in later tables.

For Tables 7.31 to 7.34, Gain = +40 dB, C1 = 33 nF

Table 7.31 +40 dB (1 kHz) C1 = 33 nF 1xE96 (1% tolerance)

Component	Desired value	Actual value	Nominal error
R0	66.696 Ω	66.5 kΩ	−0.29%
R1	66.850 kΩ	66.5 kΩ	−0.52%
R2	9.5723 kΩ	9.53 kΩ	−0.44%

Table 7.32 +40 dB (1 kHz) C1 = 33 nF 2xE24 (1% tolerance)

Component	Desired value	Actual value	Parallel part A	Parallel part B	Nominal error	Effective tolerance
R0	66.696 Ω	66.667 Ω	120 Ω	150 Ω	−0.04%	0.71%
R1	66.850 kΩ	68.571 kΩ	120 kΩ	150 kΩ	−0.27%	0.71%
R2	9.5723 kΩ	9.6000 kΩ	16 kΩ	24 kΩ	+0.29%	0.72%

Table 7.33 +40 dB (1 kHz) C1 = 33 nF 3xE24 (1% tolerance)

Component	Desired value	Actual value	Parallel part A	Parallel part B	Parallel part C	Nominal error	Effective tolerance
R0	66.696 Ω	66.666 Ω	150 Ω	200 Ω	300 Ω	−0.04%	0.60%
R1	66.850 kΩ	66.835 kΩ	160 kΩ	220 kΩ	240 kΩ	−0.02%	0.59%
R2	9.5723 kΩ	9.5758 kΩ	22 kΩ	30 kΩ	39 kΩ	+0.04%	0.59%

Table 7.34 +40 dB (1 kHz) C1 = 33 nF Capacitors (1% tolerance)

Component	Desired value	Actual value	Parallel part A	Parallel part B	Parallel part C	Parallel part D	Nominal error	Effective tolerance
C1	33 nF	33 nF	33 nF	–	–	–	0%	1%
C2	11.408 nF	11.40 nF	4.7 nF	4.7 nF	1 nF	1 nF	−0.07%	0.59%

For Tables 7.35 to 7.38, Gain = +40 dB, C1 = 40 nF

Table 7.35 +40 dB (1 kHz) C1 = 40 nF 1×E96 (1% tolerance)

Component	Desired value	Actual value	Nominal error
R0	55.025 kΩ	54.9 kΩ	−0.23%
R1	55.150 kΩ	54.9 kΩ	−0.45%
R2	7.8973 kΩ	7.87 kΩ	−0.35%

Table 7.36 +40 dB (1 kHz) C1 = 40 nF 2×E24 (1% tolerance)

Component	Desired value	Actual value	Parallel part A	Parallel part B	Nominal error	Effective tolerance
R0	55.025 Ω	55.000 Ω	110 Ω	110 Ω	−0.04%	0.71%
R1	55.150 kΩ	55.000 kΩ	110 kΩ	110 kΩ	−0.27%	0.71%
R2	7.8973 kΩ	7.8787 kΩ	13 kΩ	20 kΩ	−0.23%	0.72%

Table 7.37 +40 dB (1 kHz) C1 = 40 nF 3×E24 (1% tolerance)

Component	Desired value	Actual value	Parallel part A	Parallel part B	Parallel part C	Nominal error	Effective tolerance
R0	55.025 Ω	55.059 Ω	130 Ω	130 Ω	360 Ω	+0.06%	0.62%
R1	55.150 kΩ	55.192 kΩ	110 kΩ	160 kΩ	360 kΩ	+0.08%	0.63%
R2	7.8973 kΩ	7.8957 kΩ	18 kΩ	22 kΩ	39 kΩ	−0.02%	0.60%

Table 7.38 +40 dB (1 kHz) C1 = 40 nF Capacitors (1% tolerance)

Component	Desired value	Actual value	Parallel part A	Parallel part B	Parallel part C	Parallel part D	Nominal error	Effective tolerance
C1	40 nF	40 nF	10 nF	10 nF	10 nF	10 nF	0%	0.50%
C2	13.827 nF	13.82 nF	6.8 nF	4.7 nF	1.5 nF	820 pF	−0.05%	0.61%

Equivalent RIAA Configurations

You may be wondering if the four configurations in Figure 7.7 actually cover all possible single-stage series-feedback RIAA arrangements. The configurations can be differently arranged as if two components are in series, with nothing connected to their junction point; it does not matter in which order they occur. In Figure 7.21 the configurations A and A' are electrically identical. This looks pretty obvious, but it is perhaps a bit less so for B B', C C', and D D', in which the identical topology can be drawn in several different ways. Burkhard Vogel in his monumental book on noise, *The Sound of Silence*, describes Configuration-A as a Type-Eub network, Configuration-B' as a Fub-B network, and

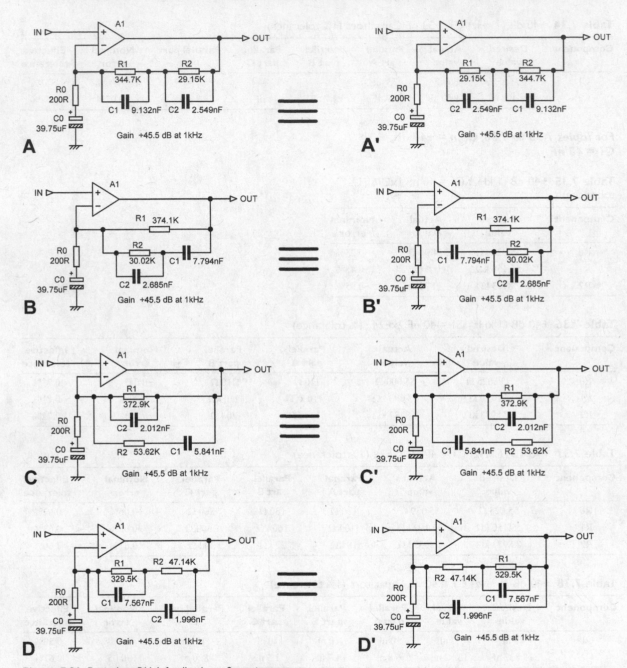

Figure 7.21 Equivalent RIAA feedback configurations

Configuration-C as a Fub-A network.[10] Configuration-D is not examined.

RIAA Components

Many of the factors affecting the choice of components for the RIAA network, such as accuracy, linearity, and cost, have already been dwelt on at length. Here are a few more points to ponder.

Resistors should be metal film for good linearity, with two or more near-equal E24 values paralleled to obtain the nonpreferred resistance values. See Chapter 2.

For close-tolerance capacitors the best solution seems to be axial polystyrene types, which are freely available at 1% tolerance up to 10 nF. Some paralleling is required, and in fact is highly desirable. This is because the sum of multiple capacitors is more accurate than a single component of the same tolerance, so long as the mean is well controlled, because the capacitances sum arithmetically but the random errors partially cancel. This is described in detail in Chapter 2. Both the preamp examples at the end of this chapter use multiple capacitors in this way.

Polystyrene capacitors have two foils, and one of them will be on the outside of the component and thus vulnerable to capacitive crosstalk and hum pickup. It is desirable that the capacitor is orientated so that the outer foil is connected to the circuit node with the lowest impedance; very often this can be arranged to be the stage output, which is at a very low impedance and immune to capacitive pickup.

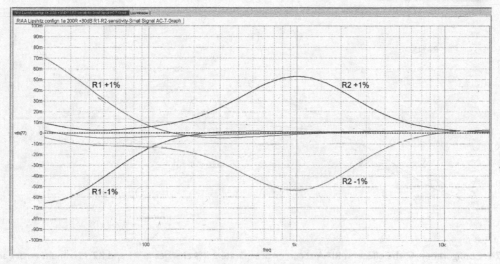

Figure 7.22 Configuration-A: the effect on RIAA accuracy of separate ± 1% changes in R1 and R2. Maximum errors are 0.070 dB for R1 and 0.053 dB for R2.

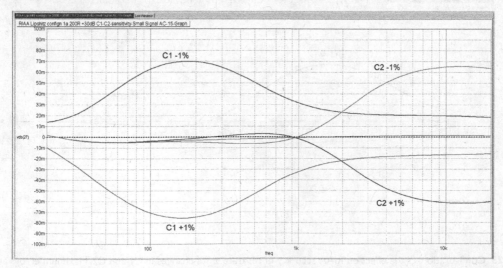

Figure 7.23 Configuration-A: the effect on RIAA accuracy of separate ± 1% changes in C1 and C2. Maximum errors are 0.070 dB for C1 and 0.065 dB for C2.

Some capacitor manufacturers mark the outer foil; for example, the outer foil of polystyrene capacitors manufactured by LCR is indicated by the mitred corner on the packaging. Inexpensive polystyrene caps may not have consistent foil placement. Other types of capacitor, such as polypropylene, have similar considerations, With axial types, the outer foil is likely to be marked by a line at one end, if indeed it is marked at all. The outer foil of a capacitor can be quickly identified with an oscilloscope. Ground one lead and put the probe on the other, and see how much hum the capacitor picks up from your fingers. Reverse the connections and repeat. When the outer foil is grounded, much less hum is picked up.

RIAA Component Sensitivity: Configuration-A

The "component sensitivity" of a circuit defines how much its response varies as a result of component value tolerances. It has nothing to do with gain or signal levels. It is

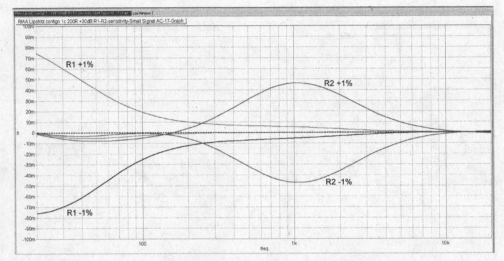

Figure 7.24 Configuration-C: the effect on RIAA accuracy of separate ± 1% changes in R1 and R2. Maximum errors are 0.075 dB for R1 and 0.047 dB for R2. Similar to Configuration-A except that R2 has very little effect below 200 Hz.

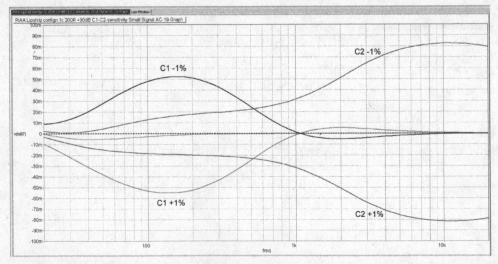

Figure 7.25 Configuration-C: the effect on RIAA accuracy of separate ± 1% changes in C1 and C2. Maximum errors are 0.052 dB for C1 and 0.082 dB for C2. Very similar to Configuration-A except that C2 has significant effect below 1 kHz.

Table 7.39 Maximum errors for 1% deviation in components values for Configurations A and C

Configuration	R1	R2	C1	C2
A	0.070 dB	0.053 dB	0.070 dB	0.065 dB
C	0.075 dB	0.047 dB	0.052 dB	0.082 dB

much affected by the way the circuit works—the higher the Q of an active filter, the greater its component sensitivity.

Components have tolerances on their value, and we need to assess what RIAA accuracy is possible without spending a fortune on precision parts; ± 1% is the best tolerance readily available for metal film resistors and polystyrene capacitors, so at first it appears anything better than ± 0.1 dB accuracy is out of the question. This is not so, for the simple reason that across the audio band more than one component determines the response. Higher precision can be obtained by using multiple components, as we have noted before.

We saw earlier that Configuration-C was more economical than Configuration-A. We might wonder whether we pay for that in increased component sensitivity for Configuration-C. Sensitivity analysis can be done by involved mathematics or more simply by making ± 1% changes in the components of a preamp simulation. The results in Figures 7.22 and 7.23 include the IEC amendment, implemented by C0. The effect of C0 tolerances on accuracy has already been examined in this chapter.

RIAA Component Sensitivity: Configuration-C

Figures 7.24 and 7.25 show the same results but for Configuration-C. The component sensitivities for both configurations are summarised in Table 7.39, which shows that there is not much to choose between them. Configuration-A is better for C2, but Configuration-C is better for R2 and C1. There is no reason here to not use Configuration-C.

Open-Loop Gain and RIAA Accuracy

There is no point in having a super-accurate RIAA network if the active element does not have enough open-loop gain to correctly render the response demanded. This was a major problem for two- and three-transistor discrete MM input stages, but one might have hoped that it would have disappeared with the advent of usable opamps. However, life is flawed, and gain problems did not wholly vanish. The TL072 was at one time widely used for MM inputs because of its affordability, even though its JFET input devices are a poor match to MM cartridge impedances and its distortion performance was not of the best. However, there was another lurking problem.

As we have seen earlier in this chapter, the appropriate gain (at 1 kHz) for an MM input is between +30 and +40 dB. The TL072 does not have enough open-loop gain to give an accurate response with a closed-loop gain of a +35 dB. Figure 7.26 shows the result of a simulation using Configuration-A RIAA with accurate values

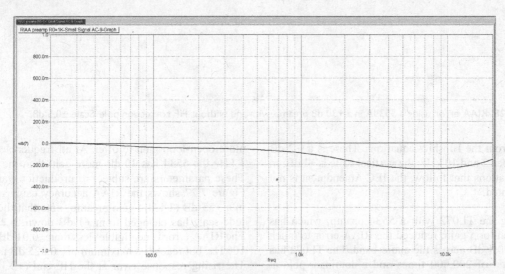

Figure 7.26 RIAA error using a TL072 in a +35 dB preamp; lack of open-loop gain causes a 0.2 dB dip between 3 and 10 kHz. Scale ±1 dB

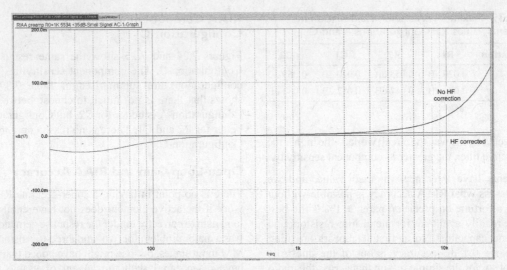

Figure 7.27 RIAA error using a 5534A in a +35 dB preamp, with and without HF correction pole. Scale ±0.2 dB.

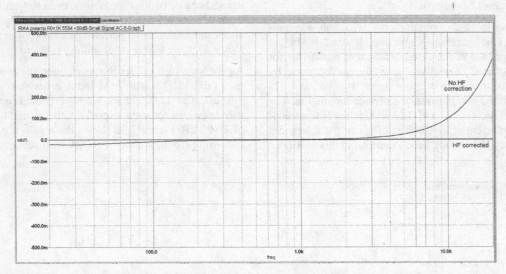

Figure 7.28 RIAA error using a 5534A in a +30 dB preamp, with and without HF correction pole. Scale ±0.5 dB.

derived from the Lipshitz equations. There is a 0.2 dB dip between 3 and 10 kHz; the vertical scale is ±1 dB. In the simulations that follow, the IEC Amendment is not implemented.

Replacing the TL072 with a 5534 opamp, which has more open-loop gain, reduced the RIAA error to much less than 0.1 dB across the audio band. The TL072 has an LF gain of 200,000 times and a dominant pole at 20 Hz (all typical specs). The 5534 has a lower LF gain of 100,000, but its pole (uncompensated) is much higher at 1 kHz, so at any frequency above 40 Hz the 5534 has

more open-loop gain to offer. At all frequencies above 1 kHz the 5534 has 30 dB more gain than the TL072. These parameters are subject to production variations. Figure 7.27 shows the RIAA accuracy using a 5534A in a +35 dB (1 kHz) gain preamp; note that the amplitude scale has changed from ±1 dB down to ±0.2 dB. The RIAA error is negligible (less than ±0.01 dB) above 100 Hz, but reaches a maximum of −0.025 dB at 30 Hz. The rising response error at the HF end (+0.12 dB at 20 kHz) is due to the gain levelling off at f6, but this is here cancelled by an HF correction pole that reduces the HF error to less than ±0.01 dB.

Figure 7.28 shows the same results for a 5534 in a +30 dB (1 kHz) preamp. HF accuracy is better, but the LF error is not much reduced, being −0.020 dB at 30 Hz, indicating it may not be due to a lack of open-loop gain. This error can be reduced to −0.005 dB by adjusting the value of R1 upwards by trial and error, no other RIAA components being altered. One wonders if there might be a small systematic error buried in the Lipshitz equations, or more likely in my application of them. The uncorrected HF response has a greater error of 0.37 dB at 20 kHz, as the gain is levelling off towards unity at a

lower f6 frequency, but this is once more fully sorted out by the HF correction pole.

The 5534 has quite a complicated internal structure with nested Miller loops for compensation. It is not obvious (to me, anyway) whether this has anything to do with the interaction of open-loop gain with RIAA accuracy. It seemed worthwhile to do a few more simulations to get a better idea of the situation. The SPICE opamp model is replaced with the simplest possible conceptual opamp. This has only two parameters—the LF open-loop gain

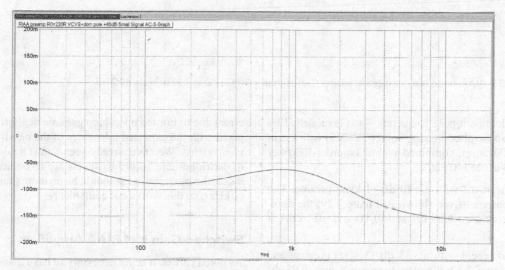

Figure 7.29 RIAA error with conceptual opamp. Closed-loop gain +40 dB at 1 kHz. Scale ±0.2 dB.

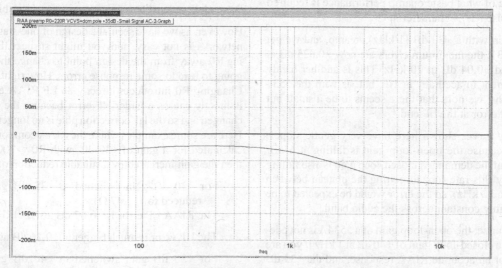

Figure 7.30 RIAA error with conceptual opamp. Closed-loop gain +35 dB at 1 kHz. Scale ±0.2 dB.

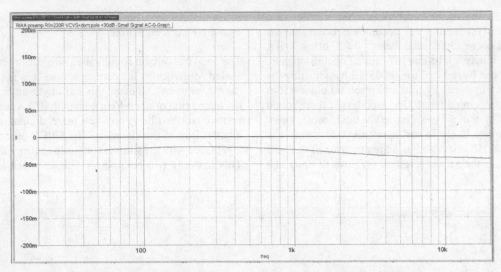

Figure 7.31 RIAA error with conceptual opamp. Closed-loop gain +30 dB at 1 kHz. Scale ±0.2 dB.

and the dominant-pole frequency—and is modelled by a voltage-controlled-voltage source (VCVS) with a gain of 100,000 times, combined with a 1st-order RC filter that is −3 dB at 100 Hz.

We will start off with a +40 dB (1 kHz) preamp. The simulation results are shown in Figure 7.29; the maximum errors are −0.09 dB at 150 Hz and −0.16 dB at 20 kHz.

Rinse and repeat with a +35 dB (1 kHz) preamp, and we get Figure 7.30; the maximum errors are −0.03 dB at 40 Hz and −0.095 dB at 20 kHz. This is a useful improvement in accuracy at both LF and HF and gives some idea of what basic opamp performance is required for a precise RIAA characteristic.

Do it again with a +30 dB (1 kHz) preamp, and we get Figure 7.31; the maximum errors are now −0.025 dB at 30 Hz and −0.04 dB at 20 kHz. This is another handy improvement in accuracy at HF, but, as with the 5534 simulations, we note that there seems to be a small but persistent error at the LF end.

The error curves shown here have very gentle slopes. This is because the open-loop gain is falling at 6 dB/octave, but the demanded closed-loop gain is also falling at roughly this rate (not forgetting the plateau between 500 Hz and 2 kHz), so the error would be expected to be very roughly constant across the audio band.

To summarise, the open-loop gain of a 5534A is not adequate for a closed-loop gain of +40 dB at 1 kHz if you are aiming for an accurate RIAA response, and the +35 dB (1 kHz) situation is marginal. For +30 dB (1 kHz) the

errors due to limited open-loop gain are negligible compared with the expected tolerances of the passive RIAA components. We have already seen that if a wide range of cartridges and recording levels are to be accommodated, the minimum gain should be no more than +30dB (1 kHz), so this works out quite nicely.

Switched-Gain MM RIAA Amplifiers

As noted earlier, it is not necessary to have a wide range of variable or stepped gain if we are only dealing with MM inputs, due to the limited spread of MM cartridge sensitivities—only about 7 dB. According to Peter Baxandall, at least two gain options are desirable.[11]

However, as we have seen, the design of one-stage RIAA networks is not easy, and you might suspect that altering R0 away from the design point to change the gain is going to lead to some response errors. How right you are. Changing R0 introduces directly an LF RIAA error and indirectly causes a larger HF error because the gain has changed and so the HF correction pole is no longer correct. Here are some examples where the RIAA components are calculated for a gain of +30 dB, with R0 = 200 Ω, and then the gain increased by a suitable reduction of R0:

For +30 dB gain switched to +35 dB gain, (R0 reduced to 112.47 Ω).
The RIAA LF error is +0.07 dB 20Hz–1kHz.

The HF error is much bigger at −0.26 dB at 20 kHz.

For +30 dB gain switched to +40 dB gain, (R0 reduced to 63.245 Ω).

The RIAA LF error is +0.10 dB 20Hz–1kHz.
The HF error is −0.335 dB at 20 kHz.

These figures include the effect of finite open-loop gain when using a 5534A as the opamp; this increases the errors for the +40 dB gain option.

Thus for real accuracy we need to switch not only R0 but also R1 in the RIAA feedback path and R3 in the HF correction pole; this will be very clumsy. If the RIAA error tolerance is ±0.1 dB, switching R1 could be omitted, but two resistors still need to be switched. This assumes that

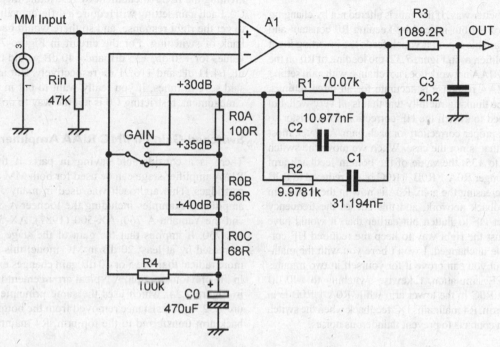

Figure 7.32 Phono amplifier with gain switchable to +30 dB, +35 dB, and +40 dB (1 kHz)

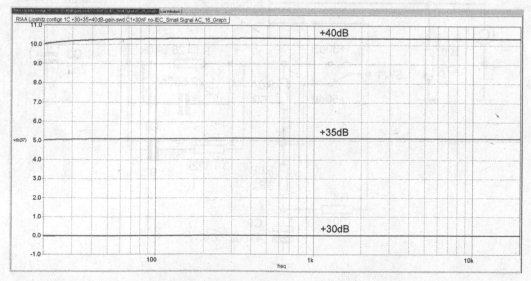

Figure 7.33 RIAA accuracy of phono amplifier switchable +30 dB, +35 dB, and +40 dB (1 kHz)

the IEC amendment is performed by a CR network after the MM stage, as described earlier; this will be unaffected by changes in R0. Otherwise, if the IEC amendment is implemented by a small value of C0, you would need to switch that component as well, to avoid gross RIAA errors below 100 Hz. All in all, switching the value of R0 is not an attractive proposition if you are looking for good accuracy.

There is a better way. If the gain is altered not by changing the value of R0 but instead by keeping R0 constant and having a variable tap on it which feeds the inverting input of the amplifier, as in Figure 7.32, the loading of R0 on the rest of the RIAA network does not change with gain setting and the RIAA response is accurate for all three settings. You may be thinking ruefully that that is all very well, but we still need to switch the HF correction pole resistor so we get the proper correction for each gain. And yet, most elegantly, that is not the case. When we move the switch from +30 to +35, the value of the bottom feedback arm R0 is no longer R0A + R0B + R0C but is reduced to R0B + R0C, increasing the gain. R0A is now in the upper arm of the feedback network, and this causes the frequency response at HF to flatten out earlier than it would have done, in just the right way to keep the required HF correction pole unchanged. I won't bore you with the mathematics, but you can prove it for yourself in two minutes with SPICE simulation. Likewise switching to +40 dB leaves just R0C in the lower arm while R0A + R0B is in the upper arm. R4 maintains DC feedback when the switch is between contacts to prevent thunderous noises.

The results can be seen in Figure 7.33. The RIAA accuracy falls off slightly at LF for the +40 dB setting because of the finite value of C0. Its value can be increased considerably if desired as it has only the offset voltage across it, and a 6V3 part will be fine.

This scheme will not work well if you insist on implementing the IEC Amendment by restricting the value of C0. Each gain setting will require a different value of C0 to get the right response, and so there would be another bank of switching. For the circuit in Figure 7.32, the values for +30 db, +35 dB, and +40 dB would be 35.49 uF, 64.11 uF, and 116.91 uF respectively. But as I have said several times, if you really want to put in the IEC Amendment, restricting C0 is not the way to do it.

Switched-Gain MM/MC RIAA Amplifiers

There is a considerable saving in parts if the same RIAA amplifier stage can be used for both MM and MC cartridges. This approach was used in many Japanese amplifiers, examples including the Pioneer A-8 (1981) and the Yamaha A-760, AX-500 (1987), AX-592, and AX-750. It implies that the gain of the stage must be increased by at least 20 dB in MC mode; this is much more radical than the 5 or 10 dB gain changes examined in the previous section. A typical arrangement is shown in Figure 7.34, which uses the same principle as Figure 7.32, with resistance removed from the bottom feedback arm transferred to the top arm. R4 maintains DC

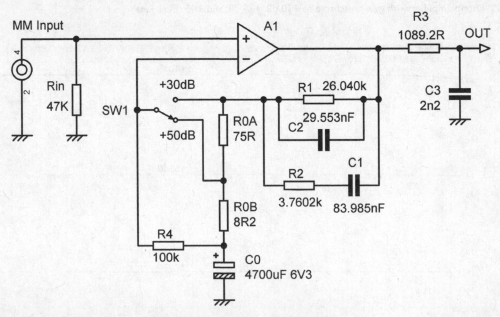

Figure 7.34 Phono amplifier with gain switchable from +30dB to +50dB (at 1 kHz)

feedback when the switch is between contacts to prevent horrible noises. SPICE simulation shows that the RIAA accuracy is well within ±0.1 dB for the +30 dB setting. This is also true down to 40 Hz for the +50dB setting, but the response then rolls off due to the finite value of C0, being 0.3 dB down at 20 Hz. C0 is already about as large as is practicable at 4700 uF/6V3; at the time of writing the smallest I found was 25mm high and 10mm diameter. Improving the +50 dB LF response, or switching to a higher gain, will require C0 to be replaced by a short circuit and DC conditions maintained by a servo. This method is described in Chapter 11.

If a BJT input device is used there is the problem that the collector current needs to be low to get low current noise, which is essential for a good MM noise performance; on the other hand the collector current needs to be high for low MC noise. The use of JFET input devices avoids this compromise because of the absence of current noise means a high drain current can be used in both cases. It is probably significant that most of the amplifiers mentioned used JFET input devices.

A significant complication is that the spread in sensitivity of MC cartridges is very much greater at about 36 dB than for MM cartridges (less than 10 dB) and having a single fixed MC gain is not very satisfactory.

Shunt-Feedback RIAA Equalisation

The shunt-feedback equivalent of the basic RIAA stage is shown in Figure 7.35. It has occasionally been advocated because it avoids the unity-gain problem, but it has the crippling disadvantage that with a real cartridge load, with its substantial inductance, it is about 14 dB noisier than the series RIAA configuration.[12] A great deal of grievous twaddle has been talked about RIAA equalisation and transient response, in perverse attempts to render the shunt RIAA configuration acceptable despite its serious noise disadvantage. Since the input resistor R0 has to be 47 kΩ to load the cartridge correctly, the RIAA network has to operate at a correspondingly high impedance and will be noisy.

A series-feedback disc stage cannot make its gain fall below one, as described earlier, while the shunt-feedback version can; however an HF correction pole solves that problem completely. Shunt feedback eliminates any possibility of common-mode distortion, but then at the signal levels we are dealing with that is not a problem, at least with bipolar input opamps. A further disadvantage is that a shunt-feedback RIAA stage gives a phase inversion that can be highly inconvenient if you are concerned to preserve absolute phase.

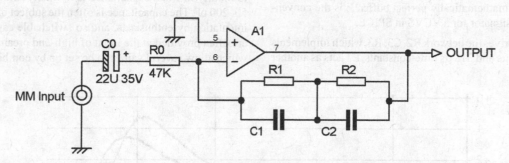

Figure 7.35 Shunt-feedback RIAA configuration. This is 14 dB noisier than the series-feedback version.

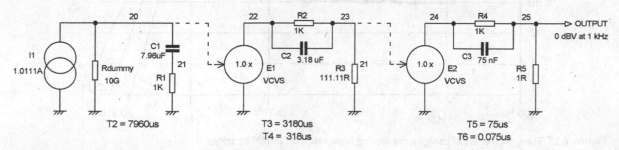

Figure 7.36 Inverse RIAA network for SPICE simulation

Simulating Inverse RIAA Equalisation

SPICE simulation is well suited to the task of checking that the RIAA component values chosen are accurate. The best way to do this is to build an inverse RIAA model to feed the RIAA preamplifier being simulated. This is much, much simpler than designing the pre-amplifier RIAA network because the time-constants can be completely decoupled from each other by using unity-gain buffers with zero-impedance outputs. The required response can be implemented in many ways, but my version is shown in Figure 7.36. The component values have nothing to do with practical circuitry and are chosen simply for ease of calculation.

The first network (C1, R1) implements the 7960 µs time-constant of the notorious IEC Amendment. Since this first network is the inverse of a bass-roll-off, its output must continue to rise indefinitely at 6 dB/octave as frequency falls, and it is therefore implemented with a current source, so that as the impedance of C1 rises the output voltage at node 20 rises indefinitely. The apparently odd value of 1.011 A for the current source is in fact cunningly chosen to give a final output of 0 dBV at 1 kHz, which simplifies SPICE output plotting. The 10-gigaohm resistor Rdummy is required, as SPICE otherwise considers node 20 to be at an undefined DC level and objects strongly. The voltage at node 20 controls the output of the VCVS (voltage-controlled voltage source) E1, which has its gain set to unity. It has zero output impedance and so acts as a mathematically perfect buffer. E is the conventional designator for a VCVS in SPICE.

E1 then drives the network R2, C3, R3, which implements the 3180 µs and 318 µs time-constants. E2 acts as another perfect buffer for the voltage at node 23 and drives R4, C3, R5, which implements the 75 µs time-constant. The very low value for R5 allows the output to go rising at 6 dB/octave to well beyond 20 kHz; the response does not level out until the T6 zero at 2.12 mHz is reached. If the IEC Amendment is not required, increase C1 to 10,000 uF so it has no effect in the audio band.

Physical Inverse RIAA Equalisation

Building a sufficiently accurate inverse RIAA network for precision measurements is not to be entered upon lightly or unadvisedly. The component values will need to have an accuracy a good deal better than 1%, and this makes sourcing components difficult and expensive. A much better alternative is to use a test system such as those by Audio Precision that allow an equalisation file to modify the generator output level during a frequency sweep.

MM Cartridge Loading and Frequency Response

The standard loading for a moving-magnet cartridge is 47 kΩ in parallel with a certain amount of capacitance, the latter usually being specified by the maker. The resulting resonance with the cartridge inductance is deliberately used by manufacturers to extend the frequency response, so it is wise to think hard before trying to modify it. Load capacitance is normally in the range 50–200 pF. The capacitance is often the subject of experimentation by enthusiasts, and so switchable capacitors are often provided at the input of high-end preamplifiers which allow several values to be set up by combinations

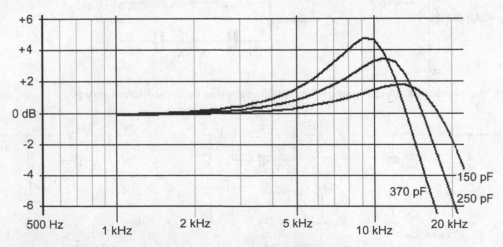

Figure 7.37 The typical effect of changing the loading capacitance on an MM cartridge

of switch positions. There is usually a minimum value of 47 or 100 pF which is always in circuit, and this component should be placed right on the input connector to maximise EMC immunity. The exact effect of altering the capacitance depends on the inductance and resistance of the cartridge, but a typical result is shown in Figure 7.37, where increasing the load capacitance lowers the resonance peak frequency and makes it more prominent and less damped. It is important to remember that it is the total capacitance, including that of the connecting leads, which counts.

Because of the high inductance of an MM cartridge, adjusting the load resistance can also have significant effects on the frequency response, and some preamplifiers allow this too to be altered, though the option is less common. The only objective way to assess the effects of these modifications is to measure the output when a special (and expensive) test disc is played.

When loading capacitance is used it should be as near to the input socket as possible so it can contribute to filtering out RF before it radiates inside the enclosure. However its effectiveness for EMC purposes is likely to be much compromised if the capacitors are switched. Normal practice is that the smallest capacitor is permanently in circuit so it can be mounted right on the rear of the input socket. A continuously variable loading capacitance could be made with an old-style tuning capacitor (two-section for stereo); looking back they were marvels of mass-produced precision engineering. The maximum value in an old medium-wave radio is often a rather convenient 500 pF. This would look well cool but naturally takes up a lot of space, and the variable-bootstrapping of a fixed capacitor would be much more compact.

The effective capacitance is the total seen by the cartridge, which includes the capacitance of the connecting lead from deck to preamplifier; this can easily be 100 pF per metre, so there is a strong incentive to avoid long cables. Ideally the phono amp should be right next to the deck, though low capacitance cables giving 30 pF/meter are available. There is also the capacitance of the leads inside the tone arm, commonly regarded as being 10–20 pF. The cable capacitance can be nullified if the screen is not grounded but driven by a low-impedance version of the signal at the phono amplifier input, provided by some sort of unity-gain buffer. The Neumann PUE 74 phono amplifier had this feature and is described in detail by Vogel.[13] Instrumentation guard technology puts another grounded screen on the outside of the guard screen to prevent radiation, but that seems unnecessary here where the signal are only a few mV; making the guard drive a very low impedance—a fraction of an ohm being easy—should prevent any possibility of crosstalk. The idea does not seem to have gained much traction.

The exact nature of the cartridge resonance does not have a consensus in the hi-fi community. There is also the possibility of what is usually called the "cantilever resonance" which is a mechanical resonance between the effective tip mass of the stylus and the compliance of the vinyl it is tracking, the latter making up the spring part of the classic mass-and-spring system. The effective tip mass of the stylus is contributed to by the mass of the diamond tip, the cantilever, and the generator element on the other end, which may be a piece of iron, a magnet, or coils; it usually ranges from 0.2–0.7 milligrams. There is also the question of the contribution of the cantilever compliance and the possibility of a torsional resonance of the cantilever.[14] You are probably thinking by now that this is a mass of electromechanical compromises that should be left alone, and you are probably right.

Not everyone agrees. A scheme for cancelling the effects of the cantilever resonance with a sophisticated active filter was put forward by Steven van Raalte in *Linear Audio* Volume 3.[15] A slightly earlier attempt, in 1953,[16] simply put a series LC circuit across the cartridge output.

MM Cartridge-Preamplifier Interaction

One often hears that there can be problems due to interaction between the impedance of the cartridge and the negative-feedback network. Most commentators are extremely vague as to what this actually means, but according to Tomlinson Holman,[17] the factual basis is that it used to be all too easy to design an RIAA stage, if you are using only two or three discrete transistors, in which the NFB factor is falling significantly with frequency in the upper reaches of the audio band. This could be due to heavy dominant-pole compensation to achieve stability when the gain is effectively unity at HF. On the other hand, the amount of NFB tends to increase with frequency as the RIAA equalisation reduces the closed-loop gain.

Assuming series feedback is used, a falling NFB factor means the input impedance will fall with frequency, which is equivalent to having a capacitive input impedance. This interacts with the cartridge inductance and allegedly can cause a resonant peak in the frequency response, in the same way that cable capacitance or a deliberately added load capacitance can do.

For this reason a flat-response buffer stage between the cartridge and the first stage performing RIAA equalisation was sometimes advocated. One design including this feature was the Cambridge Audio P50, which used a Darlington emitter-follower as a buffer; with this approach compromising the noise performance would seem to be inevitable because there is no gain to lift the signal level above the noise floor of the next stage.

MM Cartridge DC and AC Coupling

Some uninformed commentators have said that there should be no DC-blocking capacitor between the cartridge and the preamplifier. This is insane. Keep DC out of your cartridge. The signal currents are tiny (For MM cartridges 5 mV in 47 kΩ = 106 nA, while for MC ones 245 uV in 100 Ω = 2.45 uA; a good deal higher) and even a small DC bias current could interfere with linearity. I am not aware of any published work on how cartridge distortion is affected by DC bias currents, but I think it pretty clear they will not improve things and may make them very much worse. Large currents might partially demagnetise the magnet, be it moving or otherwise, ruining the cartridge. Even larger currents due to circuit faults might burn out the coils, ruining the cartridge even more effectively. You may call a lack of blocking capacitors high end, but I call it highly irresponsible.

If I had a £15,000 cartridge (and they do exist, by Koetsu and Clearaudio) I would probably put two blocking capacitors in series. Or three.

References

1. Lipshitz, S. P. "On RIAA Equalisation Networks" *Journal of Audio Engineering Society*, June 1979, p. 458, onwards.

2. Howard, K. "Cut & Thrust: RIAA LP Equalisation" *Stereophile*, Mar 2009. See also www.stereophile.com/content/cut-and-thrust-riaa-lp-equalization-page-3 Accessed June 2014.

3. Howard, Keith "Cut & Thrust: RIAA LP Equalisation" *Stereophile*, Mar 2009. See also www.stereophile.com/content/cut-and-thrust-riaa-lp-equalization-page-2 Accessed June 2014.

4. Tobey, R., and Dinsdale, J. "Transistor High-Fidelity Pre-Amplifier" *Wireless World*, Dec 1961, p. 621.

5. Carter, E., and Tharma, P. "Transistor High-Quality Pre-amplifier" *Wireless World*, WW, Aug 1963, p. 376.

6. Cordell, R. "VinylTrak—A Full-Featured MM/MC Phono Preamp" *Linear Audio*, Volume 4, Sept 2012, p. 131.

7. Tobey, R., and Dinsdale, J. "Transistor High-Quality Audio Amplifier" *Wireless World*, Jan 1965, p. 2.

8. Self, D. "A Precision Preamplifier" *Wireless World*, Oct 1983.

9. Self, D. "Elektor Preamplifier 2012" *Elektor*, Apr, May, June 2012.

10. Vogel, B. *The Sound of Silence*. 2nd edition. Springer, 2011, p. 523. ISBN 978-3-642-19773-d.

11. Baxandall, P. Letter to Editor "Comments on 'On RIAA Equalisation Networks'" *JAES*, Volume 29 #1/2, Jan/Feb 1981, pp. 47–53.

12. Walker, H. P. "Low-Noise Audio Amplifiers" *Electronics World*, May 1972, p. 233.

13. Vogel, B. "Low-Noise Audio Amplifiers" *Electronics World*, May 1972, pp. 220–224.

14. Stanley Kelly Ortofon S15T cartridge review. *Gramophone*, Oct 1966.

15. Raalte, S. "Correcting Transducer Response With an Inverse Resonance Filter" *Linear Audio*, Volume 3, Apr 2012, p. 69.

16. Russell, G. H. "Inexpensive Pickups on Long-Playing Records" *Wireless World*, July 1953, p. 299.

17. Holman, T. "New Factors in Phonograph Preamplifier Design" *Journal of Audio Engineering Society*, May 1975, p. 263.

Chapter 8

Archival and Non-Standard Equalisation

Archival Transcription

This chapter deals with the differing requirements for reproducing discs, or indeed cylinders, that are not made to the microgroove standard. The subject of archival transcription is a large and complex one, and here I concentrate on the electronics. For disc replay there are many factors relating to stylus size and shape, tracking weight, etc., which we cannot explore here.

The major electronic problem is replay equalisation. Forget about having one universally accepted equalisation, like the RIAA characteristic for microgroove discs. As you have seen in earlier chapters, implementing the RIAA both accurately and economically is quite a challenge. Early discs used a wide variety of equalisations, and so an archival phono preamplifier must be able to provide all these in an accurate manner, though there is no need for great economy in specialised equipment that will only be made in very small numbers.

Although the media involved is always monaural, a stereo cartridge is normally used for transcription, and archival preamplifiers are likewise stereo, with a facility for summing varying proportions of the two channels to create a final mono output. This is because the two walls of a mono groove are unlikely to be identical, and the best results may be obtained by summing them unequally.

The preservation and distribution of archival material in analogue formats requires that it be replayed with the highest quality attainable and converted to digital. The most comprehensive book I am aware of on the general subject is the *Manual Of Analogue Sound Restoration Techniques*[1] by Peter Copeland.[2]

Coarse-Groove Discs

The great majority of the material for archival transcription is represented by coarse-groove discs, often referred to as "78s" because they were designed to rotate at 78 rpm rather than the 33 rpm of microgroove discs. They were produced from 1898 until the mid-Fifties, and were composed of mineral powders in an organic binder ("shellac") which gave much higher surface noise than the vinyl later used for microgroove records. Playing time was limited to 3 minutes, later extended to 6. Table 8.1 summarises the radical changes in groove dimensions required to get 22 minutes playing time from microgroove discs; completely different styli are required.

Wax Cylinders

As I have made clear, I think vinyl is an obsolete technology. However, if you're going to be retro, I say no half-measures. The first medium for recording and playback was tin-foil on a cylinder, introduced by Edison in 1877, but the results were poor even by the standards of the day. Wax cylinders proved much better. A standard cylinder system was agreed by Edison Records, Columbia Phonograph, and others in the late 1880s. These cylinders were 4 inches (10 cm) long, 2¼ inches in diameter, and played for about two minutes. The grooves made 100 turns per lateral inch.

Disc records appeared in 1901, first 10-inch, then later 12-inch. These played for about 3 and 4 minutes respectively but had poor quality compared with contemporary cylinders. To meet the competition on playing time, Edison introduced the Amberol cylinder in 1909, which increased the maximum playing time to 4.5 minutes, turning at 160 rpm, by increasing the groove density to

Table 8.1 Approximate groove dimensions for coarse and microgroove discs

	Coarse groove		Microgroove	
	Micron (μm)	Mil (inch/1000)	Micron (μm)	Mil (inch/1000)
Groove width	150	6	50	2
Groove depth	75	3	25	1
Groove bottom radius	38	1.5	8	0.3
Flat intergroove spacing	100	4	38	2.5
Stylus tip radius	63	2.5	12.5–23	0.5–1

200 turns per lateral inch. Later came the beautiful Blue Amberol cylinders, with a celluloid playing surface on a core of plaster of Paris. Despite these developments, discs decisively defeated cylinders in what must have been the first audio format war, though Edison continued to produce new cylinders until October 1929.

Wax cylinders have certain advantages. The format is inherently linear-tracking, which eliminates all the problems of angular alignment and stylus side-force. A worm gear used to move the stylus in alignment with the grooves on the cylinder, whereas disc replay uses the grooves to pull the stylus across the playing surface, creating a side-force and increasing groove wear. The speed of the stylus in the groove is constant, which eliminates the end-of-side distortion on the inner grooves of discs. Around 1900, it was acknowledged that cylinders in general had significantly better audio quality than discs, but by 1910 disc technology had improved and the difference disappeared.

Probably the greatest disadvantage of cylinders compared with discs is the space they take up. The volume on the inside of the cylinder has no use beyond supporting the player surface. Discs, however, can be stacked in a compact pile with no wasted space.

Various one-off cylinder replay machines have been built for the transcription of old recordings. These include the Archéophone[3] designed by Henri Chamoux, in 1998, and a cylinder player built by BBC engineers in the early 1990s. The latter used a linear-tracking arm from a contemporary turntable and an Ortofon cartridge. It could use a wide range of different equalisations and was used to transfer archival content to DAT tapes. Cylinders can also be "replayed" by optical scanning, which has the great advantage that it can cause no groove damage.

It is impossible to give a single correct equalisation characteristic for the replay of cylinders. The recordings were made acoustically, without any standard electrical equalisation being applied, as would later be done as part of the RIAA standard, and each case must be decided on its merits. Variable equalisation preamplifiers exist but are rare and expensive. There is information on how to build them later in this chapter.

It is, as you might imagine, now extremely unusual for music to be released in cylinder format—a notable (and probably unique) exception being the release of the track "Sewer" in 2010 by the British steampunk band The Men That Will Not Be Blamed For Nothing.[4] This was a very limited edition indeed; only 40 cylinders were produced and only 30 were put on sale.

Non-Standard Replay Equalisation

The familiar RIAA curve for microgroove records has three basic corner frequencies, f3, f4 and f5, plus f2 if you count the IEC amendment. This is the Lipshitz numbering convention. Early replay equalisation was simpler than this, with only f4 specified as the "bass turnover frequency" below which the signal received bass boost at 6 dB/octave.

The RIAA curve is usually specified in terms of its time-constants. These have the Lipshitz names of T3, T4, and T5, corresponding to the turnover frequencies f3, f4, and f5, and their values are 3180 usec (f3 = 50.05 Hz) 318 uS (f4 = 500.5 Hz) 75 usec (f5 = 2112 Hz). The time-constant T2 is the IEC amendment of 7960 usec (f2 = 20.02 Hz), but this merely acts as a feeble substitute for a subsonic filter and is of no interest here, as it is only intended to be used with microgroove (LP) discs. Any preamp intended for archival transcription is almost certain to have a proper subsonic filter, often with a variable cutoff frequency. Nonstandard equalisation is more often defined in terms of the frequencies rather than the time-constants. The two are easily interconverted; see Equations 8.1 and 8.2.

$$f = \frac{1}{2\pi T} \qquad\qquad 8.1$$

$$T = \frac{1}{2\pi f} \qquad\qquad 8.2$$

All cartridges (except for very rare strain-gauge types) respond to the velocity of the stylus and not its displacement. If a recording is made to suit this, then the groove amplitude will continuously decrease from LF to HF; this is called constant-velocity recording. This causes two problems; excessively large groove amplitude at LF, which limits playing time and at some point becomes untrackable, and excessively small groove amplitude at HF, so surface noise becomes a problem. Early recording equalisation, from the mid-1920s, tackled the first problem by introducing a 6 dB/octave LF roll-off when recording, starting at a set frequency f4 and giving constant-amplitude recording below that. An example is curve A in Figure 8.1, which corresponds to the Radiofunken characteristic in Table 8.2. In the literature f4 is often called the "bass turnover frequency". To undo this at replay complementary LF boost had to be provided, starting at the same frequency. This boost had to be curtailed before subsonic frequencies were reached, or record warps and turntable rumble would be grotesquely exaggerated. Therefore replay equalisation always had another frequency, f3, where the gain flattened out; I have used 50 Hz in Figure 8.1. Since f3 was not part of any official recording characteristic until about 1940, it seems to have been left to the judgement of those designing replay amplifiers.

Later, from the mid-1930s, the second problem of excessively small groove amplitude at HF was addressed by increasing the HF content recorded. This is called pre-emphasis and relies on the fact that an audio signal has relatively low levels at HF. At replay HF cut (de-emphasis) is applied to undo the pre-emphasis, and this attenuates the surface noise. See curve B in Figure 8.1,

Table 8.2 Generally agreed replay equalisation frequencies (without de-emphasis)

Manufacturer or organisation	f4 Hz
Acoustic gramophone	0
AES standard	400
Brunswick	500
Columbia (Eng.)	250
Decca 78	150
Early 78s (mid-'30s)	500
EMI (1931)	250
HMV (1931)	250
NAB standard	500
New Records	750
Oriole	Inconsistent
Parlophone	500
Pathe	Inconsistent
Radiofunken	400

which corresponds to the Decca 1934 characteristic in Table 8.3. In the literature f5 is often called the "treble transition frequency". The HF cut curve also dropped at 6 dB/octave, for ease of implementation, except in the case of Decca's FFRR (Full Frequency Range Recording) records, which fell at a 3 dB/octave slope. I have found no information on how this was supposed to be implemented—presumably it was approximated by using extra overlapping time-constants. See my cross-over book [5] for how to do this with varying degrees of accuracy.

There were dozens of different recording characteristics used, and choosing the corresponding replay characteristic is a significant problem in itself. The history of even one record label can be complicated with different characteristics used in different years, and according to Gary Galow sometimes even on different sides of the same disc. The histories of the characteristics are not universally agreed and are subject to some debate, much of which can be found on the Internet. In Tables 8.2 and 8.3 I have done my best to give some generally accepted examples, including the extremes, so we can see what range of frequencies is required in an archival preamp. It is far from comprehensive, and much greater depth of information can be found in Peter Copeland's book. [1] Other useful references are[6] and[7]. In most of the literature the frequencies f4 and f5 are used rather than time-constants. Sometime f5 is not quoted directly but in

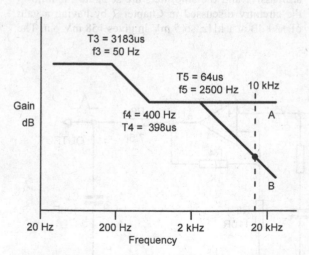

Figure 8.1 Early equalisation systems used replay curves like A; later ones like B added de-emphasis to attenuate surface noise; f5 is sometimes specified in terms of the attenuation at 10 kHz

Table 8.3 Generally agreed replay equalisation frequencies (with de-emphasis)

Manufacturer	f4 Hz	f5 Hz	10 kHz attenuation dB
Capitol (1942)	400	2500	−12
Columbia (1925)	200 (250)	5500 (5200)	−7 (−8.5)
Columbia (1938)	300 (250)	1590	−16
Decca (1934)	400	2500	−12
Decca FFRR (1949)	250	3000*	−5
London FFRR (1949)	250	3000*	−5
Mercury	400	2500	−12
MGM	500	2500	−12
Victor (1925)	200–500	5500 (5200)	−7 (−8.5)
Victor (1938–47)	500	5500 (5200)	−7 (−8.5)
Victor (1947–52)	500	2120	−12

* 3 dB/octave slope above f5

terms of the attenuation at 10 kHz with respect to 1 kHz. In difficult cases the only thing to do is to judge by ear the correct characteristic to use, and this has implications that will surface later.

The entry for "acoustic gramophone" in Table 8.2 indicates that no equalisation at all was used, giving a straight constant-velocity characteristic. This was suitable for purely mechanical reproducers such as the Orthophonic Victrola with its folded exponential horn. Note the entries marked "inconsistent".

Tables 8.2 and 8.3 show that f4 must be variable at least between 150 and 750 Hz, and Table 8.3 shows that f5 must be at least variable between 1590 and 5500 Hz. Having decided on f3, f4, and f5, you have to implement them. In Chapter 7 I stated firmly that RIAA equalisation should be done in one active series-feedback stage and demonstrated that all other approaches involving partly passive RIAA introduced serious compromises in noise or headroom, or both. The downside is that the active one-stage method is much more difficult to design because of the interacting values in the RIAA feedback network.

For archival transcription the priorities are different. If f3, f4, and f5 are combined in one network, changing one of them requires recalculating the whole network and then changing every component in it. This is obviously impractical, and the only answer is to use some form of partly passive configuration that allows the time-constants to be set independently. Increased electronic noise is not a problem because it will be well below the surface noise of ancient media. Headroom is likewise unlikely to be an issue because of the generally low recording levels and the ease of including a flat variable-gain stage in a partly-passive configuration.

Figure 8.2 shows a typical Passive-Passive equalisation circuit (so-called because both equalisation stages are passive and the amplifiers are separate). It follows the circuitry discussed in Chapter 7 by having a gain of +30 dB at 1 kHz, so 5 mV in gives 158 mV out. The

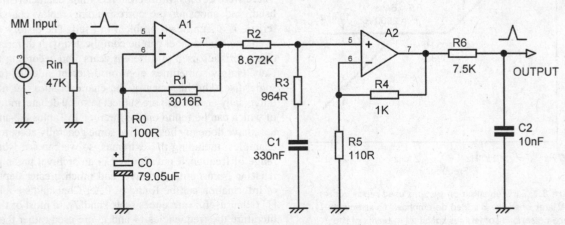

Figure 8.2 Passive-Passive equalisation configuration

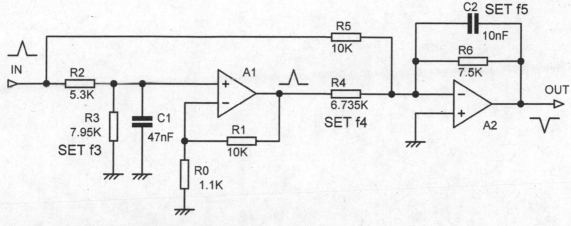

Figure 8.3 Partly passive equalisation with constant gain at 1 kHz when the time-constants are altered

values of the equalisation components are taken from Gary Galow's *Linear Audio* article,[8] for reasons that will become clear; they give an RIAA response accurate to with ±0.02 dB. As noted the IEC amendment f2 is unlikely to be used, but if required it can be implemented by making C0 equal to 79.05 uF as shown. If f2 is not required, C0 should be 220 uF or 470 uF. In this circuit f3 is set by the time-constant T3, the product of (R2 + R3) and C1. Likewise f4 is set by the time-constant T4, the product of R3 and C1. The treble turnover frequency f5 is set simply by R6 and C2; the output must not be significantly loaded, and in many cases a buffer stage will be needed. Here it is very clear that 7.5 kΩ and 10 nF give a T5 of 75 us, so f5 is 2112 Hz. If f5 is not used then R6 and C2 are simply omitted. It is normal practice to make the time-constants switchable. Making them fully variable by using ganged pots would much degrade the matching of the stereo channels.

Sometimes you have to judge by ear the "correct" characteristic to use, or at least one that gives satisfactory results. When doing this, it is very convenient if the gain stays the same at 1 kHz while the frequency extremes are altered. This does not occur with the simple circuit of Figure 8.2, and great credit goes to Gary Galow and Mike Shields for devising the circuit in Figure 8.3.[8] The values here give the standard RIAA curve, to allow easy checking of the circuit operation; when simulated using 5534s it is accurate to ±0.02 dB. The circuit works by summing the path through R5, which sets the 1 kHz gain, with the path through R4, which sets f3 and f4. Once again it is easy to see that R6 = 7.5 kΩ and C2 = 10 nF gives a T5 of 75 us and thus an f5 of 2112 Hz. This is easily adjusted by altering R6. The equations linking f3

and f4 with the related component values are more complicated but are fully explained in Galow's article.[8]

This circuit is purely an equaliser and not an amplifier; it has unity gain at 1 kHz. Be aware that adding a C0 in series with R0 will *not* give an accurate IEC amendment f2, because of the two paths.

References

1. Copeland, P. *Manual of Analogue Sound Restoration Techniques*. London: The British Library, September 2008. This can be downloaded free of charge from www.bl.uk/reshelp/findhelprestype/sound/anaudio/analoguesoundrestoration.pdf Accessed Nov 2016.

2. Wikipedia entry on Copeland, Peter. https://en.wikipedia.org/wiki/Peter_Copeland Accessed Nov 2016.

3. https://en.wikipedia.org/wiki/Arch%C3%A9ophone Accessed Nov 2016.

4. https://en.wikipedia.org/wiki/The_Men_That_Will_Not_Be_Blamed_for_Nothing Accessed Nov 2016.

5. Self, D. *The Design of Active Crossovers*. Focal Press, 2011, pp. 335–339. ISBN 978-0-240-81738-5.

6. Audacity user guide http://wiki.audacityteam.org/wiki/78rpm_playback_curves Accessed Nov 2016.

7. https://midimagic.sgc-hosting.com/mixcurve.htm Accessed Nov 2016.

8. Galow, G. "An Archival Phono Amplifier" *Linear Audio*, Volume 5, p. 77. www.linearaudio.net/ Accessed Sept 2013.

Chapter 9

Moving-Magnet Inputs

Noise and Distortion

Noise in MM RIAA Amplifiers

The subject of noise in moving-magnet (MM) RIAA preamplifiers is an involved business. An MM cartridge is a combination of resistance and a significant amount of inductance, with neither parameter having standard values, and this is combined with the complications of RIAA equalisation.[1] Burkhard Vogel has written a monumental 740-page book solely on RIAA amp noise,[2] but even this does not exhaust the subject; I will do the best I can in a chapter. The basic noise mechanisms are described in Chapter 1.

The first priority is to find out the physical limits that set how low the noise can be, with the electronics considered alone, i.e. with no groove noise. The best possible equivalent input noise (EIN) for a purely resistive source, such as a 200 Ω microphone, is easily calculated to be −129.6 dBu with a noiseless amplifier at the usual temperature and bandwidth, but the same calculation for a moving-magnet input is much harder. Real amplifiers have their own noise, and the amount by which the source-amplifier combination is noisier than the source alone is the noise figure (NF), and we want to get this as low as possible. Noise figures are rarely if ever used in audio specifications, probably because they are very revealing; an NF of 20 dB usually indicates that someone doesn't know what they're about. Manufacturers seem to have no interest at all in quoting MM noise specs in a way that would allow easy comparison.

Most of the complications in calculating theoretical noise occur only when an MM cartridge is driving an RIAA preamp directly. When an MC cartridge is in use, the RIAA stage will be driven from an MC head amp, the output impedance of which should be very low, and this makes the noise situation much simpler (see later in this chapter).

A-weighting is not used in this chapter (or any other) except where explicitly stated.

Vinyl Groove Noise

The purely electronic noise will be much lower than the noise generated by the stylus sliding along a groove, unless you have a very peculiar (and probably valve-based) phono amplifier. The *Radio Designer's Handbook* says that the groove noise (sometimes called surface noise) of vinyl is 60–62 dB below maximum recorded level, and this will be 20 dB or more above the electronic noise.[3] This is the dynamic range rather than the signal/noise ratio. The best modern reference is Burkhard Vogel's book, which devotes Chapter 11 (22 pages) to groove noise.[4] He concludes that the best direct metal master mother discs achieve a signal/noise ratio of −72 dB (A-weighted), and another 2 dB are lost in getting to the final record. Non-DMM discs will show −64 dB (A-weighted) or worse. This seems to be as good as it gets, because groove noise increases as the record wears with playing.

We are therefore looking at (or rather listening to) groove noise which is between −70 and −64 dB below nominal level, not forgetting the A-weighting. Later in this chapter you will see that a humble 5534A in a simple phono amplifier gives a signal/noise ratio of −78.7 dB (unweighted) or −81.4 dBA (A-weighted) without load synthesis, which is only 3.1 dB worse than a wholly noiseless amplifier. Therefore in the best DMM case the groove noise will be 11.4 dB above the electronic noise, and even if we managed to make the electronics completely noiseless the total noise level would only drop by 0.33 dB; this would not be detectable even in an A/B comparison.

A very experienced vinyl enthusiast (60 years plus, going back to shellac) told me that he had never, ever, known noise fail to increase when the stylus was lowered onto the run-in groove.[5]

This presents a philosophical conundrum; is it not a waste of time to strive for low electronic noise when the groove noise is much greater, and the contribution of the electronic noise negligible? If obtaining a good electronic noise performance was difficult and expensive this argument would have more force, but it is simply not so. This chapter will show how to get within about 2 dB of the lowest noise physically possible using cheap opamps and a little ingenuity.

There is also specmanship, of course. The lower the noise specification the better the sales prospects? One might hope so.

Cartridge Impedances

The impedance of the cartridge strongly influences the noise performance of an MM RIAA stage. Manufacturers do not always supply this data, and so I have had to make the best of what is available. Some of the cartridges listed in Table 9.1 are vintage, some are up to date, the collection covering from about 1972 to 2016. Resistance ranges from 430 Ω to 1550 Ω, and inductance generally from 330 to 720 mH, apart from the Grado series, which are more moving-iron than MM in operation, but given their 5 mV per cm/s output they are going to be used with an MM input. Moving-iron pickups go back a long way; see an article by Francis in *Wireless World* for 1947,[6] where a 1:100 step-up transformer was used to get the signal up to a suitable level to apply to a valve preamplifier. The Shure V15V values have been confirmed by Burkhard Vogel.[7]

The inductance of the cartridge is a very important element in determining the noise performance, as will shortly be made clear. Figure 9.1 shows the inductance of 51 cartridges, covering both historical and contemporary models. The six types in the "0–300 mH" column are the Grado series. The seven types in the rightmost two columns are DJ types with higher output and inductance than normal hi-fi cartridges; see Chapter 8.

Noise Modelling of RIAA Preamplifiers

The basic noise situation for a series-feedback RIAA stage using an opamp is shown in Figure 9.2. The cartridge is modelled as a resistance Rgen in series with a significant inductance Lgen and is loaded by the standard 47 kΩ resistor Rin; this innocent-looking component causes more mischief than you might think. The

Table 9.1 Some MM cartridge impedances, both current and historical

Type	Resistance Ω	Inductance mH
Audio-Technica AT440	Not stated	490
Audio-Technica AT15SS	500	720
CS1 "Carl Cox"	430	400
Glanz MFG-31E	900	110
Goldring 1006	660	570
Goldring 1012GX	660	570
Goldring 1042	660	570
Goldring 2044	Not stated	720
Goldring 2100	550	550
Goldring 2200, 2300	550	680
Goldring 2400, 2500	550	720
Goldring Elan	700	560
Goldring Elektra	700	560
Grado Prestige Green 1, Black 1	475	45
Grado Prestige Red 1, Blue 1	475	45
Grado Reference Sonata 1, Platinum 1	475	45
Ortofon 2M Red, Blue	Not stated	700
Ortofon 2M Bronze, Black	Not stated	630
Ortofon OM Super 20	1000	580
Shure ME75-ED Type 2	610	470
Shure ME95-ED	1500	650
Shure M97	1550	700
Shure V15V MR	815	330
Shure V15V V	815	330
Shure V15V IV	1380	500
Shure V15V III	1350	500
Shure M44G	650	650
Stanton 5000 AL-II	535	400

amplifier A1 is treated as noiseless, its voltage noise being represented by the voltage generator Vnoise, and the current noise of each input being represented by the current generators Inoise+ and Inoise−, which are uncorrelated. It does not matter to which side of Vnoise the current generator Inoise+ is connected, because

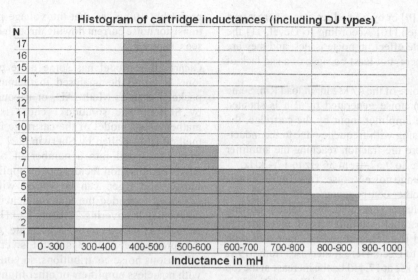

Figure 9.1 MM cartridge inductance, including DJ cartridges

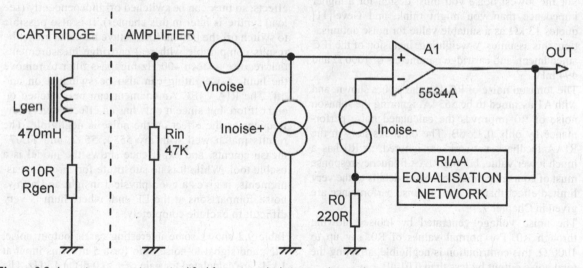

Figure 9.2 A moving-magnet input simplified for noise calculations, with typical cartridge parameter values (Shure ME75-ED2)

Vnoise has no internal resistance and the connections are equivalent. I do not consider $1/f$ noise and other low-frequency electronic disturbances in detail because there is absolutely nothing you can do about them except choose an appropriately spec'd device or opamp.

The contributions to the noise at the input of A1 are:

1) The Johnson noise of the cartridge resistance Rgen. This sets the ultimate limit to the signal/noise ratio. The proportion of noise from Rgen that reaches the amplifier input falls with frequency as the impedance

of Lgen increases. Here the fraction reaching the amplifier falls from 0.99 to 0.48, from 36 Hz to 17.4 kHz. A complication that is not visible in the diagram is that the effective value of Rgen is not simply the resistance of the coils. It increases in value with frequency (while still remaining resistive—we are not talking about inductance here) as a consequence of hysteresis and eddy current magnetic losses in the iron on which the coils are wound, and possibly skin effect.[8] These losses are sometimes modelled by a frequency-dependent resistance placed across Lgen, as by Hallgren,[9] or by a fixed resistance across part of a tapped Lgen.[10] According to Gevel,[11] the losses have little effect on noise issues, and they are not modelled here, not least because of a sad lack of data.

2) The Johnson noise of the 47 kΩ input load Rin. Some of the Johnson noise generated by Rin is shunted away from the amplifier input by the cartridge, the amount decreasing with frequency due to the inductance Lgen. Here the fraction reaching the amplifier rises from 0.013 to 0.52, from 36 Hz to 17.4 kHz.

3) The opamp voltage noise Vnoise. This contribution is unaffected by other components.

4) The noise voltage generated by Inoise+ flowing through the parallel combination of the cartridge impedance and Rin. This impedance increases with frequency due to Lgen. Here it increases from 619 Ω at 36 Hz to 24.5 kΩ at 17.4 kHz; the increase at the top end is moderated by the shunting effect of Rin. This increase has a major effect on the noise behaviour. For the lowest noise you must design for a higher impedance than you might think, and Gevel[11] quotes 12 kΩ as a suitable value for noise optimisation; this assumes A-weighting, inclusion of the IEC amendment, and cartridge parameters of 1000 Ω and 494 mH.

5) The Johnson noise of R0. For the values shown, and with A1 assumed to be 5534A, ignoring the Johnson noise of R0 improves the calculated noise performance by only 0.35 dB. The other resistors in the RIAA feedback network are ignored, as R0 has a much lower value, but the RIAA frequency response must of course be modelled. More details of the very limited effect that R0 has on noise performance are given in Chapter 7.

6) The noise voltage generated by Inoise− flowing through R0. For normal values of R0, say up to 1000 Ω, this contribution is negligible, affecting the total noise output by less than 0.01 dB.

Contributions 1, 2, and 4 are significantly affected by the rising impedance of the cartridge inductance Lgen with frequency. On top of this complicated frequency-dependent behaviour is overlaid the effect of the RIAA equalisation. This would reduce the level of white noise by 4.2 dB, but we are not dealing with white noise— the HF part of the spectrum has been accentuated by the effects of Lgen, and with the cartridge parameters given, RIAA equalisation actually reduces the noise amplitude by 10.4 dB.

The model as shown does not include the input DC-blocking capacitor Cin. This needs to be 47 uF, or preferably 100uF, so that the voltage produced by the transistor noise current flowing through it is negligible— see Chapter 7.

Clearly this model has some quite complex behaviour. It could be analysed mathematically, using a package such as MathCAD, or it could be simulated by SPICE. The solution I chose is a spreadsheet mathematical model of the cartridge input. The basic method is described by Sherwin.[12] The audio spectrum is divided into a number of octave bands so RIAA equalisation factors can be applied, and Vnoise, Inoise, and Rgen can be varied with frequency if desired. I extended the Sherwin scheme by using ten octave bands covering 22 Hz to 22 kHz; ten bands are enough to make the process accurate. An advantage of the spreadsheet method is that it is very simple to turn off various noise contributions, so you can experiment with noiseless amplifiers or other flights from physical reality. For example, the noise generated by the 47 kΩ resistor Rin is modelled separately from its loading effects, so they can be switched off independently (see load synthesis later in this chapter). It is also possible to switch off the bottom four octave bands to make the results comparable with real cartridge measurements that require a steep 400 Hz highpass filter to remove the hum. A-weighting can also be switched on and off. The RIAA IEC Amendment can be switched on and off too, but since it only has an effect on very low frequencies the effect on the noise is negligible. The results match well with my 5534, 5532, and TL072 measurements, and experience shows the model is a usable tool. While it is no substitute for careful measurements, it gives a good physical insight and allows noise comparisons at the LF end, where hum is very difficult to exclude completely.

Table 9.2 shows some interesting cases; output noise, EIN, and signal-to-noise ratio for a 5 mV rms input at 1 kHz are calculated for gain of +30.0 dB at 1 kHz. The IEC amendment is included. The cartridge parameters were set to 610 Ω + 470 mH, the measured values for the Shure M75ED 2. Bandwidth is 22 Hz–22 kHz, no A-weighting is used, and $1/f$ noise is not considered. Be aware that the 5534A is a low-noise version of the 5534, with a typical voltage noise density of 3.5 rather than 4 nV√Hz, and a typical current noise density of 0.4 rather than 0.6 pA/√Hz. There is also an A-version of the 5532, but curiously the data sheets show no noise advantage. The voltage noise and current noise densities used here are the manufacturer's "typical" figures. I am not aware of any data on how much they vary around the quoted values in practice.

Table 9.2 RIAA noise results from the MAGNOISE2 spreadsheet model under differing conditions, in order of quietness. Cases 0 to 3 assume a noiseless amplifier and are purely theoretical. Cartridge parameters 610 Ω + 470 mH. Unweighted.

Case	Amplifier type	e_n nV/rtHz	i_n pA/rtHz	Rin Ω	R0 Ω	Noise output dBu	S/N ref 5mV input dB	EIN dBu	NF ref Case 2 dB	Ref Case 6a dB
0	Noiseless amp, no Rgen	0	0	1000M	0	−136.8	−123.1	−166.8	−41.2	−44.3
1a	Noiseless amp	0	0	1000M	0	−99.4	−85.6	−129.4	−3.2	−6.3
1b	Noiseless amp	0	0	10M	0	−99.3	−85.5	−129.3		
1c	Noiseless amp	0	0	1M5	0	−98.9	−85.1	−128.9		
1d	Noiseless amp	0	0	1M	0	−98.7	−84.9	−128.7		
1e	Noiseless amp	0	0	1M5	220	−97.5	−83.7	−127.5		
1f	Noiseless amp	0	0	1M	220	−97.4	−83.6	−127.4		
2	Noiseless amp	0	0	47k	0	−95.6	−81.8	−125.6	0 dB ref	−3.1
3	Noiseless amp	0	0	47k	220	−94.9	−81.1	−124.9	0.7	−2.4
4a	2SK710 FET, Id = 2 mA	0.9	0	47k	220	−94.7	−80.9	−124.7	0.9	−2.2
4b	J310 FET, Id = 10 mA	2	0	47k	220	−94.2	−80.4	−124.2	1.4	−1.7
5a	2SB737 BJT, *Ic* =70uA	1.75	0.39	47k	220	−93.6	−79.8	v123.6	2.0	−1.1
5b	2SB737 BJT, *Ic* =100uA	1.47	0.46	47k	220	−93.4	−79.6	−123.4	2.2	−0.9
5c	2SB737 BJT, *Ic* =200uA	1.04	0.65	47k	220	−92.7	−78.9	−122.7	2.9	+0.2
6a	5534A BJT	3.5	0.4	47k	220	−92.5	−78.7	−122.5	3.1	0 dB ref
6b	5534A BJT	3.5	0.4	47k	470	−92.1	−78.3	−122.1	3.5	+0.4
7	OPA1642 JFET	5.1	0.0008	47k	220	−91.8	−78.0	−121.8	3.8	+0.7
8	5534A BJT	3.5	0.4	47k	1000	−91.4	−77.6	−121.4	4.2	+1.1
9	5532A BJT	5	0.7	47k	220	−90.5	−76.5	−120.5	5.1	+2.0
10	OPA2134 JFET	8	0.003	47k	220	−89.3	−75.5	−119.3	6.3	+3.2
11	LM4562 BJT	2.7	1.6	47k	220	−87.9	−74.1	−117.9	7.7	+4.6
12	LME49720 BJT	2.7	1.6	47k	220	−87.9	−74.1	−117.9	7.7	+4.6
13	OPA604 JFET	10	0.004	47k	220	−87.9	−74.1	−117.9	7.7	+4.6
14	OP275 BJT+JFET	6	1.5	47k	220	−87.3	−73.5	−117.3	8.3	+5.2
15	AD797 BJT	0.9	2	47k	220	−86.6	−72.8	−116.6	9.0	+5.9
16	TL072 JFET	18	0.01	47k	220	−83.4	−69.6	−113.4	12.2	+9.1
17	LM741 BJT	20	0.7 ?	47k	220	−82.4	−68.6	−112.4	13.2	+10.1

First let us see how quiet the circuit of Figure 9.2 would be if we had miraculously noise-free electronics.

Case 0: We will begin with a completely theoretical situation with no amplifier noise and an MM cartridge with no resistance Rgen. Lgen is 470 mH. Rin is set to 1000 MΩ; the significance of that will be seen shortly. The noise out is a subterranean and completely unrealistic −136.8 dBu, and that is *after* +30 dB of amplification. This noise comes wholly from Rin and can be reduced without limit if Rin is increased without limit. Thus if Rin is set to 1000 GΩ the noise out is −166.8 dBu.

You may ask why the noise is going up as the resistance goes down, whereas it is usually the other way around. This is because of the high cartridge inductance, which means the Johnson noise of Rin acts as a current rather than a voltage, and this goes up as the Rin resistance goes down.

Case 1: We now switch on the Johnson noise from Rgen (610 Ω). We will continue to completely ignore the cartridge loading requirements and leave Rin at 1000 MΩ, at which value it now has no effect on noise. The output noise with these particular cartridge parameters is then −98.8 dBu (Case 1a). This is the quietest possible condition (if you can come up with a noiseless amplifier), but you will note that right from the start the signal/noise ratio of 85 dB compares badly with the 96 dB of a CD, a situation that merits some thought. And there is, of course, no groove noise on CDs. All of this noise comes from Rgen, the resistive component of the cartridge impedance. The only way to improve on this would be to select a cartridge with a lower Rgen but the same sensitivity, or start pumping liquid nitrogen down the tone arm. (As an aside, if you *did* cool your cartridge with liquid nitrogen at −196 °C, the Johnson noise from Rgen would only be reduced by 5.8 dB, and if you are using a 5534A in the preamplifier, as in Case 6a below, the overall improvement would only be 0.75 dB. And, of course, the compliant materials would go solid and the cartridge wouldn't work at all. Hold the cryostats!)

With lower, but still high, values of Rin the noise increases; with Rin set to 10 MΩ (Case 1b) the EIN is −128.7 dBu, a bare 0.1 dB worse. With Rin set to 1 MΩ (Case 1c) the EIN is now −128.2 dBu, 0.8 dB worse than the best possible condition. (Case 1a)

Case 2: It is however a fact of life that MM cartridges need to be properly loaded, and when we set Rin to its correct value of 47 kΩ things deteriorate sharply, the EIN rising by 3.2 dB (compared with Case 1a) to −125.6 dBu. That 47 kΩ resistor is not innocent at all. This case still assumes a noiseless amplifier and appears to be the appropriate noise reference for design, so the noise figure is 0 dB. (However, see the section on load synthesis later in this chapter, which shows how the effects of noise from Rin can be reduced by some non-obvious methods.) Cases 1a,b,c therefore have negative noise figures, but this has little meaning.

Case 3: We leave the amplifier noise switched off but add in the Johnson noise from R0 and the effect of Inoise− to see if the value of 220 Ω is appropriate. The noise only worsens by 0.7 dB, so it looks like R0 is not the first thing to worry about. Its contribution is included in all the cases that follow. The noise figure is now 0.7 dB.

We will now take a deep breath and switch on the amplifier noise.

Case 4: Here we use a single J310 FET, a device often recommended for this application.[11] With the drain current Id set to 10 mA, the voltage noise is about 2 nV√Hz; the current noise is negligible, which is why it is overall slightly quieter than the 2SB737 despite having more voltage noise.

Case 5: In these cases a single discrete bipolar transistor is used as an input device, not a differential pair. This can give superior noise results to an opamp. The transistor may be part of a fully discrete RIAA stage, or the front end to an opamp. If we turn a blind eye to supply difficulties and use the remarkable 2SB737 transistor (with Rb only 2 Ω typical), then some interesting results are possible. We can decide the collector current of the device, so we can to some extent trade off voltage noise against current noise, as described in Chapter 1. We know that current noise is important with an MM input, and so we will start off with quite a low Ic of 200 uA, which gives Case 5c in Table 9.2. The result is very slightly worse than the 5534A (Case 6a). Undiscouraged, we drop Ic to 100 uA (Case 5b), and voltage noise increases but current noise decreases, the net result being that things are now 0.9 dB quieter than the 5534A. If we reduce Ic again to 70 uA (Case 5a), we gain another 0.2 dB, and we have an EIN of −123.6 and a noise figure of only 2.0 dB.

Voltage noise is now increasing fast, and there is virtually nothing to be gained by reducing the collector current further.

We therefore must conclude that even an exceptionally good single discrete BJT with appropriate support circuitry will only gain us a 1.1 dB noise advantage over the 5534A, while the J310 FET gives only a 1.7 dB advantage, and it is questionable if the extra complication is worth it. You are probably wondering why going from a single transistor to an opamp does not introduce a 3 dB noise penalty, because the opamp has a differential input with two transistors. The answer is that the second opamp transistor is connected to the NFB network and sees much more favourable noise conditions; a low and resistive source impedance in the shape of R0.

The 2SB737 is now obsolete. For information on replacements see Chapter 11.

Case 6: Here we have a 5534A as the amplifying element, and using the typical 1 kHz specs for the A-suffix part, we get an EIN of −122.5 dBu and an NF of 3.1 dB. (Case 6a with R0 = 220 Ω) Using thoroughly standard technology, and one of the cheapest opamps about, we are within 3 decibels of perfection; the only downside is that the opportunities for showing off some virtuoso circuit design with discrete transistors appear limited. Case 6a is useful as a standard for comparison with other cases, as in the rightmost column of Table 9.2.

First, how does the value of R0 affect noise? In Case 6b R0 is increased to 470 Ω, and the noise is only

0.4 dB worse; if you can live with that, the increase in the impedance of the RIAA feedback network allows significant savings in expensive precision capacitors. Reducing R0 from 220 Ω to 100 Ω is doable at some cost in capacitors but only reduces the noise output by 0.2 dBu. In Chapter 7 the value of R0 can be manipulated to get convenient capacitor values in the RIAA network, because it has only a weak effect on the noise performance.

Second, we have seen that the presence of Lgen has a big effect on the noise contributions. In Case 6a, if we reduce Lgen to zero the noise out drops from −92.5 to −94.7 dBu. Halving it gives −93.8 dBu. Minimum cartridge inductance is a good thing.

Third, what about Rgen? With the original value of Lgen, setting Rgen to zero only reduces the noise from −92.5 to −93.5 dBu; the cartridge inductance has more effect than its resistance.

Case 7: The OPA1642 is a relatively new JFET input opamp with noise densities of 5.1 nV/√Hz for voltage and a startlingly low 0.0008 pA/√Hz for current. This modern JFET technology gives another way to get low MM noise—accept a higher e_n in order to get a very low i_n. The OPA1642 gives an EIN of −121.8 dBu, beating the 5532 but not the 5534A with R0 = 220 Ω. At the time of writing the OPA1642 is something like 20 times more expensive than the 5532.

Case 8: We go back to the 5534A, with R0 now raised substantially further to 1000 Ω, and the noise is now 1.1 dB worse than the 5534A 220 Ω case. This is a good demonstration that the value of R0 is not critical.

Case 9: It is well-known that the single 5534A has somewhat better noise specs than the dual 5532A, with both e_n and i_n being significantly lower, but does this translate into a significant noise advantage in the RIAA application? Case 7 shows that on plugging in a 5532A the noise output increases by 2.0 dB, the EIN increasing to −120.5 dBu. The NF is now 5.1 dB, which looks a bit less satisfactory. If you want good performance, then the inconvenience of a single package and an external compensation capacitor are well worth putting up with. If your circuit design ends up with an odd number of half-5532s per channel, a 5534A can be placed in the MM stage, where its lower noise is best used.

Case 10: Here we try out the FET-input OPA2134, which is a good opamp when DC accuracy and low bias currents are required; we find the e_n is much higher at 8 nV/√Hz, but i_n is very low indeed at 3 fA√Hz. It looks like we might be in with a chance, but the greater voltage noise does more harm than the lower current noise does good, and the EIN goes up to −119.3 dBu. The OPA2134 is therefore 3.2 dB noisier than the 5534A and 2.5 dB noisier than the 5532A; and it is not cheap. The noise figure is now 6.3 dB, which to a practised eye would show that something had gone amiss in the design process.

Case 11: The LM4562 BJT input opamp gives significant noise improvements over the 5534/5532 when used in low-impedance circuitry, because its e_n is lower at 2.7 nV/√Hz. However, the impedances we are dealing with here are not low, and the i_n, at 1.6 pA√Hz, is four times that of the 5534A, leading us to think it will not do well here. We are sadly correct, with EIN deteriorating to −117.9 dBu and the noise figure an unimpressive 7.7 dB. The LM4562 is almost 5 dB noisier than the 5534A and at the time of writing is a lot more expensive. Measurements confirm a 5 dB disadvantage.

Case 12: The LME49720 is a recent BJT input opamp with the same voltage and current densities as the LM4562, and so gives the same EIN of −117.9 dBu, 5 dB noisier than the 5534A.

Case 13: The OPA604 is a FET-input opamp that is often recommended for MM applications by those who have not studied the subject very deeply. It has noise densities of 10 nV/√Hz for voltage and a low 0.004 pA/√Hz for current. This different balance of voltage and current noise results again in the same EIN of 117.9 dBu, 5 dB noisier than the 5534A.

Case 14: The OP275 has both BJT- and FET-input devices. Regrettably this appears to give both high voltage noise and high current noise, resulting in a discouraging EIN of −117.3 dBu and a noise figure of 8.2 dB. It is 5.2 dB noisier than a 5534A in the same circuit conditions. Ad material claims "excellent sonic characteristics", perhaps in an attempt to divert attention from the noise. It is expensive.

Case 15: The AD797 has very low voltage noise because of its large BJT input transistors, but current noise is correspondingly high, and it is noisy when used with an MM cartridge. And it is expensive, especially so since it is a single opamp with no dual version. Definitely not recommended for MM; allegedly useful in submarines.

Case 16: The TL072 with its FET-input has very high voltage noise at 18 nV√Hz but low current noise. We can expect a poor performance. We duly get it, with EIN rising to −113.4 dBu and a very indifferent noise figure of 12.2 dB. The TL072 is 9.1 dB noisier than a 5534A and 8.4 dB noisier than a 5532A. The latter figure is confirmed (within experimental error, anyway) by the data listed in the later section on noise measurements. There is now no reason to use a TL072 in an MM preamp; it must be one of the worst you could pick.

Case 17: Just for historical interest I tried out the LM741. The voltage noise measures about 20 nV√Hz. I have no figures for the current noise, but I think it's safe to assume it won't be better than a 5532, so I have used 0.7 pA√Hz. Predictably the noise is the highest yet, with an EIN of −112.4 dBu, but it is a matter for some thought that despite using a really ancient part it is only 10 dB worse than the 5534A. The noise figure is 13 dB.

Opamps with Bias Cancellation

You may be wondering what has happened to other well-known opamps, particularly the OP-27 and the LT1028. Both are sometimes recommended for audio use because of their low voltage noise density (e_n), but this ignores a serious problem. The OP-27 has a low e_n of 3 nV/√Hz and i_n of 0.4 pA/√Hz and from these figures alone gives a calculated EIN of −123.0 dB, which beats the 5532A noise, but . . . when you measure it in real life it is actually several dB noisier; I have confirmed this several times. This is due to extra noise generated by bias-current cancellation circuitry. Correlated noise currents are fed into both inputs and will only cancel if both inputs see the same impedance. In this MM RIAA application the impedances are wildly different, and the result is much increased noise. This problem with the OP-27 was originally pointed out to me by Marcel van de Gevel.[13]

The LT1028 gives a poor performance in MM applications because, while it has an appealingly low e_n of 0.85 nV√Hz, its i_n is high at 1 pA√Hz as a result of running big input BJTs at high collector currents. The EIN is by calculation −120.9 dBu, making it a shade quieter than the 5532A. But . . . the LT1028 also has bias-current cancellation circuitry, and the data sheet explicitly states: "The cancellation circuitry injects two correlated current noise components into the two inputs." According to Gevel,[11] the effective voltage noise in a typical MM application is about 39 nV/√Hz, so even the LM741 would be a better choice.

A relatively new addition to this group is the OP2227, which has promising noise densities of 3 nV/√Hz for voltage and 0.4 pA/√Hz for current, the same as the OP27, which figures give the same calculated EIN of −123.0 dB. It has however bias-current cancellation, and while I have no practical experience with this opamp, there seems no reason why it should not have the same excess noise problem as the OP27 and LT1028.

Hybrid Phono Amplifiers

In Table 9.2 the noise results are shown for single discrete devices as well as opamps. These cannot of course be used alone in a phono amplifier because of the need for both substantial open-loop gain and good load-driving ability. The discrete device can be used as the first stage of a discrete amplifier, as described in Chapter 10, but it is more convenient to combine the discrete device with an opamp, which will give both the open-loop gain and the load-driving ability required at lower cost and using less PCB area. The 5532 or 5534 is once again very suitable.

Figure 9.3 shows a basic arrangement. For optimal noise the Ic of Q1 will probably be in the range 50–200 uA, and most of this is supplied through R8. While there is always a DC path through the RIAA network because of the need to define the LF gain, trying to put all of the Ic through it would lead to an excessive voltage drop, which would appear as a big offset at the output. Instead the DC flowing through R1 is just used for fine tuning of Q1 operating conditions by negative feedback. This means that if R8 and Vbias are correctly chosen, there might be a few 100mV of offset either way at the opamp output. This is not large enough to significantly affect the output swing, but it needs blocking; C5 is shown as nonpolar to emphasise the point that the offset might go either way.

C7 gives dominant-pole compensation of the loop; the RIAA usually causes the closed-loop gain to fall to unity at high frequencies (but see the section on switched-gain phono amps in Chapter 7), and achieving HF stability may require some experimentation with its value. The two supply rails are heavily filtered by R6, C5 and R7, C6 to keep out ripple and noise; no fancy low-noise supply is needed, 78/79 series regulators work just fine.

A more sophisticated hybrid amplifier is demonstrated by the MC amplifier in Chapter 11, where R0 is 3.3 Ω and C0 would be inconveniently large at 4700 uF. To avoid this the negative feedback network has the same gain at DC as AC, and the standing output voltage is controlled by a DC servo.

Noise in Balanced MM Inputs

So far all the MM amplifiers considered have been of the usual unbalanced input type. There is a reason for this.

There is some enthusiasm out there for balanced MM inputs, on a "me-too!" basis, because they are almost universally used in professional audio for excellent reasons.

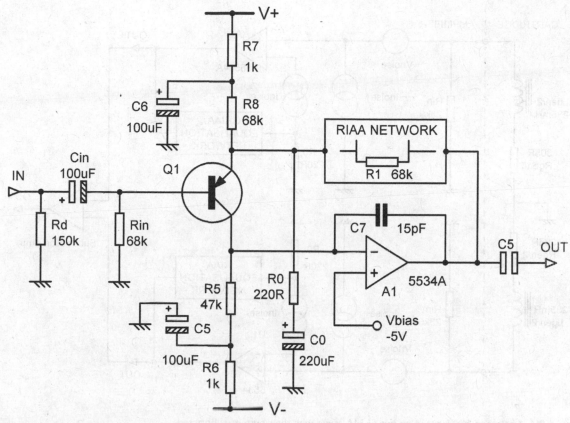

Figure 9.3 Basic arrangement of a hybrid MM phono amplifier with typical values. Note nonpolar output blocking cap.

However, an MM cartridge and its short connecting lead (short to control shunt capacitance) are nothing like the average professional connection that links two pieces of powered equipment and so is likely to have nasty currents flowing through its ground wire. The internals of an MM cartridge are shown in Figure 6.1; the coils are floating. How might common-mode interference, which is what balanced inputs reject, get into the cartridge or lead?

1) Electrical fields into the cartridge. Any sensible cartridge is electrically shielded, so balancing is not required. For electrically unscreened cartridges (there is one brand that is globally famous for humming), the coupling will not be identical for the two ends of the coil, so it won't be a true common-mode signal; I dare say you could have a "balanced" input in which you set the gain of hot and cold inputs separately so you could try to null the hum. Good luck getting that to stay nulled as the arm moves across the disc; this not an idea to pursue.

2) Magnetic fields into the cartridge. These will cause a differential voltage across the floating cart coil, just as for the signal, and will not be rejected in any way by a balanced input.

3) Electrical or magnetic coupling into the cable. Negligible with usual cable lengths and even half-sensible cable layout; i.e. keep it away from mains wiring and transformers. A balanced input is therefore not required.

These points were debated at length on DIYaudio, and no evidence was offered that they were wrong. For this reason balanced MM inputs receive only limited attention in this book.

When dealing with line inputs, a balanced input is much noisier than an unbalanced input. The conditions for an MM balanced input are quite different, but it still seems highly likely that it will be noisier because two (or more) amplifiers are used rather than one, and we don't want a noise penalty if there are no countervailing benefits. Let's find out . . .

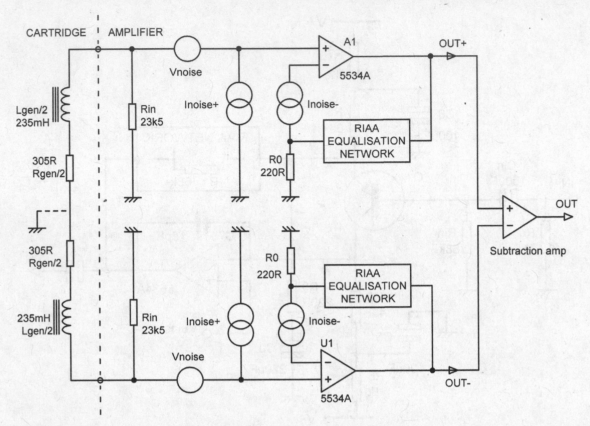

Figure 9.4 A balanced MM input using two 5534A stages with their outputs subtracted

Figure 9.4 depicts a balanced MM input made up of two 5534A stages with their outputs subtracted (or "phase-summed"). All the noise sources are shown. The equal loading on each cartridge pin makes the coil appear balanced to the amplifiers; the "ground" at the midpoint of the cartridge is purely notional, with no physical connection there.

Therefore the 47 kΩ load is split into two 23.5 kΩ resistances; these give less Johnson noise by a factor of √2, and also the voltage produced by Inoise+ will be halved. The Vnoise is unaffected, and we now have two uncorrelated sources of it.

The unweighted noise output for a standard unbalanced 5534A amplifier in a +30 dB (1 kHz) amplifier is −92.51 dBu, as in Table 9.2. Cartridge parameters are 610 Ω + 470 mH. Reducing Rin to 23.5 kΩ and changing the cartridge parameters to 305 Ω + 235 mH, as we are only dealing with half of the cartridge, reduces the noise output to −94.36 dBu. We then have to subtract the outputs of the two sides, which is equivalent to summing their noise. Neglecting the noise of the subtracting amplifier, which is quite realistic given the relatively high noise output of the input stages, the result is −91.36 dBu. This is only 1.1 dB noisier than an unbalanced input, and would be quite acceptable if a balanced input solved other problems, but as noted earlier I've yet to hear any convincing argument that it does.

The arrangement of Figure 9.4 is an illustration of principle and is not claimed to be optimal. For one thing, there are two RIAA networks, and they are very likely the most expensive part of the circuit. They will have to be accurate for accurate RIAA, and that may well be enough to give a good practical CMRR, if you can find a use for it.

Figure 9.4 uses two opamps, which effectively puts four input devices in series in the input circuit, though the two on the opamp inverting inputs see benign noise conditions in the shape of the low resistance of R0. This could be addressed by using an instrumentation amplifier IC, but the possible noise advantage is small.

Noise Weighting

The frequency response of human hearing is not flat, especially at lower listening levels. Some commentators therefore feel it is appropriate to use psychoacoustic weighting when studying noise levels. This is almost invariably ANSI A-weighting despite the fact that it is generally considered inaccurate, as it undervalues low frequencies. ANSI B-, C-, and D-weightings also exist but are not used in audio. The ITU-R ARM 468 weighting (CCIR-468) is a later development and generally considered to be much better but is only rarely used in audio (ARM stands for Average-Responding Meter). I prefer unweighted measurements, as you are one step closer to the original data.

A-Weighting

The A-weighting curve is shown in Figure 9.5. It approximately follows the Fletcher-Munson 40-phon line. It

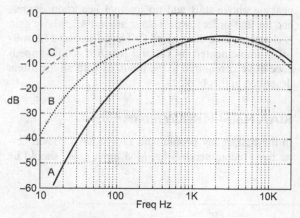

Figure 9.5 ANSI weighting curves A, B, and C. Only A-weighting is used in audio.

passes through 0 dB at 1 kHz and has a maximum gain of 1.3 dB at 2.5 kHz. The low-frequency roll-off steadily steepens as frequency falls, while the high-frequency roll-off is at 12 dB/octave (−3 dB at 12 kHz).

A-weighting is defined in the ANSI S1.42 standard. The required filter design is based on direct implementation of the filter's transfer function based on poles and zeros; no standard circuit is given. The EU version of the standard is IEC 61672−1, which does not even give the poles and zeros; it simply requires that the filter magnitudes fall within a specified error mask. This accounts for the extraordinarily wide variation in circuitry that appears if you do an image search on Google for "A-weighting schematics". Their accuracy is a matter for speculation, so I present my version in Figure 9.6; awkward component values abound, here implemented as 2xE24 resistors and E12-series capacitors. E6 capacitors could be used, but it would mean a lot more paralleling to get suitably accurate values. The filter is accurate to the A-weighting spec within ±0.1 dB over 80 Hz–20 kHz. Below 80 Hz the error slowly grows to 0.5 dB, but since this is at filter attenuations of more than 20 dB the effect on measurements will be negligible.

The wholly passive weighting network has a loss at 1 kHz of −18.37 dB, and the A1 stage simply brings the overall gain at this frequency back up to 0 dB. It is assumed that the noise being measured has already been brought up to a suitable level (say 0 dBu) and so is immune to contamination by any normal opamp circuit noise.

A-weighting reduces the level of white noise by 4.4 dB, but as noted earlier we are not dealing with white noise here, as the spectrum is altered by the effects of the cartridge Lgen and the RIAA equalisation. Useful information on this and the effects of A-weighting on the result are given by Hallgren.[9]

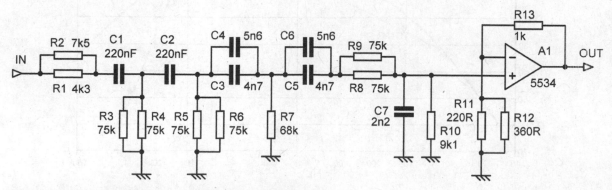

Figure 9.6 A-weighting filter using 2xE24 resistors and E12 capacitors

You may therefore be wondering how the unweighted results described earlier will be affected by the application of A-weighting. Will the order of merit in Table 9.2 be upset?

Table 9.3 shows the effects of A-weighting on selected cases from Table 9.2. The noise level drops by between 2 and 4 dB, depending on the magnitude of voltage noise compared to current noise. A-weighting does not introduce any revolutionary changes into the order of merit of the various amplifiers in Table 9.2.

Table 9.3 The effect of A-weighting on the calculated noise performances of various amplifiers

Case	Amplifier type	Unweighted noise out dBu	A-weighted noise out dBu	A-weighting difference dB
5a	2SB737 70uA	−93.6	−96.3	2.7
5b	2SB737 100uA	−93.4	v96.0	2.6
5c	2SB737 200uA	−92.7	−95.1	2.4
6a	5534A	−92.5	−95.4	2.9
7	5532A	−90.5	−93.3	2.8
8	OPA2134	−89.3	−92.8	3.5
9	LM4562	−87.9	−89.9	2.0
11	TL072	−83.4	−87.1	3.7

ITU-R Weighting

The ITU-R ARM weighting is compared with A-weighting in Figure 9.7. The specification for the ITU-R ARM 468 weighting differs in that the circuit of a standard passive filter is given; unfortunately this requires two unwelcome inductors, and every value is awkward. Not even the 600 Ω terminating resistor can be found in the E96 series. It is therefore more commonly implemented as an active filter. The LF side of the ITU curve has a 6 dB/octave slope, while the HF roll-off has a much steeper 30 dB/octave slope, indicating that a 5th-order lowpass filter is required. The ITU-R is normally implemented to give 0 dB at 2 kHz rather than 1 kHz, as this reduces the amount of gain around 5 kHz and reduces the likelihood of headroom problems. My version is shown in Figure 9.8. This is accurate to within ±0.1 dB over 10 Hz–31 kHz.

In Figure 9.8 the 5th-order lowpass is composed of the 3rd-order lowpass stage using A1, which has no gain peaking above 0dB and no internal headroom issues, and the 2nd-order lowpass stage around A2, whose gain is chosen to give the desired Q. The 6dB/octave LF roll-off is implemented by C4 and R7, R8. A low-impedance drive is required to the input, as with all the filters described in this book.

RIAA Noise Measurements

In the past, many people who should have known better have recommended that MM input noise should be

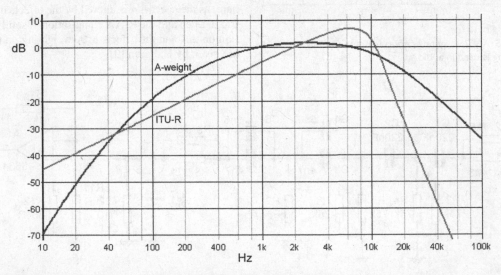

Figure 9.7 The ITU-R ARM weighting compared with ANSI weighting curve A

Noise and Distortion 165

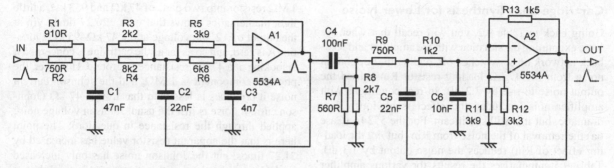

Figure 9.8 ITU-R ARM weighting filter using 2×E24 resistors and 1×E3 capacitors

measured with a 1 kΩ load, presumably thinking that this emulates the resistance Rgen, which is the only parameter in the cartridge actually generating noise—the inductance is of course noiseless. This overlooks the massive effect that the inductance has in making the impedance seen at the preamp input rise very strongly with frequency, so that at higher frequencies most of the input noise actually comes from the 47 kΩ loading resistance. I am grateful to Marcel van de Gevel for drawing my attention to some of the deeper implications of this point.[13]

The importance of using a real cartridge load is demonstrated in Table 9.4, where the noise performance of a TL072 and a 5532 are compared. The TL072 result is 0.8 dB too low, and 5532 result 4.9 dB too low—a hefty error. In general results with the 1 kΩ resistor will always be too low, by a variable amount. In this case you still get the right overall answer—i.e. you should use a 5532 for least noise—but the dB difference between the two has been exaggerated by almost a factor of two by undervaluing the 5532 current noise.

These tests were done with an amplifier gain of +29.55 dB at 1 kHz. Bandwidth was 400 Hz–22 kHz to remove hum, rms sensing, no weighting, cartridge parameters were 610 Ω + 470 mH.

Table 9.4 Measured noise performance of 5532 and TL072 with two different source impedances

Zsource	TL072	5532	5532 benefit	5532 EIN
1 kΩ resistor	−88.0	−97.2 dBu	+9.8 dB	−126.7 dBu
Shure M75ED 2	−87.2	−92.3 dBu	+5.1 dB	−121.8 dBu

The 1 kΩ recommendation was perhaps made because the obvious measurement method of loading the input with a MM cartridge has serious difficulties with hum from the ambient magnetic fields. To get useful results it is essential to enclose the cartridge completely in a grounded mumetal can—I use a can from a redundant microphone transformer, and it works very well. I suppose the ideal load would be a toroidal inductor, but it would be an expensive custom part. It is also necessary to use complete electrostatic screening of the amplifier itself. If it has a 22 uF input coupling capacitor and the input is short-circuited, the impedance downstream of the capacitor is 145 Ω at 50 Hz, which is enough to make it susceptible to electrostatic hum pickup.

RIAA Amps Driven from an MC Head Amp

All the discussion above deals with an RIAA preamplifier driven by an MM cartridge. As we noted at the start, the MM RIAA stage may also be driven from an MC head amp. The noise conditions for the RIAA amplifier are quite different, as it is now fed from a very low impedance, plus probably a series resistor in series with the MC amp output to give stability against stray capacitances. My current MC head amp design (see Chapter 12) has an EIN of −141.5 dBu with a 3.3 Ω input source resistance. Its gain is +30 dB, so the output noise is −111.5 dBu. The series output resistor is 47 Ω; its Johnson noise at −135 dBu is negligible, and likewise the effect of the RIAA stage input current noise flowing in it. The noise at the output of the RIAA stage is then −85.7 dBu, which is higher than any of the figures in Table 9.2 except those for TL072 and LM741. In this situation the value of R0 is relatively unimportant.

Cartridge Load Synthesis for Lower Noise

Going back to Table 9.2, you will recall that when we were examining the situation with the amplifier and feedback network noise switched off, adding in the Johnson noise from the 47 kΩ loading resistor Rin caused the output noise to rise by 3.2 dB. In real conditions with amplifier noise included the effect is obviously less dramatic, but it is still significant. For the 5534A (Case 6a) the removal of the noise from Rin (but *not* the loading effect of Rin) reduces the noise output by 1.3 dB. Table 9.5 summarises the results for various amplifier options; the amplifier noise is unaffected, so the noisier the technology used, the less the improvement.

This may appear to be utterly academic, because the cartridge must be loaded with 47 kΩ to get the correct response. This is true, *but it does not have to be loaded with a physical 47 kΩ resistor*. An electronic circuit that has the V/I characteristics of a 47 kΩ resistor, but lower noise, will do the job very well. Such a circuit may seem like a tall order—it will after all be connected at the very input, where noise is critical, but unusually, the task is not as difficult as it seems.

Figure 9.9a shows the basic principle. The 47 kΩ Rin is replaced with a 1 MΩ resistor whose bottom end is driven with a voltage that is phase-inverted and 20.27 times that at the top end. If we conceptually split the

1 MΩ resistor into two parts of 47 kΩ and 953 kΩ, a little light mathematics shows that with −20.27 times Vin at the output of A2, the voltage at the 47 kΩ–953 kΩ junction A is zero, and so as far as the cartridge is concerned it is looking at a 47 kΩ resistance to ground. However, the physical component is 1 MΩ, and the Johnson current noise it produces is less than that from a 47 kΩ (Johnson current noise is just the usual Johnson voltage noise applied through the resistance in question). The point here is that the apparent resistor value has increased by 21.27 times, but the Johnson noise has only increased by 4.61 times because of the square root in the Johnson equation; thus the current noise injected by Rin is also reduced by 4.61 times. The noise reduction gained with a 5534A (Case 6a) is 1.3 dB, which is very close to the 1.5 dB improvement obtained by switching off the Rin noise completely. If a resistor larger than 1 MΩ is used slightly more noise reduction can be obtained, but that would need more gain in A2, and we would soon reach the point where it would clip before A1, restricting headroom. In this case, with a gain of 21 times, we get a very good noise figure of 1.8 dB, though the lowest noise output comes from the 2SB737 at 70 uA.

The implementation made known by Gevel[11] is shown in Figure 9.9b. This ingenious circuit uses the current flowing through the feedback resistor R0 to drive the A2 shunt-feedback stage. With suitable scaling of R3 (note

Table 9.5 The noise advantages gained by load synthesis with Rin = 1 MΩ. R0 =220Ω From MAGNOISE2. NF is ref Case 1c in Table 9.2, with Rin = 1M5. (EIN = −128.9dBu) Ref was Case 2 in 2nd edition of SSAD

Case	Amplifier type	Rin =47KΩ		Rin =1M		Advantage	Rin =1M5		Advantage
		EIN dBu	NF dB	EIN dBu	NF dB	dB	EIN dBu	NF dB	dB
4a	2SK710 FET, Id = 2 mA	−124.7	3.5	−127.1	1.1	2.9	−127.3		
4b	J310 FET, Id = 10 mA	−124.2	4.0	−126.3	1.9	2.5	−126.4		
5a	2SB737 70uA	−123.6	4.6	−125.3	2.9	1.7	−125.4		
5b	2SB737 100uA	−123.4	4.8	−125.1	3.1	1.7	−125.2		
5c	2SB737 200uA	−122.7	5.5	−124.1	4.1	1.4	−124.2		
6a	5534A	−122.5	5.7	−123.8	4.4	1.3	−123.9		
7	5532A	−120.5	7.7	−121.3	6.9	0.8	−121.4		
8	OPA2134	−119.3	8.9	−119.9	8.3	0.6	−119.9		
9	LM4562	−117.9	10.3	−118.3	9.9	0.4	−118.4		
11	TL072	−113.4	14.8	−113.5	14.7	0.1	−113.5		

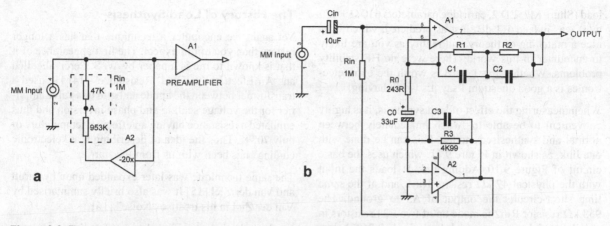

a

b

Figure 9.9 Electronic load synthesis: a) the basic principle; b) the Gevel circuit

that here it has an E96 value), the output voltage of A2 is at the right level and correctly phase-inverted. When I first saw this circuit I had reservations about connecting R0 to a virtual ground rather than a real one and thought that extra noise from A2 might find its way back up R0 into the main path. (I hasten to add that these fears may be quite unjustified, and I have not found time so far to put them to a practical test.) The inverting signal given by this circuit is amplified by 20.5 times rather than 20.27, but this has a negligible effect on the amount of noise reduction.

Because of these reservations, I tried out my version of load synthesis as shown in Figure 9.10. This uses the basic circuit of Figure 9.9a; it is important that the inverting stage A3 does not load the input with its 1 kΩ input resistor R4, so a unity-gain buffer A2 is added. The inverting signal is amplified by 20 times, not 20.27, but once again this has negligible effect on the noise reduction.

In practical measurements with a 5534A as amplifier A1, I found that the noise improvement with a real cartridge

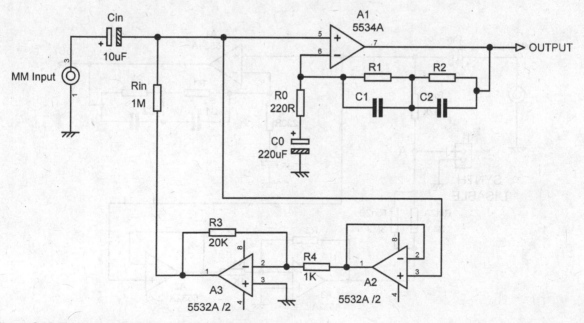

Figure 9.10 Electronic load synthesis: the Self circuit

load (Shure M75ED 2, cartridge parameters 610 Ω + 470 mH) was indeed 1.3 dB, just as predicted, which is as nice a matching of theory and reality as you are likely to encounter in this world. There were no HF stability problems. Whether the 1.3 dB is worth the extra electronics is a good question; I say it's worth having.

When measuring the effect of load synthesis, it is highly convenient to be able to switch immediately between normal and synthesised modes. This can be done with one link, as shown in Figure 9.11, which uses the basic circuit of Figure 9.10. Adding link J1 loads the input with the physical 47 kΩ resistor Rin1, and at the same time short-circuits the output of A3 to ground. The 953 kΩ resistor Rin2 is made up of two E24 resistors in parallel, which give a combined value only 0.26% below the nominal value.

This technique has been called "electronic cooling", presumably because it could be regarded as analogous to dipping the loading resistance in liquid nitrogen or whatever to reduce Johnson noise. I must admit I don't like the term, as it could be understood to mean that thermoelectric elements have been used to cool down the input stage, a technique I do not think has been used in hi-fi yet. I prefer to call it "electronic loading", "active input impedance", or "load synthesis", the last being perhaps the most explicit. It would also be useful for tape head preamplifiers, but they are a bit of a minority interest these days.

The History of Load Synthesis

Yet again we encounter a technique that has a longer history than you might expect. The first appearance of it that is known to me is a paper by W. S. Percival called an "An Electrically 'Cold' Resistance".[14] He used a transformer between the anode and grid of a valve amplifier for the voltage scaling and phase inversion and thus emulated a resistance having an effective temperature of only 70° K. Thus the idea of describing it as "electronic cooling" has been with us from the start.

The same technique was later expanded upon by Strutt and Van der Ziel.[15] It was also briefly summarised by Van der Ziel in his treatise "Noise".[16]

Load synthesis was referred to by Tomlinson Holman in his famous 1975 paper "New Factors In Phonograph Preamplifier Design"[17] (which is still well worth reading) as a recent innovation:

> Two noise reduction techniques have appeared in recent designs. One is to use quite low impedances in the RIAA feedback . . . The second involves the use of a synthesized input impedance through the use of an extra feedback loop which bootstraps the cartridge termination resistor to reduce its noise contribution. One commercial embodiment of the bootstrap method produced a signal-to-noise ratio of 85 dB re 10 mV, 1 kHz input, ANSI 'A' weighted with a cartridge input.

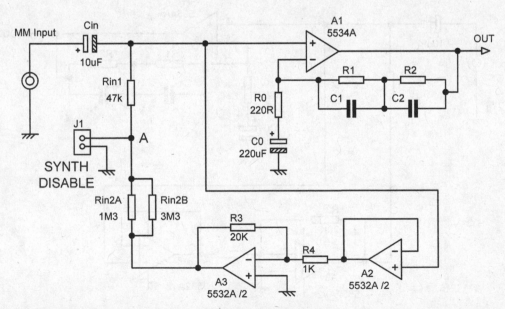

Figure 9.11 Switching between normal loading and load synthesis with one link: the Self circuit

For comparison, assuming the reference level is 10 mV rms, the 5534A in Case 6a gives an *unweighted* S/N ratio of 84.7 dB without load synthesis and 86.0 dB with it. Clearly it has a better noise performance than whatever was used in the "commercial embodiment".

Bootstrapping is not really the right term; load synthesis is more "anti-bootstrapping" in that it makes a large impedance look like a small one by applying a voltage in anti-phase, whereas conventional bootstrapping makes a small impedance look like a large one by applying a voltage in phase. I have no idea which "commercial embodiment" he was talking about, and I would be very glad to hear any suggestions. So far I have had none, and I'm sure you all can do better than that.

The technique was analysed by Hoeffelman and Meys in the JAES in 1978.[18] The only relevant patents found so far are US 4 156 859 granted to Robert L. Forward in October 1977, which describes the use of opamps and transformers for voltage scaling and phase inversion, and US 4 232 280 (same guy), which uses FET source-followers instead of opamps.

Noise Gating

We saw earlier in this chapter that there are definite physical limits to how low the noise of a preamplifier driven by an MM cartridge can be. Noise in the quiescent state is particularly noticeable when headphones are in use. So if you want less noise, there is nothing to be done? Ha! In electronics there is almost always something that can be done.

First it has to be recognised that cartridge and preamplifier noise is pretty much irrelevant while a disc is playing. Groove noise will be something like 30 dB higher. So to make an audible difference the only thing to do is reduce the noise when a disc is *not* playing. This can be done most effectively by the use of a noise gate, which cuts the signal completely when it falls below a certain threshold. You probably think this is a dreadful idea, and under most circumstances it would be, with the noise gate chopping in and out in quiet passages of music. But . . . the difference between vinyl and all other sources is that it carries subsonic disturbances that are always present, even if the audio signal disappears completely. Even disc pressings of the highest quality produce this subsonic information, at a surprisingly high level, partly due to the RIAA bass boosting. The subsonic component is often less than 20 dB below the total programme level, and this is more than sufficient to keep the gate open (unmuted) for the duration of an LP side.

(See Chapter 12.) Therefore all that is required is a noise gate with a response down to around 1 Hz.

The noise gate will have the usual fast-attack, slow-decay characteristic, so the preamplifier is unmuted as soon as the stylus touches the disc, and muted about a second after it has been raised from the run-out groove. This delay can be made short because the relative quiet at the start of the run-out groove is sensed and stored. The rumble performance of the record deck is largely irrelevant because virtually all of the subsonic information is generated by disc irregularities.

A noise gate is composed of a high-gain amplifier to bring low-level signals up to a convenient level, a peak rectifier with fast attack and slow decay, a comparator that switches at a set threshold; this is called the side-chain because signal does not go through it and come out the other end. The comparator controls a muting element. It is often convenient to use the mute relay, which prevents switch-on thumps. The signal pick-off point must clearly be before any subsonic filtering and ideally where the signal has reached its full nominal level. If this is not possible because the subsonic filtering is done at a lower level, then more gain must be built into the side-chain to compensate.

Figure 9.12 shows an early noise gate design that was used in the Advanced Preamplifier of 1976.[19] It worked flawlessly at the time (and indeed it still does) so is worth a look.

The side-chain begins with two amplifiers with gains of 101 times, one for each channel. The inputs are clamped with diodes so that main signal path can use its full voltage swing capability without damaging the opamps. You will note they are shown as the 741 types used at the time. They are elderly but more than adequate for the task here, so this is a good place to use up some vintage parts. They are powered from the +24V and 0V rails. Fed with the nominal signal voltage of 800 mV rms (+0.3 dBu), the opamp outputs move continuously between positive and negative clipping due to the high gain; this keeps the peak-rectifier capacitor C3 fully charged. In the silent-ish passages between LP tracks the subsonic signal is not normally of sufficient amplitude to cause them to clip but will usually produce at least +3 to +4 V across C3, which gives a large margin of safety against unwanted muting.

When the stylus is lowered onto a record, C3 charges rapidly through D3, and when its voltage exceeds the +0.6V reference set up by R5–D5, the output of A2 goes high, switching on Q2, energising the mute relay, and letting signals through to the preamplifier output. This happens very quickly, and you are never going to miss the first note of the music, so long as you put the needle down in the right place.

When the stylus leaves the record surface and the subsonic signals cease, C3 slowly discharges through R4 until A2 output goes low, cutting off the base drive to Q2, and so switches off the relay.

The threshold for unmuting is determined from the gain of A1 and the +0.6V reference; a level of 850 mV rms at A1 output (allowing for the diode forward drop of D3) will switch A2, corresponding to an input level of 8.5 mV rms (equal to −39 dBu). As shown, the noise gate monitors both stereo channels and selects the maximum with D3, D4, but on reflection (and I've had 40 years to think it over) monitoring just one channel should work equally well, even if it has a vaguely unsettling feeling of asymmetry about it; a useful number of parts are saved. What you should NOT do is monitor the sum of the two channels, as much of the subsonic information is in anti-phase and will tend to cancel.

The noise gate will only work properly with vinyl inputs, so it is disabled when any other input is selected. An extra wafer on the source-select switch SW1 is arranged

to provide permanent unmute when required by pulling the inverting input of A2 low so the output of A2 will stay high even when there is no signal and C3 is fully discharged. Q1 is part of the switch-on delay system—at power-up Q1 was turned on to prevent Q2 switching on; there is a bit of a trap here because it is essential to get Q1 turned on fast before the relay has time to close. The original circuit worked reliably, but it would be much safer to rearrange things so that the switch-on delay supplies an enable rather than a disable for the relay circuit. This will be essential if a microcontroller is providing the switch-on delay, as they take a little time to initialise themselves when the power is applied.

One rough edge of this design is the side-chain amplifier, which spends most of its time clipping hard. The currents drawn from the supplies and put into the ground have sharp edges, and the connections must be arranged to keep them out of the audio supplies and ground. There are also large square-wave-ish voltages on the opamp outputs which must not be allowed to couple capacitively into the audio path.

There are more sophisticated ways to deal with these problems. Figure 9.13 shows a side-chain designed originally for a signal activation system to bring a power

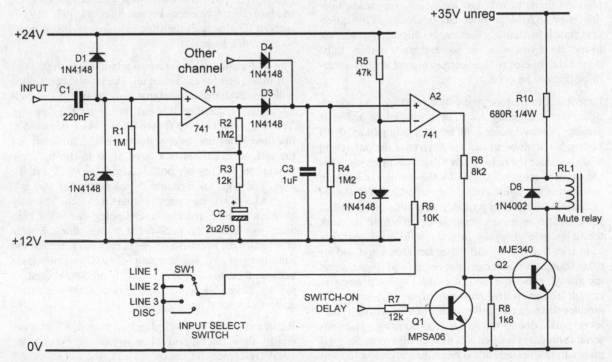

Figure 9.12 Subsonic-activated noise gate as used in the Advanced Preamplifier (1976)

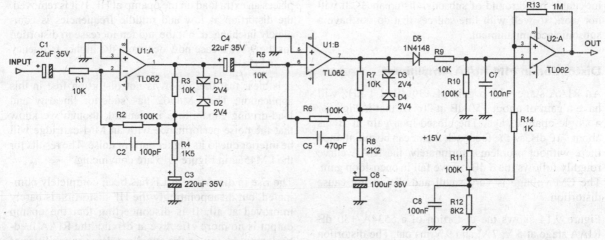

Figure 9.13 A more sophisticated subsonic-activated noise gate

amplifier out of standby when a signal appeared at the input; it has been adapted for use as a subsonics-activated noise gate by extending the LF response. The first amplifier stage U1:A has a mid-band gain of +38 dB and uses shunt feedback in the usual way to generate a virtual-earth at Pin 2. The feedback network R2, R3, R4 is in the form of a T-network to give high gain without excessively high resistor values. C2 across R2 provides an HF roll-off of −3 dB at 16 kHz to discriminate against HF noise. C3 reduces the gain to unity at DC to minimise offset voltages and gives an LF roll-off at 0.5 Hz. D1 and D2 are 2V4 Zener diodes that provide output clamping by increasing the negative feedback when the output exceeds about 3V peak in either direction; the opamp is always under feedback control and so does not generate sharp edges by clipping.

The second amplifier stage, U1:B, is a similar shunt-feedback stage with a virtual-earth at Pin 2 and a mid-band gain of +35 dB. It is fed from the first stage via capacitor C4 and input resistor R5, which create a roll-off −3 dB at 0.7 Hz. R8 and C6 give an LF roll-off of −3 dB at 0.7 Hz, while R6 and C5 give an HF roll-off −3 dB at 3.4 kHz to prevent false triggering from noise pulses. D3 and D4 are 2V4 Zener diodes that provide output clamping.

D5 and C7 are the peak rectifier, with a slow attack time set by R9 (to further discriminate against noise pulses) and the decay time set by R10. The stored voltage on C7 is applied to comparator U2:A; R13 and R14 provide a small amount of positive feedback to introduce a touch of hysteresis and so give clean comparator switching. TL062 opamps, rather than the more familiar TL072, are used because of their lower input offset voltage.

The comparator reference of +1.2 V is set by the divider R11, R12. Allowing for a 0.6 V forward drop in D5, a peak voltage of 1.8V is therefore required at the output of U1:B to trip the comparator, equivalent to +4.3 dBu. This is given by 4.3 − (38 + 35) = −68.7 dBu, which is 284 uV. This is a much lower trip level than the first design in Figure 9.12 (which is −39 dBu) because it is intended to derive its input from before any subsonic filtering. I have recommended that a phono input stage should not have a greater gain than +30 dB at 1 kHz if a really good overload margin is desired, and so the nominal output for 5 mV rms in is only 158 mV rms, (both at 1 kHz) which is −13.8 dBu. Therefore more amplification is required and two stages are used. The difference between the nominal output and the noise gate trip point is 40 dB for the original design and 55 dB for the later design, and both have worked reliably for many years.

If lower sensitivity is required for the circuitry of Figure 9.13, the gain of either or both amplifier stages can be reduced. If higher sensitivity is sought then the best way is to leave the amplifiers alone and instead reduce the reference voltage set by R11, R12. I have tested it down to +260 mV from +1.2 V, increasing the sensitivity by another 13 dB. Beyond this you are on your own, and care will be required in layout and grounding to prevent positive feedback and false triggering. This circuitry was used in a large number of amplifiers, starting about 14 years ago; it is still in production at the time of writing.

Before we leave this subject, I would like to emphasise that foolproof noise gating which never cuts off a wanted signal is feasible only for a vinyl signal, with its inevitable background of subsonic disturbances. It will not work so well with line sources that do not have a subsonic accompaniment.

Distortion in MM RIAA Amplifiers

An RIAA stage with a gain of +30 dB at 1 kHz will have a gain of about +50 dB at 20 Hz; that is high for a single-opamp stage. The closed-loop gain at HF is about +10 dB, so the feedback factor can be maintained there without problems; fortunately, the RIAA curve roughly follows the 6 dB/octave fall in open-loop gain. The CM voltage is very small and should not cause distortion.

Figure 9.14 shows the distortion of a 5534A +30 dB RIAA stage at 5 V, 7 V, and 9 Vrms out. The distortion at LF is mainly second harmonic; it was checked that this was not coming from C0; increasing it from 220uF to 1000/6V3 gave no improvement. It must be coming from the opamp. The sudden increase in distortion at about 27 kHz for the 9 Vrms case occurs when the current drawn by the RIAA network reaches the limits of the output capability of the 5534A. Note that 9 Vrms is a whole 36 dB above the nominal operating level of 150 mV rms out.

The distortion will be aggravated by external loading. Here we have just the HF correction pole, which only places an extra load on the opamp at HF. If it is removed, the distortion at low and middle frequencies is completely unchanged, but the sudden increase in distortion for the 9 Vrms case now occurs at the higher frequency of 36 kHz.

It is clear that the 5534A is not distortion free in this application. The LM4562 has superior linearity and load-driving capabilities in general, though we know that the noise performance with an MM cartridge will be inferior due to its higher current noise. The results for the LM4562 in Figure 9.15 are convincing.

The rise in distortion at LF has been completely eliminated, but disappointingly the HF distortion is barely improved at all. It is disconcerting that the opamp output is no more effective at driving the RIAA feedback network, given the excellent drive capabilities of the LM4562 into resistive loads. I suspect the reason is that the output stage uses VI limiting for overload protection, as opposed to the simple current-limiting used in the 5532, and this makes it more likely to come into action when driving a highly reactive load like an RIAA network, which above 1 kHz looks pretty much like a capacitance in series with a low-value resistor (R0) connected to ground. The use of a "helper" opamp to assist in driving the RIAA network would be likely to reduce the distortion; see Chapter 1. Figure 9.15 confirms that the LF distortion is wholly from

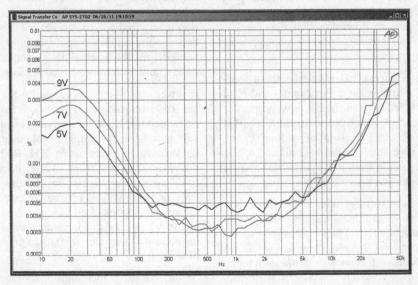

Figure 9.14 +30 dB (1 kHz) 5534A RIAA preamp THD at 5, 7, and 9 Vrms out. No IEC amendment, no external load except the HF correction pole.

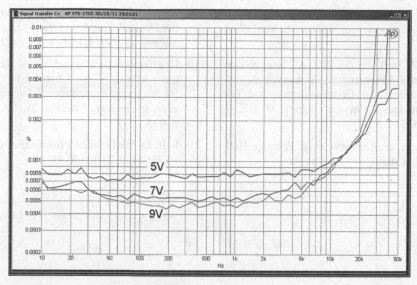

Figure 9.15 As for Figure 9.14 but using one LM4562 section

the 5534 opamp and is nothing to do with capacitor distortion.

Conclusions

While MM RIAA cartridge noise is a complicated business, the clear result is that the 5534A is the cheapest and easiest way to get within 3 dB of the theoretical best noise performance. This is because it not only has low noise in general but also a favourable balance between its voltage noise and current noise for an MM source. It is a happy chance that it is inexpensive. To get within 2 dB of perfection, another 5532 can be added to implement load synthesis.

Money can be saved by using a 5532, but you are then 5 dB from theoretical best noise, and running two channels through the same package is very likely to compromise crosstalk.

For these reasons the 5534A is used in all the opamp designs in this book.

References

1. Lipshitz, S. P. "On RIAA Equalisation Networks" *Journal of Audio Engineering Society*, June 1979, p. 458, onwards.

2. Vogel, B. *The Sound of Silence*. 2nd edition. Springer, 2011, p. 523. ISBN 978-3-642-19773-d.

3. Langford-Smith, F. *Radio Designer's Handbook*. 1953, Newnes reprint 1999, Chapter 17, p. 705. ISBN 0 7506 3635 1.

4. Vogel, B. *Radio Designer's Handbook*. 1953, pp. 201–224.

5. Crossley, D. Personal Communication, Nov 2016.

6. Francis, E. H. "Moving Iron Pickups" *Wireless World*, Aug 1947, p. 285.

7. Vogel, B. "Adventure: Noise" (Calculating RIAA noise) *Electronics World*, May 2005, p. 28.

8. Al-Asadi et al. "A Simple Formula for Calculating the Frequency-Dependent Resistance of a Round Wire" *Microwave and Optical Technology Letters*, Volume 19, No. 2, Oct 5 1998, pp. 84–87.

9. Hallgren, B. "On the Noise Performance of a Magnetic Phonograph Pickup" *Journal of Audio Engineering Society*, Sept 1975, p. 546.

10. Elliot, Rod Elliott Sound Products: Sound.whsites.net/articles/cartridge-loading.html Accessed Nov 2016.

11. de Gevel van, M. "Noise and Moving-Magnet Cartridges" *Electronics World*, Oct 2003, p. 38.

12. Sherwin, J. "Noise Specs Confusing?" National Semiconductor Application Note AN-104" *Linear Applications Handbook*, 1991.

13. de Gevel van, M. Private Communication, Feb 1996.

14. Percival, W. S. "An Electrically 'Cold' Resistance" *Wireless Engineering*, Volume 16, May 1939, pp. 237–240.

15. Strutt, M. J. O., and Van der Ziel, A. "Suppression of Spontaneous Fluctuations in Amplifiers and Receiver for Electrical Communication and For Measuring Devices," *Physica*, Volume 9, No. 6, June 1942, pp. 513–527.

16. Van der Ziel, A. *Noise*. New York: Prentice-Hall, 1954, pp. 262 et seq.

17. Holman, T. "New Factors in Phonograph Preamplifier Design" *Journal of Audio Engineering Society*, Volume 24, No. 4, May 1975, p. 263.

18. Hoeffelman, J. M., and Meys, Rene P. "Improvement of the Noise Characteristics of Amplifiers for Magnetic Transducers" *Journal of Audio Engineering Society*, Volume 26, No. 12, Dec 1978, p. 935.

19. Self, D. "Advanced Preamplifier Design" *Wireless World*, Nov 1976, pp. 41–46.

Chapter 10

Moving-Magnet Inputs

Discrete Circuitry

Discrete MM Input Stages

Moving-magnet (MM) input stages constructed from discrete transistors have certain advantages. They are not limited to opamp supply rail voltages. They do not produce complex distortion products or crossover distortion like some opamps. The circuit operation is completely under the designer's control, and every parameter, such as each transistor's collector current, can be chosen by the designer. This chapter examines how to use that freedom of design and at the same time gives an historical overview of MM phono stages.

Discrete moving-magnet (MM) input amplifiers were almost universal until the early 1970s. For a long time opamps had, quite deservedly, a poor reputation for noise when used in this application.

When the first bipolar transistor MM inputs were designed, active components were still expensive, and adding another transistor to a circuit was not something to be done lightly. The circuitry from that era therefore looks to us very much cut-to-the-bone, but before disrespecting it we need to remember that it was designed under very different economic constraints.

A major problem with early discrete MM amplifiers was a simple lack of open-loop gain to give an accurate RIAA response network in the low-frequency region, even if the RIAA network was accurate, which it rarely was. Another problem was that an RIAA feedback network, particularly one designed for low closed-loop gain, and/or a relatively low RIAA network impedance to reduce noise, presents a heavy load at high frequencies because the impedance of the capacitors becomes low. Heavy loading at HF was commonly a major cause of increased distortion and headroom-limitation in discrete RIAA stages that had either common-collector or emitter-follower output topologies with asymmetrical

clipping behaviour; an NPN emitter-follower is much better at sourcing current than sinking it. The 20 kHz output capability, and thus the overload margin, was often brought down by 6 dB or even more. Replacing the emitter resistor of an emitter-follower with a current source gives a much better HF output current capability, and this can be further doubled for the same quiescent dissipation by using a simple push-pull Class-A output structure.

One-Transistor MM Input Stages

I will say at once that the one-transistor MM input stage is purely an historical curiosity. Its performance cannot be expected to be anything other than dreadful by today's standards. Nevertheless, the idea is worth looking at. Figure 10.1b shows a one-transistor MM input stage designed by Jack Dinsdale (of whom more later) in 1961.[1] At that time transistors were *very* expensive, and using two in a single stage would have been thought highly extravagant.

If you have but one transistor to play with, there are only three possible configurations; common-collector, (i.e. emitter-follower) common-base, and common-emitter. The first gives no voltage gain, and the second has a low input impedance that looks unpromising. That leaves the common-emitter configuration, as in both halves of Figure 10.1. Since it inherently inverts, the only possibility is shunt feedback, which as we saw in Chapters 7 and 9 is inherently much noisier than its series-feedback equivalent. The standard approach at the time, derived from valve designs, was that in Figure 10.1a, which had an input resistor R1 of 47 kΩ to give the correct cartridge loading, with the RIAA equalisation performed by the negative feedback network C1, R2, C2 in conjunction with the impedance of the collector load R3.

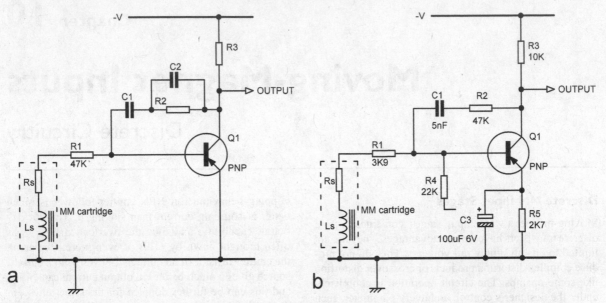

Figure 10.1 a) Basic one-transistor shunt-feedback MM amplifier; b) using the cartridge inductance to perform the LF part of the RIAA equalisation (Dinsdale 1961)

This arrangement is bound to give a poor noise performance. Dinsdale's solution, in Figure 10.1b, was to make the input impedance low and implement the LF part of the RIAA equalisation by the interaction of the cartridge inductance with it, giving a 6 dB/octave slope. As frequency falls, the impedance of the inductance falls and the current into the input increases. He described this method as more "efficient", which presumably means a greater transfer of energy through R1 and hence a better signal/noise ratio. Components R5 and C3 set the DC conditions, with base bias provided through R4.

The idea of loading the MM cartridge inductance with a low input resistance to achieve the LF boost section of the RIAA equalisation has come back to haunt us many times since then. The terrible snag is that since the LF equalisation is set by the cartridge inductance, changing the cartridge type almost certainly means you have to change the loading resistor too. You can of course add a control marked "cartridge inductance", but this assumes you actually know the cartridge inductance, and know it precisely. Inaccuracy in the inductance setting will give errors in the RIAA response. You will have to rely on the manufacturer's technical specification for the value of the inductance—which is usually given in suspiciously round figures—unless you plan to measure it yourself. That means acquiring an expensive precision component bridge and taking great care that you do not apply

an excessive test signal. The two channels are unlikely to be identical, so you will need custom values for each channel. This is not a route many people are going to want to take, and for this reason, throughout this book you will find that the idea of using the cartridge inductance as a critical part of the RIAA equalisation receives very little sympathy.

Two-Transistor MM Input Stages

Figure 10.2 shows a typical two-transistor MM input amplifier from the late '60s. The configuration is generally considered to have been introduced by Jack Dinsdale in 1965, in a classic preamplifier design[2] that was one of the first to deal effectively with the new RIAA equalisation requirements for microgroove records. It is a two-stage series-feedback amplifier composed of two common-emitter stages. R3 and C1 make up an RF filter; note R3 is 3k9; this is considerably greater than the DC resistance of most MM cartridges and looks like it would introduce unnecessary Johnson noise and unhelpfully convert the current noise of Q1 into voltage noise. The RIAA network is R6, R7, C4, C5, and it has a high impedance to reduce loading on the stage output. Since R6 has the high value of 1M8, the RIAA network cannot be used for the DC feedback that is required to set the quiescent conditions. There is a separate DC feedback network comprising R1, R4, R5, R10, and C3 which

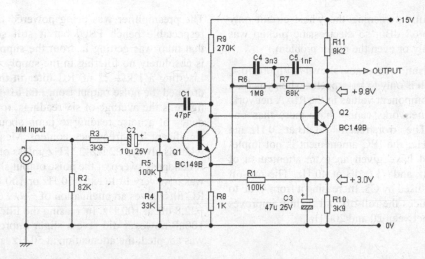

Figure 10.2 A typical two-transistor MM amplifier as commonly used in the 1960s and early '70s. Gain +39 dB at 1 kHz.

establishes the appropriate voltage across R10. C3 keeps signal frequencies out of this path. The RIAA network is in Configuration-A (see Chapter 7). No attempt is made to implement the IEC Amendment, as it was not introduced until 1976.

Because of its simplicity, this stage inevitably contains compromises. The second collector resistor, R11, needs to be high in value to maximise open-loop gain but low to adequately drive the RIAA network and any external loading.

An MM preamp has to deliver a maximum low-frequency boost of nearly 20 dB, on top of the gain required to get the desired output level at 1 kHz. If the cartridge output is taken as 5 mV rms at 1 kHz and the amplifier output is 150 mV rms (which is about as low as you could hope to get away with then if you were sending this signal to the outside world), then a total closed-loop gain of 20 + 34 = 54 dB is required at low frequencies. The open-loop gain obviously needs to be considerably higher than this, for a decent feedback factor is required not only to reduce distortion but also to ensure that the RIAA equalisation is accurately rendered by the feedback network. By 1970 it had become clear that the two-transistor configuration was really not up to the job, and more sophisticated circuits using three transistors or more were developed, aided by falling semiconductor cost. While the two-transistor MM preamplifier must now be regarded as of purely historical interest, it is highly instructive to see just what can be done with it by modification.

The circuit shown in Figure 10.2 was deliberately chosen as representative of contemporary practice in its era, and

it has not been modified or optimised in any way. It is closely based on a small RIAA preamplifier PCB called the "Lenco VV7", which was intended for upgrading systems to use MM cartridges where the amplifier had only a ceramic pickup input; see Figure 10.3. It was a Swiss product distributed in Britain by Goldring in the early 1970s. It had an integral mains PSU (see the tiny transformer on the left) with half-wave rectification and RC smoothing. What the proximity of that transformer did to the hum levels I do not know, but it looks awfully close to the preamp, which is in the screening can to the right. You will note that the single-rail supply is by modern standards low, at +15V; opamp-based preamplifiers today normally run from ±15V or ±17V, giving them a 6 dB headroom advantage at once. The gain is +39 dB at 1 kHz.

I built up Figure 10.2 with BC184 transistors, using an external DC supply. I found that the first-stage (Q1) collector current was 42 uA, and the second-stage (Q2) collector current was 0.63 mA. On measuring it I was not exactly surprised that the performance was mediocre. There was a high level of hum at the output: −66 dBu

Figure 10.3 The Lenco phono preamplifier

at 50 Hz. Carefully screening the whole circuit only reduced this to −68 dBu, so electrostatic pickup was clearly not the only or even the major problem.

The RIAA equalisation accuracy, shown in Figure 10.4, is not good, which is only to be expected when you look at the standard component values in the RIAA network. Accurate RIAA networks cannot have more than one preferred value. The errors reach +2.3 dB at 20 Hz and +0.7 dB at 20 kHz; the IEC amendment is not implemented; it would have given an extra attenuation of −3.0 dB at 20 Hz and −1.0 dB at 40 Hz. The roll-off below 20 Hz is caused by C3. Increasing it from 47uF to 100uF much reduces the roll-off and slightly improves RIAA accuracy between 20 and 200 Hz.

The preamplifier was being powered from a perfectly respectable bench PSU, but it still seemed possible that hum was getting in from the supply rail, as there is absolutely no filtering in the supply to Q1 collector. Inserting a 1 kΩ–22 uF RC filter in the supply to R9 dropped the noise output from −68 to −73.4 dBu. (This figure is the average of six readings, to reduce the tendency of a noise reading to jump about when there is significant low-frequency content. Measurement bandwidth is always 22 Hz–22 kHz unless otherwise stated.) A bandpass sweep of the noise output showed that there was now very little extra 50 Hz or 100 Hz content. The RC filter gives an attenuation of −16.9 dB at 50 Hz and −22.8 dB at 100 Hz. Increasing the filter capacitance to 100uF however did give a slight improvement, so this was adopted; the attenuation at 50 Hz is now −29.9 dB.

Maximum output with a +15V supply rail was 3.4 Vrms at 1 kHz, (1% THD) so the maximum input was 38 mV rms, giving an overload margin of only 17.6 dB. It is noticeable that clipping is not symmetrical, occurring first on the positive peaks. When this clipping does occur, there is a shift in the DC conditions of the circuit due to the way the biasing works through the filtering action of C3.

The THD at 1 Vrms out (1 kHz) was 0.010%, which by modern standards is a lot for such a low level. Figure 10.5 shows the distortion performance with a +15V rail, at 1, 2, and 3 Vrms out. The input signal was inverse RIAA equalised so that the output level remains constant with frequency. It was necessary to use the 100 Hz filter

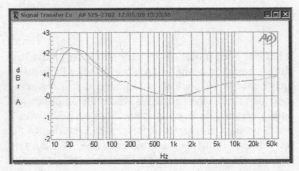

Figure 10.4 The Lenco RIAA errors are pretty gross by today's standards. The LF flatness is improved somewhat by increasing C3 to 100 uF.

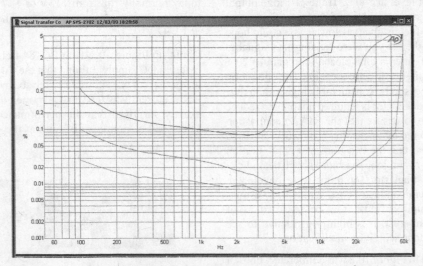

Figure 10.5 Two-transistor MM amplifier THD with a +15V rail, at 1, 2, and 3 Vrms out. Bandwidth 100 Hz–80 kHz.

on the AP to get consistent results, despite having got rid of the 50 Hz problem with the RC filter, as there is still a large LF noise component due to the RIAA LF boost.

You can see that for 1 Vrms, the mid-band distortion is around 0.01%, but there is a steady rise below 1 kHz. This is caused by the falling negative feedback factor as the RIAA curve demands more gain at lower frequencies. The other area of concern is at high frequencies; at 1 Vrms nothing too bad happens in the audio band, though THD has reached 0.02% at 20 kHz.

At the higher output level of 2 Vrms, the mid-band THD is tripled. The output stage starts to clip around 15 kHz, as Q2 can no longer drive the RIAA network, which has a falling impedance at high frequencies. Things are pretty gross at 3 Vrms out, with THD around 0.1% mid-band and HF clipping starting at 4 kHz.

Clearly this historical RIAA preamp could use a bit of improvement, starting with linearity and headroom. Let's see what can be done with it; the process will reveal a lot about how discrete circuitry works.

Two Transistors: Increasing Supply Voltage to +24 V

A pretty sure bet for improving both the linearity and headroom of a discrete amplifier is simply to increase the supply voltage. We will start by turning it up to +24 V, a voltage that can conveniently be obtained from a 7824 IC regulator. Figure 10.6 shows the results; distortion is somewhat reduced overall, and the HF overload

problem has been pushed to slightly higher frequencies, but the effect is not as dramatic as we might have hoped. The maximum output has only increased to 3.8 Vrms at 1 kHz (1% THD), which is not much of a return for increasing the supply voltage by 60%.

Casting a suspicious eye over the circuit, it's clear that it is still clipping asymmetrically. There is +18.4V on Q2 collector, whereas for a symmetrical output swing we would expect something more like +12 V. Improving the bias conditions by changing R10 from 3k9 to 2k4 reduces Q2 collector volts to +15.0V and gives much more output voltage swing capability, as well as increasing the standing current in Q2, which improves load-driving capability. The maximum output is now 6.0 Vrms at 1 kHz, an improvement of +4 dB. The input overload margin is raised to 23 dB. While this rebiasing does not give exact symmetry of clipping, it does seem to be close to optimal biasing for linearity. A good indication of this is that the distortion residual at 1 kHz is third harmonic, which suggests that some cancellation of second-harmonic distortion is going on.

The distortion performance is transformed; 3 Vrms out (1 kHz) gave 0.06% in Figure 10.6. After rebiasing it has fallen to 0.014%, as in Figure 10.7. The HF overload effect has also been pushed out to above 20 kHz, even for the 3 Vrms case. Not bad for modifications that essentially cost nothing.

The distortion improvement at lower output voltages in the mid-band is barely visible even with 100 Hz AP

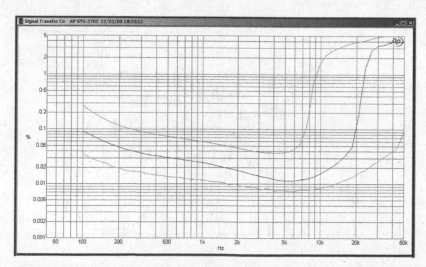

Figure 10.6 Two-transistor MM amplifier THD with a +24V rail, at 1, 2, and 3 Vrms out. Bandwidth 100 Hz–80 kHz.

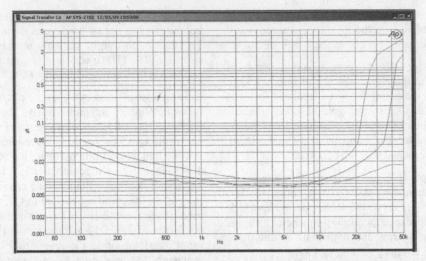

Figure 10.7 Two-transistor MM amplifier THD with a +24 V rail, at 1, 2, and 3 Vrms out, after rebiasing. Bandwidth 100 Hz–80 kHz.

filtering because of the high noise output from a circuit with +39 dB of gain at 1 kHz.

The modified circuit, with the added RC filter for the first stage, supply increased to +24 V, and biasing adjusted by changing R10, is shown in Figure 10.8. The Ic of Q1 is now 75 uA, and the Ic of Q2 is 1.1 mA.

Two Transistors: Increasing Supply Voltage to +30V

Since increasing the supply to +24 V gave considerable benefits (after rebiasing), we will increase it further to

+30 V. This increases the maximum output to 6.8 Vrms at 1 kHz (1% THD), which is 6 dB up on the original circuit. The input overload margin is raised to 24 dB. R10 has been changed again to 2k2 to optimise the biasing. The results are seen in Figure 10.9. The THD for the 1 Vrms case is now completely submerged in low-frequency noise, so I used the 400 Hz AP filter, which shows the THD at 1 Vrms (1 kHz) is about 0.0055%. This number however still contains a significant amount of noise.

HF overload behaviour has improved again, but HF distortion is somewhat worse at all three output levels.

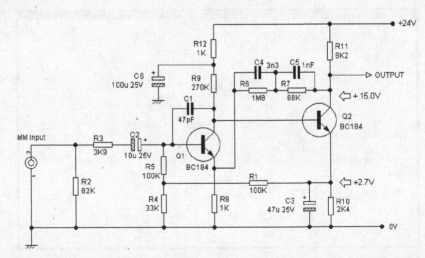

Figure 10.8 The two-transistor MM amplifier using a +24 V rail, with the RC filter R12, C6 added to the collector of Q1, and after rebiasing by altering R10

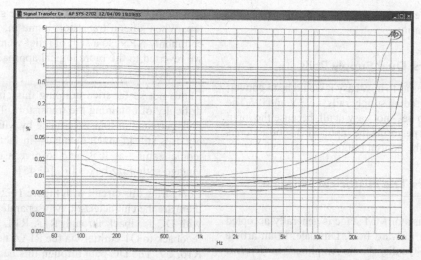

Figure 10.9 THD with a +30V rail, at 1, 2, and 3 Vrms out, rebiased again. Bandwidth 100 Hz–80 kHz for 3 and 2 Vrms, 400 Hz–80 kHz for 1 Vrms.

The LF distortion is notably improved, being more than halved.

As a side issue, we might consider how to generate the supply rail required. The 7824 IC regulator will accept a maximum input of 40 V, so it is feasible to use that with the ADJ pin elevated by some means, as described in Chapter 16. For output voltages above +30V this does not leave enough regulator headroom, and we might need to use the TL783 high-voltage regulator. This is a favourite device for generating +48 V supplies for microphone phantom power and can definitely be relied on up to this voltage.

Two Transistors: Gain Distribution

At this point I began wondering what else could be done to reduce the distortion. There are practical limits to raising the supply voltage; power dissipation increases, and there is a danger that the circuit could generate turn-on or turn-off transients that would damage stages downstream.

At this point it's worth considering what the sources of nonlinearity are. The two transistors, obviously, but the RIAA capacitors could also be contributing, as I used ordinary polyester types, and if you've read the chapter on components, you will know that these are not wholly linear, and in this application there is a significant signal voltage across them. However, the distortion generated by polyester caps is typically of the order of 0.001% at 10 Vrms, and the signal levels we are using here are much lower than that, so the capacitor

contribution is almost certainly negligible. At the time of writing I haven't got round to proving the point by substituting polypropylene capacitors, which *are* linear.

The configuration is made up of two cascaded voltage amplifiers, and it seemed to be a good idea to find out how the open-loop gain is distributed between them. The high value of the Q1 collector load suggests that it is intended to give a high-voltage gain.

Measurement showed that the signal on Q1 collector was −49 dB with reference to that on the output at Q2 collector, at 1 kHz. This was confirmed by SPICE simulation, which gave −45 dB on Q1 collector between 100 Hz and 10 kHz. Note however that the emitter of Q2 is connected to AC ground via C3, which suggests that the second stage has a low input impedance and perhaps the first stage is working as a transconductance stage, feeding a current into the base of Q1 rather than a voltage, if you see what I mean. SPICE gives the error voltage, i.e. that between the base and emitter of Q1, as 38 dB below the output voltage, so the voltage on Q1 collector is less than that going into the first stage, and this indicates that Q1 is indeed feeding a current to Q2. This is an important finding as it means that the open-loop gain, and hence the feedback factor, cannot be increased by bootstrapping the collector load of Q1, which was the idea I had at the back of my mind all along.

The low impedance at Q2 base will be further reduced by Miller feedback through the Cbc of Q2, though how significant that is is uncertain at present. Since Cbc is a function of collector voltage, this is another potential source of nonlinearity.

Two Transistors: Dual Supply Rails

The question arises as to how easy it would be to convert this stage to run off dual supply rails, i.e. ±V and 0V. The answer appears to be—not easy at all, because the input transistor Q1 that performs the input-NFB subtraction is sitting very near the bottom rail.

Two Transistors: The Historical Dinsdale MM Circuit

The first two-transistor MM stage is generally accepted to have been put forward by J. Dinsdale, in an article "Transistor High-Quality Audio Amplifier" in *Wireless World* for January 1965. This article may be 45 years old, but it is still worth reading if you can got hold of it, not least because it discusses how to make an MM preamplifier using just *one* transistor; see the start of this chapter to find how to do it and why not to do it.

You will note at once from Figure 10.10 that the circuit is upside-down to modern eyes, with a negative supply rail at the top. I find the electrolytics particularly unsettling. This was common in circuits of the era and stemmed from the fact that most germanium transistors were PNP, so if you drew the emitter at the bottom (which is where people were used to drawing valve cathodes) then inevitably you end up with a negative supply rail at the top. When silicon transistors came in, they were more commonly NPN, so a sigh of relief went up all round as we reverted to the more logical approach of having the most positive rail at the top.

I have not so far tried building this circuit, due to the difficulty of obtaining the transistors. The OC44 was a PNP germanium transistor made by Mullard. Remarkably, it is still in much demand—in fact it has achieved iconic status—as it is held to give a unique sound in vintage-style fuzz boxes.[3], [4], [5]

Comparing this circuit with Figure 10.2 you can see that there are the characteristic two separate feedback loops, with DC feedback through R13, R14, and R7 and AC feedback to Q1 emitter via the RIAA network R15, C8, R16, C7. The DC path through this network is blocked by C3. The RIAA network is in Configuration-B (see Chapter 7). There is no IEC Amendment.

C3 bootstraps R6 to raise the input impedance high enough to give 47 kΩ in conjunction with loading resistor R2; disconcertingly Dinsdale refers to this as "feedback" in his article, which it is not.

The supply rail voltage is not precisely known, as in the complete preamplifier circuit the MM stage is fed through a network of RC filters that leave the final voltage in some doubt. It clearly is not greater than −20V, judging by the rating of C13, and it seems pretty safe

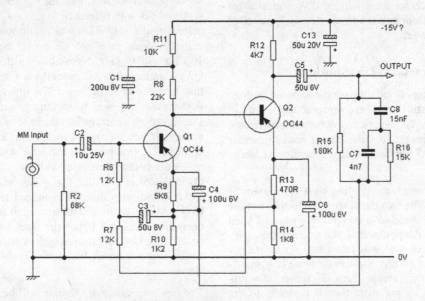

Figure 10.10 The circuit of the original Dinsdale MM stage: 1965

to assume it was around −15V. However the OC44 had a collector breakdown voltage of only 15V, so the rail might have been lower—perhaps −12V.

This circuit uses more parts than the two-transistor circuit of Figure 10.2, largely as a result of the different DC bias arrangements. Apart from the transistors, it has twelve resistors and five electrolytic capacitors. Figure 10.2 has (ignoring EMC filtering) ten resistors and two electrolytic capacitors and is the more economical solution.

Three-Transistor MM Input Stages

Adding an extra transistor improves the possible performance remarkably. An early three-transistor configuration was introduced by Arthur Bailey in 1966,[6] though his version had a rather awkward level-shift between the first and second transistors. Figure 10.11 shows a much improved version from the early '70s, designed by H. P. Walker[7] and later enhanced by my own good self.[8] It consists of two voltage amplifier stages as before, but an emitter-follower Q3 is added to buffer the collector of the second transistor from the load of the RIAA network and any external load.

This means that the second transistor collector can be operated at a much higher impedance, generating more open-loop gain.

The original Walker design had a simple 22 kΩ resistor as a collector load for Q2; when I was using this configuration[8] I split this into two 12 kΩ resistors, bootstrapping their central point from the emitter of Q3, as shown in Figure 10.11; this further increased the open-loop gain and reduced the distortion by a factor of three. With the wisdom of hindsight I strongly suspect that C6 should be increased in size; its LF roll-off in conjunction with R10 (R12 is not involved) is at 6 Hz, but that misses the point. For bootstrapping to work properly the signal voltage at the top capacitor terminal must be identical to that at the bottom, which needs a very low roll-off frequency. If C6 is increased to at least 47uF, the LF distortion should be reduced. Dominant-pole compensation is applied to Q2 by C1. Once again, the RIAA network has a high impedance, and a separate path for DC feedback must be provided by R1; there is in fact no DC feedback at all through the RIAA network because it is connected to the outside of C9. R3 and C8 are the input RF filter. The supply to the first stage is heavily filtered by R8 and C5.

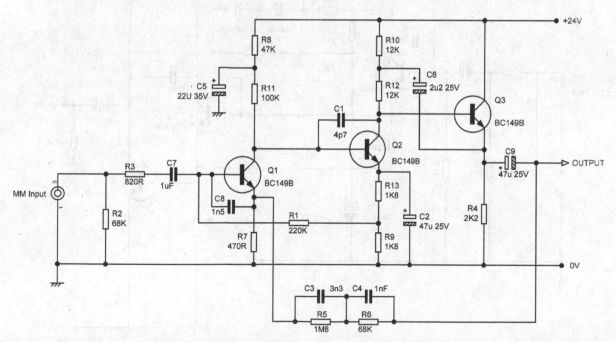

Figure 10.11 A typical three-transistor MM amplifier as commonly used in the 1960s and early '70s. The original design was by H. P. Walker; the bootstrapping of Q2 collector was added by me. I suggest increasing C6 to 47 uF.

The RIAA network is in Configuration-A. There is no IEC Amendment, as it was not introduced until 1976.

The added emitter-follower Q3, running at a much higher collector current than Q2 (5 mA versus 500 uA), much increases the output voltage swing into a load and so improves the input overload margin. However, a simple emitter-follower output stage has an asymmetrical output current capability, and so this is less effective at high frequencies where the impedance of the RIAA network is falling. Low-gain versions of this circuit may have the overload margin compromised by several dB at 20 kHz. This can be overcome by making the output stage more sophisticated—replacing the emitter resistor R4 with a current source greatly improves matters, and using a push-pull Class-A output doubles the output current capability again.

While this stage is a great improvement on the two-transistor configuration, it also is not well-adapted to dual supply rails, for the same reason; Q1 is still referenced to the bottom rail.

Figure 10.12 demonstrates another way to use three transistors in an RIAA amplifier; this configuration consists of a voltage amplifier stage, an emitter-follower, and then another voltage amplifier stage. The RIAA network is in Configuration-A. The first transistor is now a PNP type, its collector bootstrapped for increased gain by connecting the lower end of collector load R6 to the emitter-follower Q2. R2 and C8 make up an RF filter, and once more there is a high value of series resistance. Q2 drives the output transistor Q3, which is now a common-emitter voltage amplifier with collector load R8. Note the asymmetrical supply voltages; the positive rail is 8 V greater than the negative rail, and this is almost certainly intended to increase the positive-swing capabilities of the Q3 stage. Once again note the high impedances in the RIAA network to reduce the loading on the output and the consequent need for a separate DC feedback path via R10, the AC content being filtered out by C7. Versions of this configuration were used by Pioneer and Sonab.

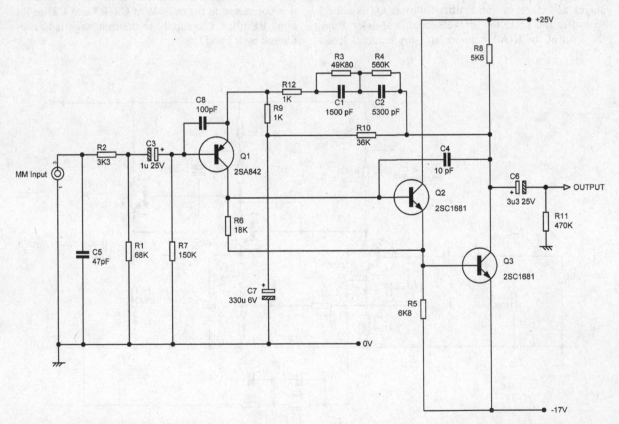

Figure 10.12 Another three-transistor MM amplifier configuration with a quite different structure. This is a simplified version of a circuit used by both Pioneer and Sonab.

Four-Transistor MM Input Stages

All of the amplifiers described so far have conventional output stages and so have trouble driving their RIAA network at high frequencies. Going from three transistors to four gives much greater freedom of design and allows us to use a markedly more capable output stage. Since Q1 is no longer connected to the bottom rail, the use of dual supplies is possible. The design in Figure 10.13 is closely based on the MM input stage of my "High-Performance Preamplifier" published in *Wireless World* in 1979.[9] By my own internal labelling system this was the MRP4. A slightly modified version (the MRP6) was used a year later in a preamplifier design for a consultancy client.

The MRP4 was the first "conventional" preamplifier design, insofar as my designs are ever conventional, that I published. It was my own reaction to the considerable complexity of the Advanced Preamplifier of 1976,[8] with its multiple discrete opamps. Back then

the available IC opamps were looked at with entirely justified suspicion; they were relatively noisy and prone to crossover distortion in their output stages. Crossover might be inescapable in power amplifiers of the day, but it was definitely not acceptable in a preamplifier. Thus discrete Class-A circuitry was used throughout the preamplifier. (The 5534 opamp was just becoming available at the time but was horribly expensive.)

This design was intended to showcase a very large overload margin, and so its gain at 1 kHz was only 10 times (+20 dB), and the output was only 50 mV rms nominal. Obviously, further downstream amplification (of variable gain) was required to get the signal up to a level that could be applied to a power amplifier. The configuration is much the same as many power amplifiers of the day, and if the singleton input transistor was replaced by a differential pair, it would look very much like the "model amplifiers" used so extensively in my book *Audio Power Amplifier Design*.[10]

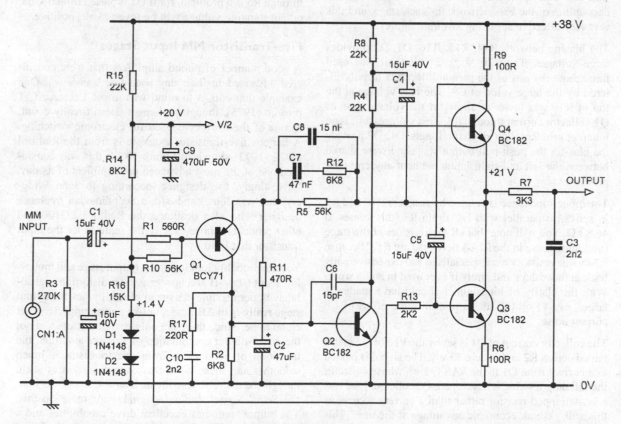

Figure 10.13 My four-transistor MM amplifier configuration with push-pull Class-A output structure. Based on the MM stage of my "High-Performance Preamplifier" published in *Wireless World* in 1979.

I used a single-transistor input because I was worried about the noise contribution from the second transistor in a differential pair. That this might create a relatively large amount of second-harmonic distortion was considered of less importance, given the 50 mV rms nominal output of the stage. The RIAA network is an example of Configuration-C, which is preferred because of its lower capacitor values for the same gain and impedance, as described in Chapter 7. It has capacitor values similar to those used in the +30 dB gain opamp MM inputs in that chapter, but R0 (which in Figure 10.13 is actually labelled R11) is larger, so the gain is 10 dB less. The stage could accept an input of 1.1 V rms at 1 kHz, an overload margin of no less than 47 dB, and 3.8 Vrms at 10 kHz. The low gain made an HF correction pole (R7, C3) essential. No attempt was made to implement the IEC Amendment.

The whole preamplifier ran off a single +38 V rail that was not regulated; instead it had a post-reservoir RC filter that reduced the supply ripple to about 50 mV rms. This low-cost approach was combined with heavy RC decoupling of the bias network for each stage, and this was very effective at getting low hum figures.

The biasing network R14, R15, R16, D1, D2 provides three voltages. The +20 V "V/2" bias rail was used throughout the rest of the preamplifier; it is heavily filtered by the large value of C9. The bias voltage at the top of R16 was lower to allow for the voltage drop of Q1 collector current through R5. This voltage-shift is an inherent problem with singleton inputs. D1, D2 provide the bias for the push-pull output stage and were shared between the left and right inputs without any crosstalk problems.

The input impedance is defined by resistors R10 and R1 in series, in parallel with DC drain R3; this comes to 46.8 kΩ. You will note that all the resistors in the stage (including those in the RIAA network) are E12, because E24 value resistors were specialised and expensive parts back in those days and rarely if ever used in audio work. With the clarity of hindsight, I should have made C1 larger, say 47 or 100 uF, to minimise the effect of LF current noise.

The collector current of Q1 is set by the Vbe of Q2 maintained across R2 and is here 53 uA. The signal is passed as a current from Q1 to the VAS Q2, on whose collector the full output swing is developed. Q2 collector load was a bootstrapped resistor rather than a current source, as this still gave an economic advantage at the time. This is then buffered by the push-pull emitter-follower Q4

with its driven current source Q3; R9 senses the current through Q4, and when it increases the voltage on Q3, base is reduced via C5. Since this is effectively a negative-feedback loop with a gain of unity and 100% feedback, the current variations are halved for a given load, and so the peak output current is doubled.

Stability is an issue here because the amplifier is working with a closed-loop gain close to unity at HF. The Miller dominant-pole capacitor, C6, is made as small as possible to maximise the slew rate, and stability is assisted by the lead-lag network R17, C10 across Q1 collector resistor. This stage was followed by a 3rd-order subsonic filter using a current-source emitter-follower, and the only THD figure I have to hand is for both stages together. THD was below 0.004 % at 6 Vrms out, from 1 kHz to 10 kHz, the input signal being inverse RIAA equalised to give constant output with frequency.

This design could be relatively easily converted to run off dual supply rails, as the first transistor is not referenced to the bottom rail. However, the voltage drop through R5 is a problem, for if Q1 is biased from 0V, the output standing voltage will be several volts positive.

Five-Transistor MM Input Stages

A good number of phono amplifiers that were considered advanced in their day used five transistors. One example that comes to mind was in the Lecson AC-1 preamp (1975), though perhaps it doesn't really count, as one of the devices is used for electronic switching. A famous five-transistor example is from the Radford ZD22 (1973) shown in Figure 10.14; this was considered one of the most advanced preamplifiers of its day. Interestingly, the designer (according to John Widgery, that was Jens Landvard, a Scandinavian freelance designer who also designed the Radford ZD100 and other products) chose exactly the same Ic for the input transistor that I did.

You are probably looking at the output stage and muttering about Class-B and how crossover distortion is absolutely impermissible in a preamplifier. In fact, the output stage really *is* in AB mode, unlike power amplifiers that claim to be. Since there are only two transistors and not the driver-output combination of a power amplifier, the transition of conduction from one transistor is much smoother and there is no crossover distortion as such, though the stage is less linear than a Class-A version; see *Small Signal Audio Design*[11] for more on this. The output stage has excellent drive capabilities and a low-impedance RIAA network in the capacitor-efficient

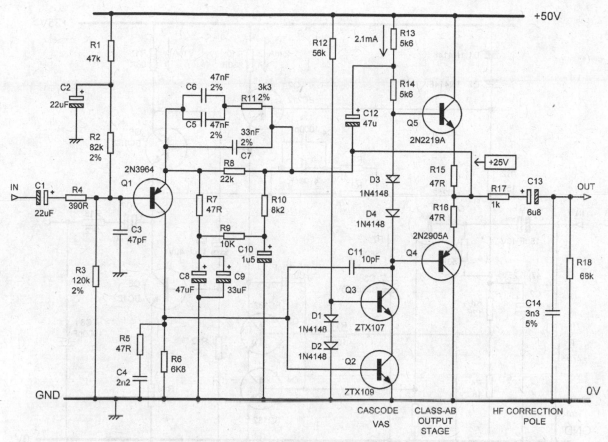

Figure 10.14 Five-transistor MM amplifier configuration as used in Radford ZD22 preamp

Configuration-C is used, with precision components (for their day) used in the critical positions.

The single +50V supply rail permits a maximum output of about 17 Vrms. The gain is +33.9 dB (1kHz), which gives an excellent maximum input of 357 mV rms (1kHz) and a very healthy overload margin of +36 dB. At this relatively low gain an HF correction pole R17, C14 is essential to keep the RIAA accurate at the HF end. In simulation the RIAA accuracy is very flat in the middle but −0.1 dB at 200 Hz and 10 kHz. The LF end rolls off due to the subsonic filter implemented by C8, C9, R9, C10, R10; this does not give any known filter characteristic but shows a −0.4 dB shelf between 50 and 100 Hz, followed by a roll-off at roughly −9 dB/octave that is −10 dB at 20 Hz and −18 dB at 10 Hz. The HF end rolls off because (from my simulations) the HF correction pole has the wrong value; changing C14 to 2n2

pushes the −0.1 dB point out to 20 kHz. Aside from that, it's a fine piece of design work.

Six-Transistor MM Input Stages

If we permit ourselves one more transistor, bringing the total to six, the MM stage of Figure 10.13 can be significantly improved in its distortion performance. The MM preamplifier in Figure 10.15 was used in an experimental preamplifier which according to my own numbering system was the MRP2; it was not published. It was designed with less regard to economy than the MRP4 and was therefore somewhat more sophisticated despite it coming earlier in the series.

In this preamplifier, as in the MRP1, the RIAA equalisation was performed in two stages, with the HF part of the RIAA implemented by R8, R9, C4, C5 in the first stage to give

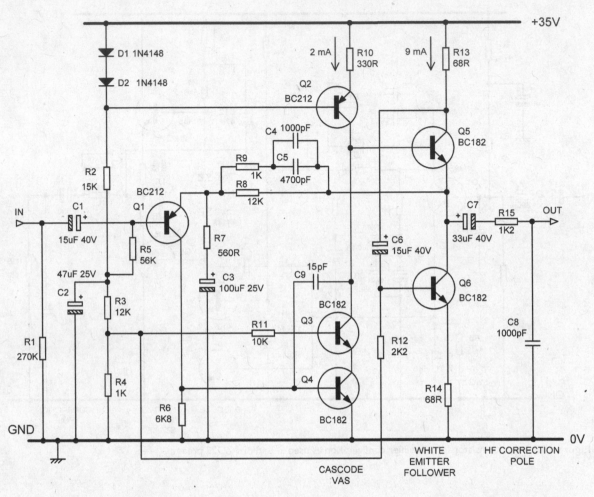

Figure 10.15 My six-transistor MM input stage for the MRP2, with single input transistor, cascode VAS with current source, White push-pull output stage, and HF correction pole. Only the HF cut of the RIAA is implemented in this stage.

good headroom at high frequencies. The gain is +26 dB (1kHz) and the maximum output of 12.4 V rms, giving a huge maximum input of 620 mV rms and an overload margin of 42 dB. The LF boost part of the RIAA was performed by the normalisation amplifier downstream, which raised all the inputs to the same nominal level. The LF boost components were switched out of the feedback loop when a line input was in use and a flat response required.

The configuration is again essentially that of the solid-state power amplifiers of the day, which tended to use a single input transistor to perform the feedback subtraction, the great advantages of the differential pair for low distortion as well as DC precision not being much appreciated at this point. A single input transistor was used here with the aim of minimising noise, working on the assumption that two transistors must be noisier than one, though in this case not 3 dB worse, as the NFB network has a much lower impedance than the cartridge. The input transistor has its collector current defined at about 88 uA by the 0.6V established across R6; this apparently low value gives a better noise performance than higher currents, with the highly inductive source resistance of an MM cartridge. Q1 passes its output current to the voltage amplifier stage, or VAS, and its transconductance combined with the value of the dominant-pole Miller capacitor C9 sets the open-loop gain at high frequencies. As in Figure 10.14, C1 would have been better made 47 or 100 uF.

The main sources of distortion in the VAS are Early effect and the nonlinear variation of Cbc with Vce. A cascode VAS reduces both effectively; the collector voltage of Q4 is kept constant by Q3, and so there is no Early effect. Likewise, since there is no signal on the collector of Q4 there can be no local negative feedback through its nonlinear Cbc, and the compensation feedback all goes through C9. Similar and perhaps slightly better results would be obtained by omitting the cascode transistor Q3 and putting an emitter-follower inside the C9 Miller loop. The first version of this circuit had a simple current-source emitter-follower output, running at the same quiescent current of 9 mA; it showed premature negative clipping at high frequencies due to the heavy loading of the HF RIAA capacitor in the feedback network, and also loading by the HF correction pole R15-C8; this problem was almost totally eliminated by converting it to the same push-pull Class-A White output structure as used in Figure 10.13.

Everything is biased from a single divider D1-D2-R2-R3-R4, which is heavily filtered by C2 to remove supply rail noise. The voltage across R4 is 1.2 V and biases both the cascode transistor Q3 and the push-pull output current-source Q6. So far as I'm concerned, this configuration may have six transistors, but it does not count as a "discrete opamp" because there is no differential input. As for the four-transistor version, there should be no difficulty in converting it to dual-rail operation, because the biasing system is not referenced to the bottom rail. Because a single input transistor does not have the DC precision of a differential pair, and there is a voltage drop across feedback resistor R8, there will be a significant offset voltage at the output, which will need to be DC blocked by a series capacitor.

The input impedance is set by the parallel combination of R1 and R5, which is 46.4 kΩ. This is a little low compared with the nominal value of 47 kΩ, representing an error of 1.3%, but it is the best you can do with two E12 resistor values; bear in mind that R5 should not be too big (say 100 kΩ or less) as it carries the base current for Q1 and there will be a voltage dropped across it. At the time of the design E24 resistors were relatively rare and expensive, and no consideration was given to using them to get parameters like the input impedance more accurate. Using E24 resistors R1 = 300 kΩ and R5 = 56 kΩ gives an input impedance of 47.2 kΩ, an error of only 0.4% in the nominal value; note this error does not include the effects of resistor tolerances.

Even greater accuracy could be achieved by using three E24 resistors in parallel. For example R1 = 560 kΩ in parallel with 620 kΩ and R5 = 56 kΩ reduces the error in nominal value to 0.10% high, which is likely to be much less than the resistor tolerances. There are many combinations of three resistors that give a combined value very close to 47 kΩ, and adding the extra constraint that they should be as nearly equal as possible allows significant improvements in accuracy when tolerances are taken into account, as random errors partially cancel; this is explained in detail in Chapter 2. The three resistors 560 kΩ, 620 kΩ, and 56 kΩ show this effect to some extent; if 1% resistors are used the effective tolerance of the combination is slightly improved to 0.85%. A better choice would be R1= 160 kΩ in parallel with 200 kΩ and R5 = 100 kΩ, which has a nominal value only 0.12% high and an effective tolerance of 0.60%, which is significantly better.

More Complex Discrete-Transistor MM Input Stages

The record for the highest supply voltage to an RIAA stage was set in 1974 by the Technics SU9600, which used seven transistors and employed ±24 V rails and a third rail at a staggering +136 V. This gives a whole new meaning to the phrase "third-rail electrification". To the best of my knowledge the record still stands. The general configuration is shown in Figure 10.16; seven transistors are used, in three cascaded differential voltage amplifiers, followed by an emitter-follower output buffer. The final voltage amplifier and the output stage work on asymmetrical supplies, running between +136 V and −24 V. The output sits at +56 V to allow a symmetrical output swing, which accounts for the DC-blocking capacitor C11. Note that several resistors around the output stage are high-wattage types. RV1 allowed gain adjustment, while RV2 and RV3 were for setting the DC conditions. My information is that the maximum input was 900 mV (frequency unstated, but presumably at 1 kHz), the THD was 0.08% (frequency and level unstated), the S/N ratio was 73 dB with reference to 2 mV, and the RIAA accuracy was ±0.3 dB.

The output device dissipation is of course enormous for a preamp stage, and the use of a constant-current source, or better still a push-pull Class-A output stage, would have allowed this to be much reduced; one can only speculate as to why those techniques were not used. There would probably have been some fearsome transients at the output on switch-on, and it is notable that an output muting relay was required, probably not so much

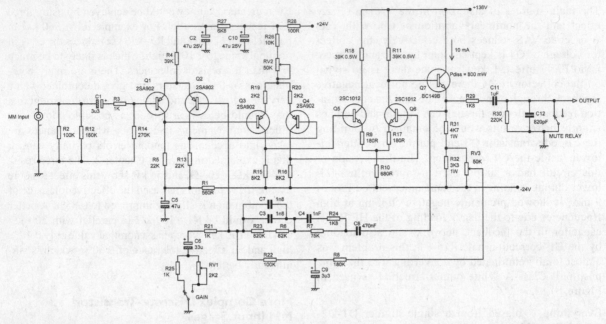

Figure 10.16 A simplified schematic of the Technics SU9600 RIAA stage, with its +136V supply rail

for reducing audible noise as to give the later stages in the preamplifier a chance of survival.

In the original circuit small capacitors were freely sprinkled over the diagram, leading me to suspect that HF stability was a serious issue during development.

References

1. Tobey, R., and Dinsdale, J. "Transistor High-Fidelity Pre-Amplifier" *Wireless World*, Dec 1961, p. 621.

2. Dinsdale, J. "Transistor High-Quality Audio Amplifier" *Wireless World*, Jan 1965, p. 2.

3. www.californiavalveworks.com/Mullard.html Accessed June 2013.

4. http://swartamps.com/oc44_transistor_dallas_rangemaster.htm Accessed Nov 2016.

5. www.williamsaudio.co.uk/Williams-OC44-Ranger. html Accessed Nov 2016.

6. Bailey, A. R. "High Performance Transistor Amplifier" *Wireless World*, Dec 1966, p. 598.

7. Walker, H. P. "Low-Noise Audio Amplifiers" *Electronics World*, May 1972, p. 233.

8. Self, D. "An Advanced Preamplifier Design" *Wireless World*, Nov 1976.

9. Self, D. "High-performance Preamplifier" *Wireless World*, Feb 1979, p. 40.

10. Self, D. *Audio Power Amplifier Design*. 6th edition. Newnes, 2013, p. 190. ISBN 978-0-240-52613-3.

11. Self, D. *Small Signal Audio Design*. 2nd edition. Focal Press, 2015, pp. 562–568. ISBN: 978-0-415-70974-3 (hbk) ISBN: 978-0-415-70973-6 (pbk) ISBN: 978-1-315-88537-7 (ebk).

Chapter 11

Moving-Coil Head Amplifiers

Moving-coil cartridges are generally accepted to have a better tracking performance than moving-magnet (MM) cartridges because the moving element is a set of light-weight coils rather than a magnet, which is inevitably made of a relatively dense alloy. Because the coils must be light, they consist of relatively few turns and the output voltage is very low, typically in the range 100–700 uV rms at a velocity of 5 cm/sec, compared with 5 mV rms from the average MM cartridge. Fortunately this low output comes from a very low impedance, which, by various technical means, allows an acceptable signal-to-noise performance to be obtained. Apart from the low output, a further complication is that the output voltage varies over a very wide range between different brands.

Moving-Coil Cartridge Characteristics

There is much greater variation in impedance and output across the available range of MC cartridges than for MM cartridges. The range quoted earlier is already much wider, but including the extremes currently on the market (2009) the output range is from 40 to 2500 uV, a remarkably wide span of 62 times or 36 dB. This is illustrated in Figure 11.1, which shows the results of a survey of 85 different MC cartridges. (Note that the ranges for the columns are wider at the right side of the diagram.) When I first became involved in designing MC amplifiers in 1986, I compiled a similar chart,[1] and it is interesting to note that the same features occurred—there are two separate clusters around 100–300 uV and 500–700 uV, and the lowest output is still 40 uV from an Audio Note cartridge (the loLtd model). The highest output of 2.5 mV comes from the Benz Micro H2, and this is only 6 dB below an MM cartridge.

Assuming that a conventional MM input stage is being used to raise 5 mV to nominal internal level and perform the RIAA equalisation, the Audio Note cartridge requires a gain of 125 times or +42 dB. The cartridge cluster around 200 uV output needs 25 times or +28 dB,

while the 500 uV cluster requires 10 times or +20 dB. If an amplifier is to cover the whole range of MC cartridges available, some form of gain switching is highly desirable.

Cartridge impedances also vary markedly, over a range from 1 Ω (Audio Note loLtd) to 160 Ω, (Denon DL-110 and DL160), with impedance increasing with output level, as you would expect—there are more turns of wire in the coils. The inductance of MC cartridges is very low, and their source impedance is normally treated as purely resistive. The recommended load impedances are also resistive (unlike the R-C combinations often used with MM cartridges) and are usually quoted as a minimum load resistance. Once more the variation is wide, from 3 Ω (Audio Note loLtd again) to 47 kΩ (Denon DL-110 and DL160 again), but a 100 Ω input load will be high enough for most of the cartridges surveyed, and 500 Ω will work for almost all of them. The Audio Note loLtd cartridge is unusual in another way—its magnetic field is produced not by permanent magnets but a DC-powered electromagnet, which presumably requires a very pure supply indeed. The manufacturers whose cartridges were included in the survey are listed in Table 11.1.

The Limits on MC Noise Performance

Because MC cartridges can be modelled for noise purposes simply as their coil resistance, it is straightforward to calculate the best signal/noise ratio possible. Even if we assume a noiseless amplifier, the Johnson noise from the coil resistance sets an inescapable limit; comparing this with the cartridge output gives the maximum signal/noise ratio. This was done for all the cartridges used to compile Figure 11.1, using the manufacturers' specs, and the answers varied from 63.9 to 90.8 dB, which is a pretty big spread. (This does not include RIAA equalisation.)

In practice things will be worse. Even if we carry on assuming a noiseless amplifier, there is resistance in the

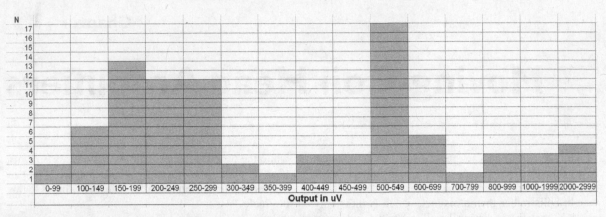

Figure 11.1 The output levels for 85 MC cartridges at 5 cm/sec (2009)

Table 11.1 MC cartridge manufacturers whose product data was used to compile Figure 11.1

Audio Note	Immutable Music
Benz Micro	Koetsu
Cardas	Lyra
Clearaudio	Miyabi
Denon	Ortofon
Dynavector	Shelter
Goldring	Sumiko
Grado	van den Hul

tone arm wiring, which has to be very thin for flexibility, and a bit more in the cable connecting turntable to preamp. Calculating the same figures for MM cartridges is a good deal more complicated because of the significant cartridge inductance; see Chapter 9.

Amplification Strategies

There are two ways to achieve the high gains required for these low-output cartridges. In the most common method a standard MM input stage with RIAA equalisation can be switched to either accept an MC input directly or the output of a specialised MC input stage, which gives the extra gain needed; this may be either a step-up transformer or an amplifier configured to work well with very low source resistances. The large amount of gain required is split between two stages, which makes it easier to achieve. Alternatively, a single stage can be used with switched gain; but this is not too hot an idea:

1) Switchable gain makes accurate RIAA equalisation much harder.

2) For good noise performance, the input device operating current needs to be low for MM use (where it sees a high impedance) and high for MC use (where it sees a very low impedance). Making this operating current switchable would be a complicated business.

3) Achieving the very high gain required for MC operation together with low distortion and adequate bandwidth will be a challenge. It is unlikely to be possible with a single opamp, and so there is little likelihood of any saving on parts.

Moving-Coil Transformers

If you have a very low output voltage and very low impedance, an obvious way to deal with this is by using a step-up transformer to raise the voltage to the level where it can be appropriately applied to an MM amplifier stage. In some ways a step-up transformer to get the signal up to MM level is the most elegant solution, as no power supply is required, but such a transformer has most of the usual disadvantages such as frequency response problems and cost, though for hi-fi use the weight is not a difficulty, and nonlinearity should not be an issue because of the very low signal levels.

In this application the cost is increased by the need for very high immunity to hum fields. While it is relatively straightforward to make transformers that have high immunity to external magnetic fields, particularly if they are toroidal in construction, it is not cheap, because the mumetal cans that are required for the sort of immunity necessary are difficult to manufacture. The root of the problem is that the signal being handled is so very small. The transformer is usually working in a modern house, which can have surprisingly large hum fields and generally presents a hostile environment for very-low-signal transformers, so very good immunity indeed is required; some manufacturer use two nested screening cans, separately grounded, to achieve this, as shown in Figure 11.2. An inter-winding electrostatic screen is usually fitted. A stereo MC input naturally requires *two* of these costly transformers.

MC transformers are designed with low primary winding resistances, typically 2 or 3 Ω, to minimise the Johnson noise contribution from the transformer. Some transformers have windings made of silver rather than copper wire, but the conductivity of silver is only 5% higher than that of copper, and the increase in cost is startling. At the time of writing, one brand of silver-wound transformer costs more than £1200—each, not for a pair. To put this in perspective, using copper and going from one American wire gauge to the next larger will drop the resistance of a winding by 12%; a rather better deal.

Because of the great variation in cartridge output levels and impedances, some manufacturers (e.g. Jensen) offer transformers with two or three primary windings, which can be connected in series, parallel, or series-parallel to accommodate a wide variety of cartridges.

A transformer secondary must be correctly loaded to give the flattest possible frequency response, and this usually means that a Zobel R-C network must be connected across it. This is Rd and Cd in Figure 11.2, where they have typical values. The values required depend not only on the transformer design but also somewhat on the

cartridge impedance, and some manufacturers such as Jensen are praiseworthily thorough in giving secondary loading recommendations for a wide range of cartridge impedances.

The very wide variation in cartridge outputs means that the step-up ratio of the transformer must be matched to the cartridge to get an output around 5 mV that is suitable for an MM input amplifier. For example, Jensen offer basic step-up ratios from 1:8 to 1:37. The maximum ratio is limited not only by transformer design issues but by the fact that the loading on the secondary is, as with all transformers, transferred to the primary divided by the square of the turns ratio. A 1:37 transformer connected to the 47 kΩ input impedance of an MM stage will have an impedance looking into the primary of only 34 Ω; such a transformer would however only be used with a very low-impedance low-output cartridge, which would be quite happy with such a loading. It is of course possible to arrange the input switching so the 47 kΩ input load is on the MM socket side of the MC/MM switch; the MM amplifier can then have a substantially higher input impedance.

Moving-Coil Input Amplifiers

The high cost of transformers means that there is a strong incentive to come up with an electronic solution to the amplification problem. The only thing that makes it possible to achieve a reasonable signal-to-noise ratio

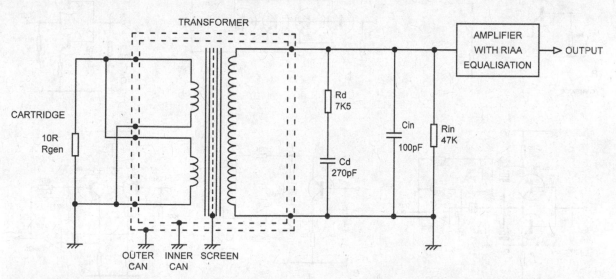

Figure 11.2 A typical MC step-up transformer circuit with twin primaries wired in parallel, dual screening cans, and Zobel network Rd, Cd across the secondary. Rin, Cin are the usual MM input loading components.

is that the very small signal comes from a very low source impedance.

MC head amplifiers come in many forms, but almost all in use today can be classified into one of the topologies shown in Figure 11.3, all of which use series feedback. Representative component values are given. The configuration in Figure 11.3a is a complementary-feedback pair using a single input transistor chosen to have a low base series resistance, r_{bb}. The feedback network must also have a very low impedance to prevent its Johnson noise from dominating the overall noise output, and this puts a heavy load on the second transistor. Typically a gain of around 46 times (+33 dB) will be implemented, with an upper feedback resistor of 100 Ω and a lower resistor of 2.2 Ω, a total load on the amplifier output of 102 Ω. The combination of limited open-loop gain and the heavy load of the feedback network means that both linearity and maximum output level tend to be uninspiring, and the distortion performance is only acceptable because the signals are so small. An amplifier of this type is analysed in Reference [2].

Figure 11.3b shows a classic configuration where multiple transistors are operated in parallel so that their gains add but their uncorrelated noise partly cancels. Using two

transistors gives a 3 dB improvement, four transistors gives 6 dB, and so on. The gain block A is traditionally one or two discrete devices, which again have difficulty in driving the low-impedance feedback network. Attention is usually paid to ensuring proper current-sharing between the input devices. This can be done by adding low-value emitter resistors, say 1 Ω, to deal with Vbe variations; they are effectively in series with the input path, and therefore degrade the noise performance unless each resistor is individually decoupled with a large electrolytic. Alternatively, each transistor can be given its own DC feedback loop to set up its collector current. For examples of this kind of circuitry see Reference [3].

Figure 11.3c shows the series-pair configuration. This simple arrangement uses two complementary input transistors to achieve a 3 dB noise improvement without current-sharing problems because essentially the same collector current goes through each device. The collector signal currents are summed in R_c, which must be reasonably low in value to absorb collector current imbalances. There is no feedback so linearity is poor. The biasing arrangements are not shown.

Figure 11.3d is a development of Figure 11.3c, closing a negative-feedback loop around the input devices. This

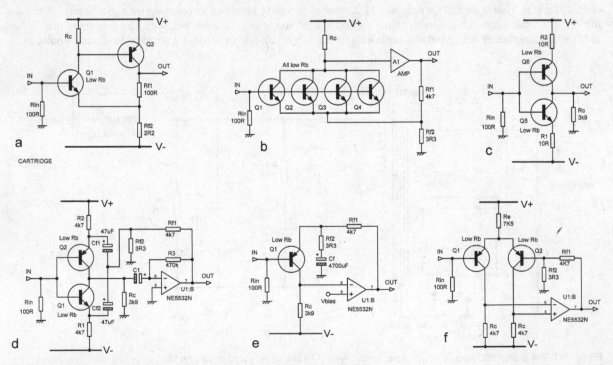

Figure 11.3 Some popular MC amplifier configurations, with typical values

must be applied to the emitters of both transistors, and so two DC-blocking capacitors Cf1,Cf2 are needed. This configuration was used in the Quad 44 preamp.

Figure 11.3e is a development of Figure 11.3a, with the input transistor inverted in polarity and the spadework of providing open-loop gain and output drive capability now entrusted to an opamp. I describe this as a hybrid amplifier because it combines a discrete device with an opamp. The much increased feedback gives excellent linearity, and less than 0.002% THD at full output may be confidently expected. However, problems remain. Rf_2 must be very low in value, as it is effectively in series with the input and will degrade the noise performance accordingly. If Rf_2 is 10 Ω (which is on the high side), Cf must be very large, for example 1000 uF, to limit the LF roll-off to −1 dB at 30 Hz. Adopting a quieter 3.3 Ω for the Rf_2 position gives significantly lower noise but demands 4700 uF to give −3 dB at 10 Hz; this is not elegant and leads to doubts as to whether for once the ESR of a capacitor might cause trouble. Cf is essential to reduce the gain to unity at DC because there is +0.6 V on the input device emitter, and we don't want to amplify that by fifty times. Negative feedback will drive the inverting input of the opamp to Vbias, and this, with the value of Rc, defines the collector current of Q1. Vbias

must be negative to give Q1 enough Vce to work in, and will be in the range −5 to −10 V.

The +0.6V offset can be eliminated by the use of a differential pair, as in Figure 11.3f. This cancels out the Vbe of the input transistor TR1, at the cost of some degradation in noise performance. The pious hope is that the DC offset is so much smaller that if Cf is omitted, and the offset is amplified by the full AC gain, the output voltage swing will not be seriously reduced. The noise degradation incurred by using a differential pair was measured at about 2.8 dB. Another objection to this circuit is that the offset at the output is still nonnegligible, about 1 V, mostly due to the base bias current flowing through R10. A DC-blocking capacitor on the output is essential.

An Effective MC Amplifier Configuration

Finding none of these configurations satisfactory, I evolved the configuration shown as a block diagram in Figure 11.4. There is no Cf in the feedback loop, and indeed no overall DC feedback at all. The input transistor and the opamp each have their own DC feedback systems. The transistor relies on simple shunt negative feedback via DC loop 1; the opamp has its output held precisely to a DC level of 0 V by the integrator A2, which

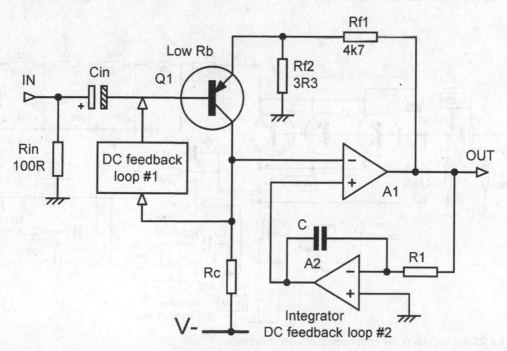

Figure 11.4 Block diagram of the MC preamplifier, showing the two DC feedback loops

acts as DC loop 2. This senses the mean output level, and sets up a voltage on the non-inverting input of A1 that is very close to that at Q1 collector, such that the output stays firmly at zero. The time-constant is made large enough to ensure that an ample amount of open-loop gain exists at the lowest audio frequencies. Too short a time-constant will give a rapid rise in distortion as frequency falls. Any changes in the direct voltage on Q1 collector are completely uncoupled from the output, but AC feedback passes through Rf1 as usual and ensures that the overall linearity is near-perfect, as is often the case with transistor opamp hybrid circuits. Due to the high open-loop gain of A the AC signal on Q1 collector is very small, and so shunt AC feedback through DC loop 1 does not significantly reduce the input impedance of the overall amplifier, which is about 8 kΩ.

As we have seen, MC cartridges vary greatly in their output, and different amplifier gain settings are highly desirable. Usually it would be simple enough to alter Rf1 or Rf2, but here it is not quite so simple. The resistance Rf_2 is not amenable to alteration, as it is kept to the low value of 3.3 Ω by noise considerations, while Rf1 must be kept up to a reasonable value so that it can be driven to a full voltage swing by an opamp output. This means a minimum of 500 Ω for the 5534/2. It is intriguing that amplifiers whose output is measured in millivolts are required to handle so much current.

Table 11.2 Gain options and maximum outputs

Gain	Gain (dB)	Max output (rms)
10 ×	+20 dB	480 mV
20 ×	+26 dB	960 mV
50 ×	+34 dB	2.4 V
100 ×	+40 dB	4.6 V
200 ×	+46 dB	10 V

These two values fix a minimum closed-loop gain of about 44 dB, which is much too high for all but the most insensitive cartridges. My solution was to use a ladder output attenuator to reduce the overall gain; this would be anathema in a conventional signal path, because of the loss of headroom involved, but since even an output of 300 mV rms would be enough to overload virtually all MM amplifiers, we can afford to be prodigal with it. If the gain of the head amplifier is set to be a convenient 200 times (+46 dB), then adding output attenuation to reduce the overall gain to a more useful +20 dB still allows a maximum output of 480 mV rms. Lesser degrees of attenuation to give intermediate gains allow greater outputs, and these are summarised in Table 11.2. For testing, an Ortofon MC10 was used with +26 dB of gain, giving similar output levels to MM cartridges. This highly successful cartridge was in production for

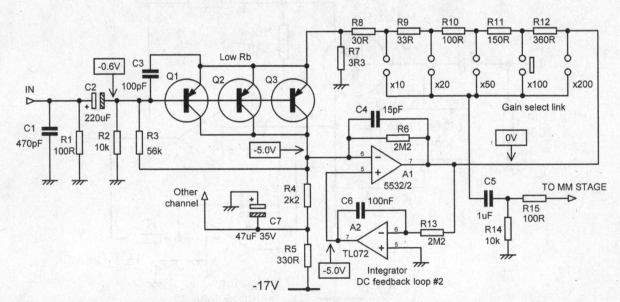

Figure 11.5 Circuit diagram of the MC preamplifier, with DC voltages

30 years and has only recently been superseded; its internal impedance is 3.3 Ω.

A final constraint on the attenuator is the need for low output impedances so that the succeeding MM input stage can give a good noise performance. The MM input should have been optimised to give its best noise figure with relatively high source impedances, but a low source impedance will still reduce its actual noise output. This means that an output attenuator will need low resistor values, imposing yet more loading on the unfortunate opamp. This problem was solved by making the attenuator ladder an integral part of the AC feedback loop, as shown in Figure 11.5. This is practicable because it is known that the input impedance of the following MM stage will be too high, at 47 kΩ, to cause significant gain variations.

The Practical Circuit

This is shown in Figure 11.5, and closely follows Figure 11.4, though you will note that the input devices have suddenly multiplied themselves by three. Capacitor C1 is soldered on the back of the MC input phono sockets and is intended for EMC immunity rather than cartridge response modification. If the need for more capacitive or resistive loading is felt, then extra components may be freely connected in parallel with R1. If R1 is raised in value, then load resistances of 5 kΩ or more are possible, as the impedance looking into C_2 is about 8 kΩ. Capacitor C2 is large to give the input devices the full benefit of the low source impedance, and its value should not be altered. Resistors R2, R3 make up DC loop 1, setting the operating conditions of Q1, Q2, Q3, while R4 is the collector load, decoupled from the supply rail by C9 and R5, which are shared between two stereo channels. Opamp IC1 is a half of a 5532, providing most of the AC open-loop gain, and is stabilised at HF by C4. R6 has no real effect on normal operation but is included to give IC1 a modicum of DC negative feedback and hence tidy behaviour at power-up, which would otherwise be slow due to the charging time of C2. IC2, half of a TL072, is the integrator that forms DC feedback loop 2, its time-constant carefully chosen to give ample open-loop gain from IC1 at low frequencies while avoiding peaking in the LF response that could occur due to the second time-constant of C2.

The ladder resistors R8–R12 make up the combined feedback network and output attenuator, the gain being selected by a push-on link in the prototype. A rotary switch could be used instead, but this should *not* be operated with the system volume up as this will cause loud clicks, due to the emitter current (about 4 mA) of Q1–Q3 flowing through R7, which causes voltage drops down the divider chain. Note that the current through R7 flows down the ground connection back to the PSU. Output resistor R15 ensures stability when driving screened cables, and C5 is included to eliminate any trace of DC offset from the output.

The power supply rails do not need to be especially quiet, and a normal opamp supply is quite adequate.

The Performance

The performance is summarised in Table 11.3. Careful grounding is needed if the noise and crosstalk performance quoted is to be obtained.

When connected to a RIAA-equalised MM stage as described in Chapter 7, the noise output from the MM stage is −93.9 dBu at 10 times MC gain and −85.8 dBu at 50 times. In the 10 times case the MC noise is actually 1.7 dB lower than for MM mode.

Transistors for MC Amplifiers

The input transistor originally chosen was the 2N4403, a type that was acknowledged as superior for this kind of application for some years due to its relatively low R_b of about 20 Ω. The conventional wisdom was that PNP devices gave slightly lower noise than their NPN equivalents because of "lower surface recombination noise".

Table 11.3 MC head amp performance figures.

Input overload level	48 mV rms
Equivalent input noise.	−141.0 dBu, unweighted, without RIAA equalisation. (with 3.3 Ω source res)
Noise figure	6.4 dB (with 3.3 Ω source res)
THD	Less than 0.002% at 7 Vrms out (maximum gain) at 1 kHz
	Less than 0.004% 40 Hz—20 kHz
Frequency response	+0, −2 dB, 20 Hz–20 kHz
Crosstalk	Less than −90 dB, 1 kHz– 20 kHz (layout dependent)
Power consumption	20 mA at ±15 V, for two channels

A single device used in the circuit of Figure 11.5 gives an EIN of −138 dB with a 4 mA collector current and a 3.3 Ω source resistance. The Johnson noise from 3.3 Ω is −147.4 dBu, so we have a noise figure of 9.4 dB.

It was then consistently found that putting devices in parallel without any current-sharing precautions whatever always resulted in a significant improvement in noise performance. On average, adding a second transistor reduced noise by 1.2 dB, and adding a third reduced it by another 0.5 dB, giving an EIN of −139.7 dBu and an NF of 7.7 dB. Beyond this, further multiplication was judged unprofitable, so a triple-device input was settled on. The current-sharing under these conditions was checked by measuring the voltage across 100 Ω resistors temporarily inserted in the collector paths. With 3.4 mA as the total current for the array it was found, after much device-swapping, that the worst case of imbalance was 0.97 mA in one transistor and 1.26 mA in another. The transistors were not all from the same batch. It appears that, for this device at least, matching is good enough to make simple paralleling practical.

A superior device for low source impedances was the purpose-designed 2SB737, with a stunningly low Rb of 2 Ω. Three of them improved the EIN to −141.0 dBu and the NF to 6.4 dB, albeit at significant cost. Sadly it is now obsolete. It was a device with unique properties, and since MC cartridges show no sign of going away, you would think there would be a secure if not enormous market for it. It can still be obtained from specialised suppliers such as the Signal Transfer Company.[4] They are a strictly limited resource, though there are probably more of them out there than there are nuvistors. The 2SB737 had an NPN complement called the 2SD786 which was almost as good but on measurement had a slightly higher r_{bb}.

This design was revisited for use in the Elektor preamplifier, and the availability-challenged 2SB737 was replaced by four 2SA1085 transistors, which lowered the noise by about 1.0 dB for 3.3 Ω and 10 Ω source resistances. This in turn has now been declared obsolete.

Why this slaughter of the innocents? Why stop making such excellent and useful parts? The only plausible reason I have heard advanced is that none of these transistors had surface-mount versions and therefore were not suitable for modern manufacturing. This raises the question of why they could not have been repackaged as SMT.

Going back to the elderly 2N4403 is not an attractive option. Paralleling ten 2N4403's to get the effective $_{bb}$ down to the level of one single 2SB737 is not exactly

elegant design, and you might run into trouble with the build-up of device capacitances. Later versions of my MC preamp used three 2SB737 in parallel, giving a very handy noise reduction. Using thirty 2N4403's to try and emulate this is not really practical politics, though you could argue anything goes in the wonderful world of hi-end hi-fi.

A most excellent survey of the available low-r_{bb} and low-noise transistors is given in Horovitz and Hill,[5] and there is no point in trying to duplicate it here. The best they found, after a heroic series of measurements, were the NPN ZTX851 (r_{bb} 1.7 Ω) and the PNP ZTX 951 (r_{bb} 1.2 Ω). The snag is that they are not designed as low-noise amplifiers; they are medium-power devices with very low saturation voltage for emergency lighting control. Not surprisingly, there is no mention of noise on the data sheet. Using transistors whose noise performance is not specified in production is a dangerous business.

It looks like it might be time to explore further the use of low-noise JFETs in MC headamps. The voltage noise is higher but the current noise is lower, so they are normally thought of as being best matched to medium-impedance sources rather than the very low values seen in MC use. However, one of the quietest amplifiers I know of is a design by Samuel Groner which uses eight JFETs in parallel to obtain a noise density of 0.39 nV√Hz.[6]

Horovitz and Hill also have a very good section on low-noise JFETs.[7]

Possible Improvements

You will have spotted that R7, at 3.3 Ω, generates as much Johnson noise as the source impedance; this only degrades the noise figure by 1.4 dB rather than 3 dB, as in this case most of the noise comes from the transistors. Reducing R7 will require the impedance of the entire negative feedback to be reduced in proportion, which will put an excessive load on A1 output. "Mother's little helper", described in Chapter 1, is likely to be helpful here.

It would be instructive to compare this design with other MC preamplifiers, but it is not at all easy as their noise performance is specified in so many different ways it is virtually impossible to reduce them all to a similar form, particularly without knowing the spectral distribution of the noise. (This chapter has dealt until now with unweighted noise referred to the input over a 400 Hz–20 kHz bandwidth, and with RIAA equalisation *not* taken into account.) Nonetheless, I suggest that this design is quieter than most, being within almost 6 dB of the theoretical minimum, with clearly limited scope for improvement. Burkhard Vogel has written an excellent article on the calculation and comparison of MC signal-to-noise ratios.[8]

References

1. Self, D. "Design of Moving-Coil Head Amplifiers" *Electronics & Wireless World*, Dec 1987, p. 1206.

2. Nordholt, E. H., and Van Vierzen, R. M. "Ultra Low Noise Preamp for Moving-Coil Phono Cartridges" *JAES*, Apr 1980, pp. 219–223.

3. Barleycorn, J. (a.k.a. S. Curtis). *HiFi for Pleasure*, Aug 1978, pp. 105–106.

4. www.signaltransfer.freeuk.com/

5. Horowitz, P., and Hill, W. *The Art of Electronics*. 3rd edition, 2015, pp. 500–505. ISBN 978-0-521-80926-9 Hbk (Note: *must* be the 3rd edition).

6. Groner, S. "A Low-Noise Laboratory-Grade Measurement Preamplifier" *Linear Audio*, Volume 3, Apr 2012, p. 143.

7. Horowitz, P., and Hill, W. "A Low-Noise Laboratory-Grade Measurement Preamplifier" *Linear Audio*, Volume 3, Apr 2012, pp. 509–519.

8. Vogel, B. "The Sound of Silence" (Calculating MC preamp noise). *Electronics World*, Oct 2006, p. 28.

Chapter 12

Subsonic Filtering

The Need for Filtering

The problem of wobbling bass-cones seems to be worse than it's ever been, and I surmise that many users changed from infinite baffle speakers to reflex (ported) designs with extended bass when it became clear that CDs gave clean LF with no subsonic rubbish. Going back to vinyl and retaining the same speakers sets the cones flapping, promising severe intermodulation distortion, not to mention Doppler effects. Intermodulation in a good solid-state amplifier should be negligible, but dreadful things can happen in valve amplifiers.

Rumble is actually not a very accurate description of the problem. The subsonic signals are not audible in themselves, but the intermodulation they cause in bass units certainly is. Audible rumble, typically from motor noise, is at higher frequencies from 50 Hz up and cannot be filtered out without losing the music. The worst subsonic disturbances are in the region 8–12 Hz, where record warps are accentuated by resonance between the cartridge vertical compliance and the effective arm mass; see Figure 12.1. Reviewing work by Happ and Karlov,[1] Bruel and Kjaer[2] Holman,[3],[4] Taylor,[5] and Hannes Allmaier[6] suggests that in bad cases the disturbances are only 20–30 dB below maximum signal velocities and that the cart/arm resonance frequency should be attenuated by at least 40 dB to reduce its effects below the general level of groove noise. These points suggested to me that a notch in the frequency response at about 10 Hz would not only greatly reduce the cart/arm resonance problem but also steepen the initial filter roll-off so it intrudes less into the audio band. This chapter will use the term "subsonic filter" rather than "rumble filter".

Vinyl sources also produce ultrasonic spurious signals caused by the stylus mistracking the groove, and clicks produced by surface scratches can exceed the amplitude of the audio. Ultrasonic filtering is very desirable and is dealt with at the end of this chapter.

Designing Filters

The design of filters is a complicated business once you get beyond simple 2nd-order Butterworths, and there is no space to go into it in detail here. I am going to assume you know your stopband from your passband and your Butterworth from your Bessel, but there will be no pole-zero diagrams and no complex algebra. I suspect that would immediately cut my audience by about 99%. I have designed many filters in my time, but I have never yet, not even with elliptical filters, had to resort to placing poles and zeros on the s-plane or wrestling with foot-long complex equations. As I have done before in my *Active Crossover* book,[7] I will present here finished designs that can be simply scaled for different frequencies.

I have used the terms *pole* and *zero* because they are hard to avoid without getting verbose. Basically, a pole is a 1st-order 6dB/octave roll-off, either lowpass or highpass. A zero is a zero-amplitude value at a given frequency in the response, in the form of a notch that is in theory infinitely deep.

Subsonic Filters

Filters to remove the subsonic disturbances are highpass filters with a cutoff frequency typically around 20 Hz. That frequency is a compromise; a higher cutoff frequency will remove the subsonics more effectively but also intrude more on musical signals you want to keep. A lower frequency keeps the music intact but has less ability to remove the subsonics. This is not the only parameter; the order of the filter (2nd, 3rd, etc.) determines how quickly the response rolls off as frequency falls, and a higher order gives better discrimination between music and subsonics, at the cost of greater complexity and greater sensitivity to component value errors. Some people at least will be concerned about the audibility of LF phase-shifts and suggest the cutoff frequency should be lower than is implied by just looking

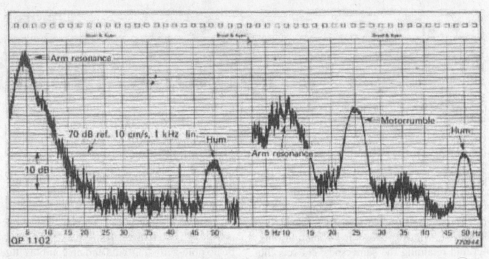

Figure 12.1 Typical noise spectra of vinyl at low frequencies

at the frequency response. This is not the case; there are mountains of evidence that phase-shifts are not audible, especially at very low frequencies. Accepting this means that the filter design can concentrate on the frequency response, and for more advanced subsonic filters that is hard enough. There is nothing approaching a consensus on this, so it can be a wise move to configure the subsonic filter so it can be switched out, possibly by a rear-panel switch, as it is not likely to be operated very often.

A differing philosophy states that the subsonic filter should always be in circuit, to minimise speaker damage if the stylus get dropped onto the vinyl. As with other subsonic disturbances, the consequences of this for ported speakers design are particularly serious.

One of the most useful filter configurations is the well-known Sallen and Key; all the filters in this chapter use it, including the elliptical filters, which combine it with a Bainter notch filter. The Sallen and Key has a hidden drawback when used as a lowpass handling high frequencies, as the response comes back up again at RF due to the nonzero opamp output impedance. For highpass filtering this issue is not relevant, and Sallen and Key works very well for our purposes here.

Before going further, it is as well to put down a formal specification of what we want from a rumble filter.

A Subsonic Filter Specification

- To give adequate attenuation of subsonic disturbances. "Adequate" is a matter of definition. At the

start of the chapter it was noted that subsonics in the cart/arm resonance area, say 8–12 Hz, would need to be attenuated by 40 dB to get them down to the wideband groove noise level. This is perhaps a counsel of perfection, and requires a serious technical effort.

- To be maximally flat in the passband, expressed as:
  - −3 dB point to be 20 Hz or lower. The RIAA specification stops at 20 Hz, so there is some justification for doing what you like at lower frequencies than that.
  - −0.1 dB error-criterion frequency to be as low as possible, ideally ruler-flat to 30 Hz.
  - Absolutely no ripples in the passband. No Chebyshev filters.
- To be unity gain with absolute phase preserved (i.e. noninverting), to allow simple insertion in existing signal paths, and to permit the filter to be switched in/out. The preservation of phase will follow directly from the filter configurations chosen.
- To have low component sensitivity; in other words errors due to component tolerances are magnified as little as possible. Expensive high-precision parts must not be required. This leads to:
- To be reasonably economical in parts cost.
- It is assumed the filter will be fed from a low impedance, such as an opamp output, and that ordinary loads are to be driven from its output. In other words, a final opamp stage may be needed to buffer a high-impedance RC network from the outside world and its loading. When required they are included in the schematics in this chapter.

Filter Types

There are many kinds of filter, and there are various ways of classifying them. The most basic division is into the familiar lowpass, highpass, bandpass, and bandstop (notch) filters. There are also allpass filters, which have flat frequency response, and if you are wondering what the point of that is, the point is that the phase-shift does vary with frequency, and this gives a way of compensating for other unwanted phase-shifts. A brick-wall filter is either a lowpass or highpass filter with a very steep, near-vertical roll-off. They do not filter out actual brick walls; if you have unexpected masonry in your equipment then the electronics is probably not the problem. Finally, I feel compelled to mention the all-stop filter, which prevents any signal at all getting through. This is as near to a joke as you are likely to get in the serious business of filter design. For an example found in the wild, see Reference[8].

In this chapter we will mainly be using lowpass and highpass filters, though notch filters will come in when we deal with elliptical highpass filters.

Filters Are Either All-Pole or Non-All-Pole

An all-pole filter is one that keeps on going down (i.e. is monotonic) once the final roll-off has begun. The best-known examples are Butterworth, Bessel, and Chebyshev filters, with the Butterworth being the most popular because of its maximally flat frequency response and simple design procedure. The Chebyshevs are all-pole despite the ripples in the passband, as once the *final* roll-off begins, the response never comes back up again. All-pole filters do not have notches (zeros) anywhere in their response.

Non-all-pole filters do have notches in the frequency response. Examples are Inverse Chebyshev filters, which have a maximally flat passband and a notch or notches in the stopband, and elliptical filters, which can simultaneously have a notch or notches in the stopband and also a ripple or ripples in the passband like the Chebyshev filter.

Filters Are Either Monotonic or Nonmonotonic

Monotonic means that there is a single trend to the frequency response; a Butterworth highpass filter is monotonic because as the frequency drops the gain is reduced, at first very slowly in the passband, then faster and faster as the roll-off point is reached, eventually falling at 6 dB/octave times the filter order in the stopband. The slope

continuously increases until it asymptotes to its final value in the stopband. Likewise Bessel and linear-phase filters are monotonic, and all-pole as well. Chebyshev filters are all-pole, but not monotonic, because they have ripple in the passband, and they are not exactly helpful if you're looking for a ruler-flat RIAA response to 30Hz or so.

There are other less common filter types such as Papoulis-Legendre, parabolic, etc., which are both all-pole and monotonic. They give different trade-offs between passband flatness, steepness of roll-off, and phase and group delay behaviour, but none of them are as flat as the Butterworth. Papoulis-Legendre and parabolic filters are not pursued further here as they roll-off more slowly than Butterworth and have no special advantages. There is more on exotic filter types in my book *The Design of Active Crossovers*.[7]

This chapter focuses on Butterworth filters because the emphasis is on flatness of frequency response, but the design tables also give full component values for Bessel filters and one form of linear-phase filter.

Butterworth Filters

The Butterworth filter has a maximally flat frequency response. It keeps as close to 0 dB as possible for as long as possible and then dives down as quickly as possible. The roll-off is as fast as it can be made while maintaining maximum flatness before the roll-off. The Butterworth is invariably designed for −3 dB at the cutoff frequency. See Figure 12.2, which shows Butterworth responses from 2nd-order up to 6th-order, with a 20 Hz cutoff frequency. Since in subsonic filtering it is primarily the amplitude response we care about, it is in many ways the natural choice of filter. For those unconvinced of this, Bessel and linear-phase filters are also described in this chapter, and all the component values for the three types given.

The component values for Butterworth filters from 2nd to 6th order are given in the filter tables here, with 2xE24 resistor combinations to make up the awkward values.

Bessel Filters

The Bessel filter characteristic approximates to linear-phase, i.e. has a flat group delay with frequency but a slower roll-off than the Butterworth characteristic. Discussions on filters always remark that the Bessel alignment has a slower roll-off but often fail to emphasise that it is a *much* slower roll-off. As stated at the beginning of this chapter, phase is not audible, especially at low frequencies, so the choice of a slower

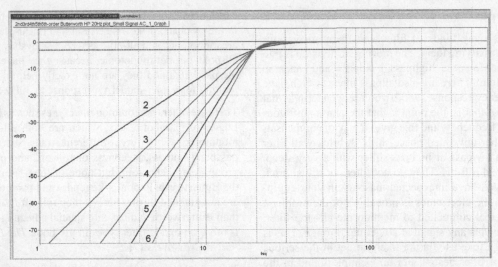

Figure 12.2 Frequency response of 2nd to 6th-order Butterworth highpass filters, cutoff −3 dB at 20 Hz

roll-off is not a good one. You may feel differently, and so the component values for Bessel filters from 2nd order to 6th order are given in the filter tables here, with 2xE24 resistor combinations to make up the awkward values.

The first man to use Bessel functions in filters was W. E. Thomson,[9] and so Bessel filters are sometimes called Thomson filters or Bessel-Thomson filters. It is important to remember that the term Bessel-Thomson does *not* refer to a hybrid between Bessel and Thomson filters, because they are the same thing.

The Bessel filter gives the closest approach to constant group delay; in other words the group delay curve is maximally flat. As a result the time response is very good, and the overshoot of a step function is only 0.43% of the input amplitude. This is often not visible on plots and has led some people to think that there is no overshoot at all; this is not true. It is certainly very small and unlikely to cause trouble in most applications. The price of these time-domain features is that the amplitude response roll-off is slow—actually very slow compared with a Butterworth filter. See Figure 12.3, and compare it with 12.2.

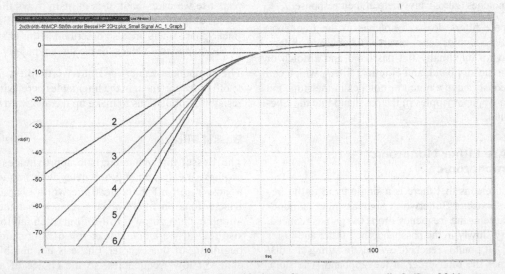

Figure 12.3 Frequency response of 2nd- to 6th-order Bessel highpass filter responses; cutoff −3 dB at 20 Hz

The cutoff frequency is defined here as the −3-dB point, but if you attempt to design a Bessel filter for a 20 Hz cutoff using this frequency, you will actually get −4.9 dB at 20 Hz. For a 2nd-order Bessel filter you must use a frequency scaling factor (FSF) of 1.2736, i.e. design for 1.2736 kHz and you will get −3 dB at 1.00 kHz. There is more on this later, where the FSFs are shown for many filter orders and types.

A simple test that the 4th-order Bessel filter is working correctly is to check for −25 dB at three times the cutoff frequency.

Linear-Phase Filters

Linear-phase filters are a compromise between Butterworth and Bessel filters. As we have seen, Bessel filters have a maximally flat group delay, avoiding the delay peak that you get with a Butterworth filter. If the requirement for a maximally flat delay is relaxed to allow equi-ripple group delay of a specified amount, then the amplitude roll-off can be much faster. The equi-ripple group delay characteristic is more efficient in that the group delay remains flat (within the set limits) further into the stopband. This is very much the same sort of compromise as in the Chebyshev filter, where the rate of amplitude roll-off is increased by tolerating a certain amount of ripple in the passband amplitude response. Linear-phase filters are all-pole filters.

Filters with of this kind are frequently referred to as "linear-phase" filters, but are sometimes (and perhaps more accurately) called Butterworth-Thomson filters. For some reason they never seem to be called Butterworth-Bessel filters, though this means exactly the same thing. The parameter m describes the move from Thomson to Butterworth, with $m = 0$ meaning pure Butterworth and $m = 1$ meaning pure Bessel. Any intermediate value of m yields a valid transitional filter. Linear-phase filters can alternatively be characterised by an angle parameter, so you can have a linear-phase 0.05 degree filter or a linear-phase 0.5° filter. This refers to the amount of deviation in the filter phase characteristics caused by allowing ripples in the group delay curve. Alternatively again, linear-phase filters can be described in % by the amount of ripple in the group delay, so an example might be described as a linear-phase 5% filter. There are therefore an infinite variety of linear-phase filters. The component values given in this chapter are for linear-phase 0.5° filters, chosen because their amplitude response is about halfway between the Butterworth and Bessel filters. These filters are here designed for −3 dB at the cutoff frequency. The frequency responses for orders 2 to 6 are given in Figure 12.4.

The component values for linear-phase 0.5° filters from 2nd order to 6th order are given in the filter tables in this chapter, with 2xE24 resistor combinations to make up the awkward values.

Figure 12.5 compares the amplitude responses of 4th-order Butterworth, linear-phase 0.5°, and Bessel highpass filters, showing how the linear-phase filter provides an intermediate solution between the Bessel and Butterworth characteristics. Note however that the linear-phase filter has a slow roll-off, which is only fractionally better than that of the Bessel filter.

Looking at Figure 12.5 gives insight into why the Butterworth filter is the type best adapted to our subsonic

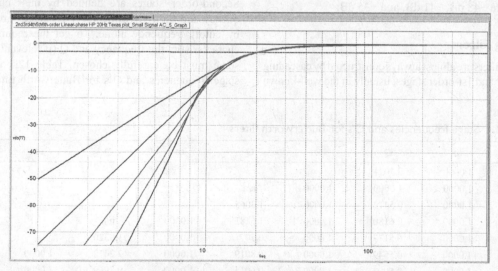

Figure 12.4 Frequency response of 2nd- to 6th-order linear-phase highpass filter responses; cutoff −3dB at 20 Hz

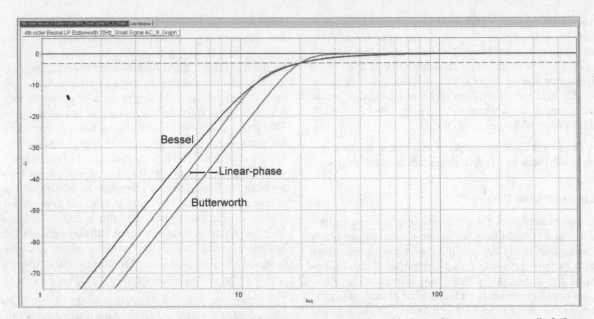

Figure 12.5 Frequency response of 4th-order Butterworth, linear-phase, and Bessel highpass filter responses; cutoff −3dB at 20 Hz

filtering application. All three responses eventually fall at −24 dB/octave, as they must do in any 4th-order filter, but the slope is established later in the linear-phase as frequency falls, and still later in the Bessel. Famously the Butterworth is much better than the other two at maintaining passband flatness before the roll-off and then falls steeply, so it gives −24 dB of attenuation at 10 Hz. The linear-phase gives only −15 dB, and the Bessel only −13.5dB at 10 Hz. At 5 Hz the respective attenuations are −48 dB, −41 dB, and −35 dB.

Building Filters

All-pole filters are almost always constructed by cascading 2nd-order and 1st-order stages, usually in the well-known

Sallen and Key configuration. Thus a 4th-order filter can be made by cascading two 2nd-order filters, and a 5th-order filter can be made by cascading two 2nd-order filters and a 1st-order filter, and so on. The order of the stages is often important to prevent excessive levels internal to the filter. In some cases filters of order greater than two can be made in a single stage, promising economy of active components, and these are also described later.

Second-order stages are defined by their cutoff frequency and their Q, while 1st-order stages are defined by cutoff frequency alone. To get the required filter characteristic, the frequency and Q (if relevant) of each stage must be carefully chosen. Table 12.1 gives the stage frequencies and Q's for Butterworth filters up to

Table 12.1 Stage frequencies and Q's for Butterworth filters

Order	ω	Q	ω	Q	ω	Q	ω	Q
2	1.0000	0.7071						
3	1.0000	1.0000	1.0000	n/a				
4	1.0000	0.5412	1.0000	1.3065				
5	1.0000	0.6180	1.0000	1.6181	1.0000	n/a		
6	1.0000	0.5177	1.0000	0.7071	1.0000	1.9320		
7	1.0000	0.5549	1.0000	0.8019	1.0000	2.2472	1.0000	n/a
8	1.0000	0.5098	1.0000	0.6013	1.0000	0.8999	1.0000	2.5628

Table 12.2 Stage frequencies and Q's for linear-phase filters

Order	freq	Q	freq	Q	freq	Q
2	1.107	0.640				
3	1.329	0.950	0.826	n/a		
4	0.905	0.610	1.615	1.340		
5	1.159	0.860	1.904	1.760	0.706	n/a
6	0.785	0.600	1.436	1.160	2.154	2.200

the 8th-order, Table 12.2 gives the stage frequencies and Q's for linear-phase filters up to the 8th-order, and Table 12.3 gives the stage frequencies and Q's for Bessel filters up to the 8th-order. The frequency columns give the factors that must be multiplied by the desired cutoff frequency. Thus to get a 1 kHz 2nd-order Bessel filter, you would design the stage for a cutoff frequency of 1.2736 kHz. The tables for linear-phase (Table 12.2) and Butterworth filters work the same way, except that for the Butterworth case all the frequency scaling factors are unity, and so design is simpler.

First-Order Lowpass and Highpass Filter Design

The usual 1st-order filter is just a resistor and capacitor. Figure 12.6a shows the normal (noninverting) lowpass

version in all its beautiful simplicity. To give the calculated 1st-order response it must be driven from a very low impedance and see a very high impedance looking in to the next stage. It is frequently driven from an opamp output, which provides the low driving impedance but very often requires a unity-gain stage to buffer its output, as shown. The highpass version is shown in Figure 12.6b.

Here are the design and analysis equations for the 1st-order filters. The design equations give the component values required for given cutoff frequency; the analysis equations give the cutoff frequency when the existing component values are plugged in. Analysis equations are useful for diagnosing why a filter is not doing what you planned. The design and analysis equations are the same for the lowpass and highpass versions.

For both versions the design equation is: **and the analysis equation is:**

$$R = \frac{1}{2\pi f_0 C} \quad \text{(choose C)} \qquad f_0 = \frac{1}{2\pi RC} \qquad \text{12.1, 12.2}$$

There are no issues with component sensitivities in a 1st-order filter; the cutoff frequency is inversely proportional to the product of R and C, so the sensitivity for either component is always 1.0. In other words a change of 1% in either component will give a 1% change in the cutoff frequency.

Table 12.3 Stage frequencies and Q's for Bessel filters

Order	freq	Q	freq	Q	freq	Q	freq	Q
2	1.2736	0.5773						
3	1.4524	0.6910	1.3270	n/a				
4	1.4192	0.5219	1.5912	0.8055				
5	1.5611	0.5635	1.7607	0.9165	1.5069	n/a		
6	1.6060	0.5103	1.6913	0.6112	1.9071	1.0234		
7	1.7174	0.5324	1.8235	0.6608	2.0507	1.1262	1.6853	n/a
8	1.7837	0.5060	1.8376	0.5596	1.9591	0.7109	2.1953	1.2258

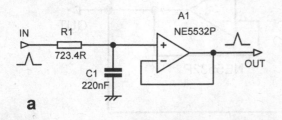 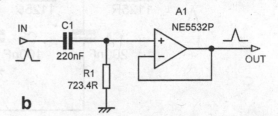

Figure 12.6 1st-order filters: a) lowpass; b) highpass. Cutoff frequency (−3 dB) is 1 kHz in both cases.

Second-Order Lowpass Filter Design

Here are the design and analysis equations for the 2nd-order unity-gain Sallen and Key lowpass filter; in other words they cover all characteristics, Butterworth, Bessel, Chebyshev, etc. While lowpass filters are of no use for subsonic filtering, the basic design principles are described here in this chapter because the comparison with the design of the equivalent highpass filter is highly instructive.

The design equations give the component values required for given cutoff frequency and Q; the analysis equations give the cutoff frequency and Q when the existing component values are fed in. Figure 12.7 shows a 2nd-order Butterworth lowpass filter with a 1 kHz cutoff frequency.

Lowpass Design Equations:
Begin by choosing a value for C2 , then:

$$R = \frac{1}{2Q(2\pi f_0)C_2} \quad C_1 = 4Q^2 C_2 \qquad \text{12.3, 12.4}$$

Lowpass analysis equations:

$$f_0 = \frac{1}{2\pi R\sqrt{C_1 C_2}} \quad Q = \frac{1}{2}\sqrt{\frac{C_1}{C_2}} \qquad \text{12.5, 12.6}$$

Table 12.4 gives the Qs and capacitor ratios required for the well-known 2nd-order filter characteristics.

Second-Order Highpass Filter Design

Sallen and Key highpass filters are the same as the lowpass filters, with the R's and the C's swapped over. Now the capacitors are equal while the resistors have a ratio that defines the Q. Figure 12.8 shows a 2nd-order Butterworth highpass with a 1 kHz cutoff frequency.

Table 12.4 2nd-order Sallen and Key unity-gain lowpass. Frequency Shift Factors, Qs and capacitor ratios for various filter types

Type	FSF	Q	C1/C2 ratio
Linkwitz-Riley	1.578	0.500	1.000
Bessel	1.2736	0.5773	1.336
Linear-phase 0.05°	1.210	0.600	1.440
Linear-phase 0.5°	1.107	0.640	1.638
Butterworth	1.000	0.707	2.000
0.5dB-Chebyshev	1.2313	0.8637	2.986
1.0dB-Chebyshev	1.0500	0.9565	3.663
2.0dB-Chebyshev	0.9072	1.1286	5.098
3.0dB-Chebyshev	0.8414	1.3049	6.812

Note that these FSFs apply *only* to 2nd-order filters. Higher-order filters in general have different design frequencies for each stage. These are given in the Tables 12.1, 12.2 and 12.3 earlier in the chapter.

Highpass Design Equations

Choose R2, then:

$$C = \frac{2Q}{(2\pi f_0)R_2} \quad R_1 = \frac{R_2}{4Q^2} \qquad \text{12.7, 12.8}$$

Highpass analysis equations:

$$f_0 = \frac{1}{2\pi C\sqrt{R_1 R_2}} \quad Q = \frac{1}{2}\sqrt{\frac{R_2}{R_2}} \qquad \text{12.9, 12.10}$$

The IEC amendment (1st-order)
The most elementary, and least effective, protection against subsonic disturbances is the IEC Amendment to the RIAA equalisation standard; see Chapter 7 for more details. The Amendment adds a single highpass (HP)

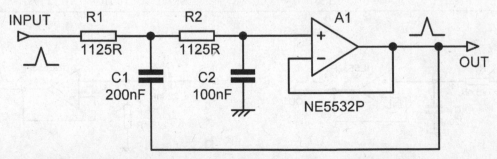

Figure 12.7 2nd-order lowpass Sallen and Key filter. Cutoff frequency is 1 kHz. Q= 0.7071 for a Butterworth response

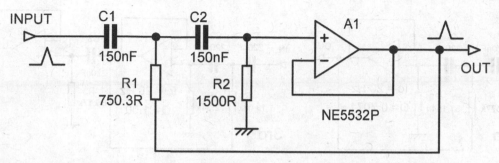

Figure 12.8 2nd-order highpass Sallen and Key filter. Cutoff frequency is 1 kHz. Q = 0.7071 for a Butterworth response.

pole at 20.02 Hz and so acts as a 1st-order "filter". It gives only −7 dB of attenuation at 10 Hz and −20 dB at 1 Hz; see Table 12.5. In any case it is often omitted by the manufacturer or switched out by the user. It is no solution to subsonic difficulties.

The various filter attenuations for the IEC and 2nd- to 6th-order Butterworth filters are summarised in Table 12.5.

The IEC figures assume the RIAA preamp has a C0 of 470uF; the −0.1 dB frequency is for the difference between IEC and non-IEC RIAA equalisation.

Here I have defined "ruler-flat" as an error of less than 0.1 dB, as in the last column of Table 12.5.

Second-order Butterworth Filter

A 2nd-order Butterworth filter is shown in Figure 12.9. It is only 12.3 dB down at 10 Hz, which gives little protection against cartridge/arm resonance problems. It is a more useful 24.0 dB down at 5 Hz, by which time the 12 dB/octave slope is well established and we are well protected against disc warps. Above the −3 dB cutoff frequency the response is still −0.8 dB down at 30 Hz, which is intruding too much into the sort of frequencies

we want to keep. The 2nd-order filter really does not bifurcate the condiment. It does however have the merit of being very easy to design; in the Sallen and Key version the two series capacitors C1 and C2 are made equal and R2 is made twice the value of R1. Other roll-off frequencies can be obtained simply by scaling the component values while keeping C1 equal to C2 and R2 twice R1. There are six resistor pairs in a 1:2 ratio in the E24 series, but that is not much help here; the nearest pair to 25kΩ–50kΩ is 18kΩ–36kΩ, which will give a cut-off frequency of 28.4 Hz with 220nF, rather adrift from what we want. If the capacitors are available in the E6 series then C1 and C2 could be made 330 nF, giving a cutoff frequency of 18.95 Hz with 18kΩ–36kΩ.

We really need more accuracy than this, so the capacitors are standardised at 220 nF, and the awkward resistance values are obtained with two E24 resistors in parallel. This method is used here and in many other places in this book wherever nonstandard values are required. In the tables here the error in the nominal value is held to less than ±0.4%. See Table 12.6. There are various integer ratios in the E24 series (see Chapter 2), but this does not help with filters of higher order because the resistor values are not in simple integer ratios.

Table 12.5 Attenuation of IEC amendment and 2nd- to 6th-order Butterworth highpass filters, all −3 dB at 20 Hz

Freq	1 Hz	5 Hz	10 Hz	20 Hz	30 Hz	−0.1 dB freq
IEC amendment	−20 dB	−12 dB	−7 dB	−3.0 dB	−1.6 dB	146 Hz
2nd-order	−52 dB	−24 dB	−12 dB	−3.0 dB	−0.79 dB	50 Hz
3rd-order	−78 dB	−36 dB	−18 dB	−3.0 dB	−0.37 dB	37 Hz
4th-order	−104 dB	−48 dB	−24 dB	−3.0 dB	−0.16 dB	32 Hz
4th-order MCP	−96 dB	−41 dB	−18 dB	−3.0 dB	−0.68 dB	49 Hz
5th-order	−126 dB	−60 dB	−30 dB	−3.0 dB	−0.08 dB	29 Hz
6th-order	−156 dB	−72 dB	−36 dB	−3.0 dB	−0.04 dB	27 Hz

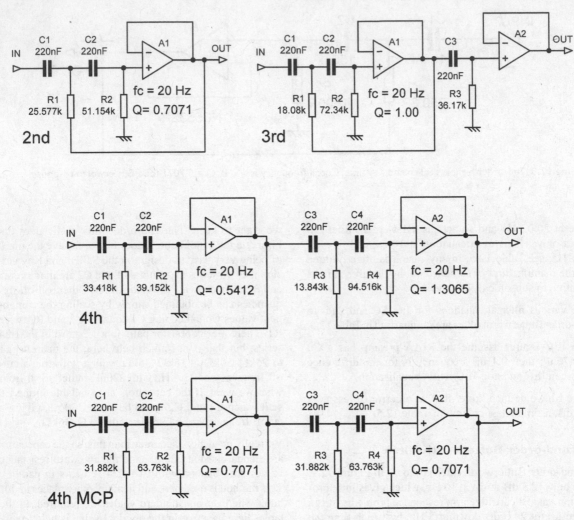

Figure 12.9 Butterworth highpass subsonic filters: 2nd-order, 3rd-order, 4th-order, and 4th-order MCP. All have −3 dB cutoff at 20 Hz.

Table 12.6 Resistor values for 2nd-order filters in 2xE24 format. Error refers to deviation from nominal value and excludes component tolerances C1=C2=220 nF.

Type	R1 Ω	R1a	R1b	Error %	R2 Ω	R2a	R2b	Error %
Bessel	39.900k	47k	270k	+0.33	53.190k	82k	150k	−0.32
Linear-phase 0.5°	31.283k	39k	160k	+0.24	51.254k	75k	160k	−0.18
Butterworth	25.577k	56k	47k	−0.09	51.154k	75k	160k	−0.18

Third-Order Butterworth Filter: Two Stage

The 3rd-order Butterworth, whose response is plotted in Figure 12.2 and circuit shown in Figure 12.9, is 18 dB down at 10 Hz, still a long way from −40 dB, but it is handily 36dB down at 5 Hz and a thumping great −78 dB

at 1 Hz. Low-frequency disc warp spurii will be well and truly suppressed. Hence my decision that a 3rd-order Butterworth was adequate at the time—as noted earlier, the wider use of reflex loudspeakers with vinyl sources makes this much less certain today. The 30 Hz response is however down by a not insignificant −0.37 dB, and the error does not fall below the 0.1dB criterion until we get up to 37 Hz. Note the output buffer after C3–R3; this point is at a high impedance and must not be significantly loaded, so in many cases a unity-gain buffer will be required. Note also that $R1 = 4 \cdot R2 = 2 \cdot R3$, though this is not likely to be of any help in selecting resistors if you want a precise cutoff frequency. It is possible to make a 3rd-order filter in one stage, as described in the next section, but this will give somewhat worse component sensitivities and may compromise the distortion performance.

All my published preamplifier designs have used a 3rd-order maximally flat Butterworth filter with −3dB cut-off at 20 Hz (with the exception of the first preamp,[10] which had no subsonic filter at all). See[11], [12], [13]. The thinking behind this was simply that a 2nd-order Butterworth was inadequate, as stated earlier, but a 3rd-order Butterworth would do the job and had the advantage that it could be implemented in one stage; more on that in the next section. A 4th-order Butterworth would have required another amplifier, as I did not at the time know how to design it in one stage. I do now (see later), but it remains uncertain if it is a good idea.

The 3rd-order filter Butterworth had the further advantage that, by suitable adjustment of its frequency response, it could also implement the IEC amendment quite closely.

It therefore did not have a classic Butterworth filter characteristic, and the filters in these preamps should not be copied if that is what you want; instead use the filter designs in this chapter. All the filters described here do just the filtering job, and it is assumed that the IEC Amendment is implemented elsewhere, if at all.

The schematics for these filters are shown in Figure 12.2. The frequency responses are plotted in Figures 12.3 and 12.4. Warp frequencies go down to 0.55 Hz for a disc rotating at 33 1/3 rpm; the plots here however stop at 1 Hz to make the notch area clearer. In every case here nothing new happens below 1 Hz or above 500 Hz. The exact resistor values for each filter, with the 2xE24 parallel values to make them up, are shown in Table 12.7.

Third-Order Butterworth Filter: Single Stage

The 3rd-order Butterworth in Figure 12.9 uses an output buffer A2 to prevent C3–R3 from being loaded by downstream circuitry, which is unlikely to have a very high input impedance. If the 3rd-order filter can be made in one stage, then A2 can be omitted, saving PCB space and power consumption. My preamplifiers [11] [12] used this technology, and I have used it in many other applications.

Figure 12.10 shows a 3rd-order Butterworth filter with unity gain and a −3 dB cutoff at 20 Hz. A complex pair of

Table 12.7 Resistor values for two-stage 3rd-order filters in 2xE24 format. Error refers to deviation from nominal value and excludes component tolerances.

Type	R1 Ω	R1a	R1b	Error %	R2 Ω	R2a	R2b	Error %	R3 Ω	R3a	R3b	Error %
Bessel	38.014k	51k	150k	+0.12	72.604k	91k	360k	+0.05	48.000k	56k	330k	−0.26
Linear-phase 0.5°	25.301k	43k	62k	+0.35	91.337k	120k	390k	+0.46	29.878k	47k	82k	−0.01
Butterworth	18.086k	22k	100k	−0.29	72.343k	82k	620k	+0.11	36.172k	39k	510k	+0.16

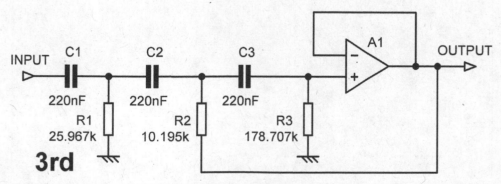

Figure 12.10 Butterworth 3rd-order subsonic filter in one stage; cutoff −3 dB at 20 Hz

poles and a single pole are implemented in a single stage. The resistor value ratios are 2.53:1.00:17.55. Table 12.8 gives the 2xE24 resistor values for one-stage 3rd-order Butterworth, linear-phase, and Bessel filters. As before, other roll-off frequencies can be had by scaling the component values while keeping the resistor ratios the same.

There is no straightforward way to design these filters. You have to solve three simultaneous equations rather than just plugging in values, and the process is described as "somewhat laborious" in the *Electronic Filter Design Handbook*.[14]

Fourth-Order Butterworth Filters: Two Stage

Fourth-order rumble filters are vanishingly rare in the hi-fi marketplace, perhaps due to worries about the audibility of rapid phase changes at the very bottom of the audio spectrum, but I suspect the fact that they are harder to design has a lot to do with it. To ease that problem at least, Figure 12.9 shows a classical 4th-order Butterworth made up of two cascaded 2nd-order stages,

the first with a Q of 0.5412 and the second with a Q of 1.3065. It is possible to make a 4th-order filter in one stage, as demonstrated in the next section, but it gives somewhat worse component sensitivities and is likely to compromise the distortion performance.[15] The 4th-order filter is −24 dB at 10 Hz, still 16 dB short of the 40 dB required to reduce bad cartridge/arm resonances to the level of groove noise. The loss at 30 Hz is only 0.16 dB, which I think many of us could live with, and we get within the 0.1 dB error criterion at 32 Hz.

The 4th-order Butterworth uses two stages with Q's that are very different—this is typical for "classical" filter designs, for want of a better word. For suitably high-order filters there is an alternative design strategy called "multiple critical poles" (MCP) which reduces the spread of Q's used.[16] Since it is the maximum Q in the filter that causes the most problems with component sensitivity, use of MCP can give a cheaper filter, but there are compromises affecting the response. Here the MCP version of the 4th-order Butterworth uses two stages with identical cutoff frequencies and identical Q's of 0.7071, so it is equivalent to two 2nd-order Butterworth

Table 12.8 Resistor values for one-stage 3rd-order filters in 2xE24 format. Error refers to deviation from nominal value and excludes component tolerances.

Type	R1 Ω	R1a	R1b	Error %	R2 Ω	R2a	R2b	Error %	R3 Ω	R3a	R3b	Error %
Bessel	36.542k	51k	130k	+0.49%	25.362k	43k	62k	+0.11%	142.20k	270k	300k	−0.07%
Linear-phase	26.097k	30k	200k	−0.04%	15.111k	20k	62k	+0.07%	182.99k	300k	470k	+0.07%
Butterworth	25.967k	33k	120k	−0.33%	10.195k	13k	47k	−0.11%	178.71k	330k	390k	+0.02%

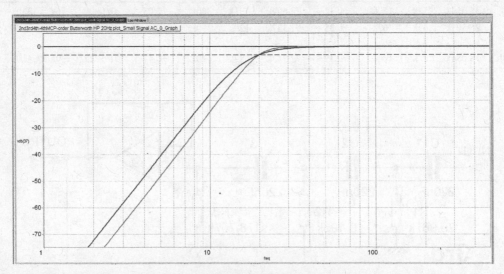

Figure 12.11 Frequency response of 4th-order classical and MCP Butterworth subsonic filters, both −3 dB at 20 Hz

Table 12.9 Resistor values for two-stage 4th-order classical filters in 2xE24 format, with error in the nominal value

Type	R1 Ω	R1a	R1b	Error %	R2 Ω	R2a	R2b	Error %	R3 Ω	R3a	R3b	Error %
Bessel	49.181k	62k	240k	+0.18	53.583k	91k	130k	−0.10	35.727k	75k	68k	−0.17
Linear-phase 0.5°	26.832k	47k	62k	−0.36	39.937k	47k	270k	+0.24	21.797k	24k	240k	+0.10
Butterworth	33.418k	43k	150k	+0.005	39.152k	75k	82k	+0.05	13.843k	24k	33k	+0.37

Type	R4 Ω	R4a	R4b	Error%
Bessel	92.723k	160k	220k	−0.10
Linear-phase 0.5°	156.56k	300k	330k	+0.37
Butterworth	94.156k	120k	430k	−0.36

filters cascaded; this is equivalent to a 4th-order Linkwitz-Riley filter, as used in active crossover design. The maximum Q used has been reduced from 1.00 to 0.7071; the −3dB frequency is still 20 Hz, and the ultimate roll-off is still at 24 dB/octave. The compromise is that the turnover is more gradual, and there is 6.5 dB less attenuation at 10 Hz and all lower frequencies; see Figure 12.11. Likewise the 0.1 dB error frequency moves right up to 49 Hz. A Q of 1.3065 is low by active filter standards and will cause no design problems, so MCP is not a good choice here; it does however demonstrate clearly a process we will make use of later for elliptical filters, where it is much more complicated.

The 2xE24 values for the classic 4th-order filters are given in Table 12.9 and for the MCP version in Table 12.10.

Fourth-Order Butterworth Filter: Single Stage

Making a 4th-order filter in a single stage is, as you might expect, rather more difficult than making a 3rd-order filter. One solution is shown in Figure 12.12, with a passband gain of 1.01 times. You will note that the ratio between R3 and R4 is now a rather large 1:243. Be aware that a gain of 1.00 times is *not* near enough; it will shift the cutoff frequency from 20 Hz to 26 Hz,

which is a big alteration for such a tiny change in gain. This should alert you to the possibility of this filter being tricky to use in practice. Investigations are continuing, but it looks like the two-stage 4th-order filter is always going to be a safer bet.

Fifth-Order Butterworth Filter: Three Stages

A 5th-order filter is usually built by cascading two 2nd-order stages and one 1st-order stage. For the Butterworth version all the three cutoff frequencies are the same, but one 2nd-order stage has a Q of 0.620 and the other a Q of 1.620. The latter is an increase on the highest Q in a 4th-order Butterworth, which is 1.3065. The Butterworth version is shown in Figure 12.13.

The first stage has a Q less than 0.7071 ($1/\sqrt{2}$) and so shows no gain peaking. The second stage has a gain peak of +4.63 dB, but with the stage order shown this is not an issue because the first low-Q stage attenuates the signal before it reaches the second stage. If the 2nd-order stages were interchanged there would be a serious loss of headroom in the region of 1 kHz–2 kHz.

Resistor values for Butterworth, linear-phase and Bessel three-stage 5th-order highpass filter are given in exact values and in 2xE24 format in Table 12.12. All capacitors are the preferred value of 220 nF, giving nonpreferred values for the resistors. The capacitors are chosen to prevent any of the resistors from becoming too small and so presenting excessive loading. The amount of gain peaking is different for each type, increasing as the filter moves from Bessel to Butterworth characteristic. The component values were calculated using the tables of stage frequency and Q given earlier in this chapter and

Table 12.10 Resistor values for two-stage 4th-order MCP filter in 2xE24 format, with error in the nominal value

Type	R1=R3 Ω	R1a=R3a	R1b=R3b	Error %	R2=R4 Ω	R2a=R4a	R2b=R4b	Error %
Butterworth	31.882k	47k	100k	+0.28	63.763k	75k	430k	+0.15

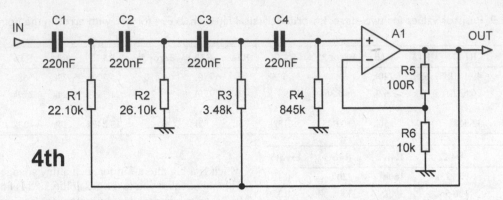

Figure 12.12 Butterworth highpass 4th-order subsonic filter in one stage. Cutoff −3 dB at 20 Hz

Table 12.11 Resistor values for single-stage 4th-order filters in 2xE24 format

Type	R1 Ω	R1a	R1b	Error %	R2 Ω	R2a	R2b	Error %	R3 Ω	R3a	R3b	Error %
Bessel	37.57k	75k	75k	−0.19	37.57k	75k	75k	−0.19	12.785k	13k	750k	−0.05
Linear-phase*	22.703k	36k	62k	+0.32	29.534k	56k	62k	−0.38	4.1387k	5k6	16k	+0.23
Butterworth	22.10k	39k	51k	0.00	26.10k	30k	200k	−0.05	3.480k	5k6	9k1	−0.38

Type	R4 Ω	R4a	R4b	Error%
Bessel	501.27k	1M	1M	−0.25
Linear-phase*	1376.2k	1M6	10M	+0.23
Butterworth	845.0k	1M6	1M8	+0.24

* Designed for 5% delay ripple

then checked by simulation. The maximum Q of 1.62 occurs in the 2nd-order stage of the Butterworth filter; note the large ratio between R3 and R4.

Fifth-order Butterworth filter: Two Stages

A 5th-order filter is commonly built in three stages, as in Figure 12.13. This requires three amplifiers, as the single pole at the end almost always needs buffering. However,

with suitable cunning you can combine a pair of complex poles and a single pole in a single, stage as around A1 in Figure 12.14. This is also done in the 3rd-order filter of Figure 12.10. The second stage need not have the same capacitor values but must have the same CR products. There is no gain peaking at the first-stage output, and so there are no internal headroom problems.

The second stage is straightforward to design, using a cutoff frequency of 20 Hz and a Q of 1.62 as in Figure 12.13 and Table 12.1. To design the first stage from scratch you have to solve three simultaneous equations to create a complex pole pair plus a single pole; but really all you need is the resistor ratios given here. I am afraid designs for linear-phase and Bessel filters are not available.

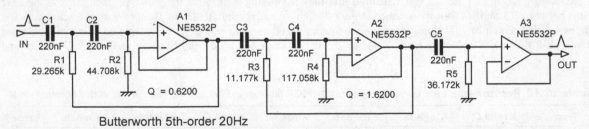

Figure 12.13 5th-order Sallen and Key Butterworth highpass filter, implemented in three stages. Cutoff −3 dB at 20 Hz.

Table 12.12 5th-order three-stage Sallen and Key highpass resistor values in 2xE24 format

Type	R1 Ω	R1a	R1b	Error %	R2 Ω	R2a	R2b	Error %	R3 Ω	R3a	R3b	Error %
Bessel	50.104k	100k	100k	−0.21	63.639k	110k	150k	−0.28	34.745k	56k	91k	−0.22
Linear-phase 0.5°	24.374k	47k	51k	+0.35	72.107k	120k	180k	−0.15	19.566k	36k	43k	+0.20
Butterworth	29.265k	43k	91k	−0.22	44.708k	75k	110k	−0.25	11.177k	13k	82k	+0.39

Type	R4 Ω	R4a	R4b	Error%	R5 Ω	R5a	R5b	Error %
Bessel	116.74k	180k	330k	−0.23	54.507k	100k	120k	+0.07
Linear-phase 0.5°	242.425k	430k	560k	+0.33	25.537k	47k	56k	+0.06
Butterworth	117.06k	160k	430k	−0.38	36.172k	39k	510k	+0.16

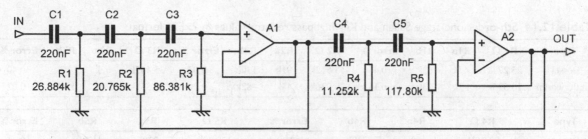

Figure 12.14 5th-order Sallen and Key Butterworth highpass filter in two stages. Cutoff −3 dB at 20 Hz.

Table 12.13 5th-order two-stage Sallen and Key highpass resistor values in 2xE24 format

Type	R1 Ω	R1a	R1b	Error %	R2 Ω	R2a	R2b	Error %	R3 Ω	R3a	R3b	Error %
Butterworth	26.884k	30k	270k	+0.43	20.765k	33k	56k	0.01	86.381k	100k	620k	−0.31

Type	R4 Ω	R4a	R4b	Error%	R5 Ω	R5a	R5b	Error %
Butterworth	11.252k	18k	30k	−0.02	117.80k	150k	560k	+0.43

Fifth-Order Butterworth Filter: Single Stage

A 5th-order highpass filter can be built in a single stage. The capacitor ratios were derived from a paper by Aikens and Kerwin[17] and transformed into resistor values by taking the reciprocal and then appropriately scaling. Aikens and Kerwin only give values for a gain of +6 dB, and this may or may not be convenient in a preamplifier gain structure. The circuit is shown in Figure 12.15, and the resistors in 2xE24 format are given in Table 12.14. The component sensitivity will be inferior to that of a three-stage 5th-order filter. I regret that a linear-phase design is not available for this filter configuration.

Sixth-Order Butterworth Filter: Three Stages

A 6th-order filter is usually built by cascading three 2nd-order stages. For the Butterworth version the three cutoff frequencies are the same, but the first 2nd-order stage has a Q of 0.5177, the next a Q of 0.707, and the final a Q of 1.9320, as in Figure 12.16. The latter is a high Q for a Sallen and Key configuration, giving a very large ratio of 14.9 times between R5 and R6. It is therefore desirable to use 47 nF for C5 and C6, to prevent R5 being too small and excessively loading amplifier A2.

There is no gain peaking in either the first or second stage, but significant peaking of +6.02 dB in the third stage.

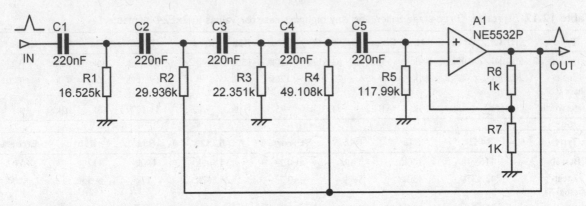

Figure 12.15 5th-order Sallen and Key Butterworth highpass filter, implemented as single stage, with gain = 2 and exact values. Cutoff −3dB at 20 Hz.

Table 12.14 5th-order one-stage Sallen and Key highpass resistor values in 2xE24 format

Type	R1 Ω	R1a	R1b	Error %	R2 Ω	R2a	R2b	Error %	R3 Ω	R3a	R3b	Error %
Bessel	25.277k	43k	62k	+0.45	53.512k	91k	130k	+0.03	35.447k	62k	82k	−0.40
Butterworth	16.525k	33k	33k	−0.15	29.936k	47k	82k	−0.20	22.351k	27k	130k	+0.02

Type	R4 Ω	R4a	R4b	Error %	R5 Ω	R5a	R5b	Error %
Bessel	82.261k	110k	330k	+0.29	170.79k	270k	470k	+0.41
Butterworth	49.108k	62k	240k	+0.33	117.9k	150k	560k	+0.35

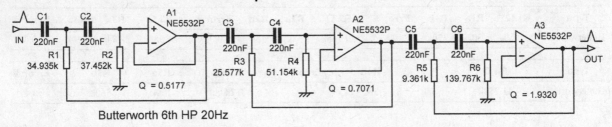

Figure 12.16 6th-order Sallen and Key Butterworth highpass filter, implemented in three stages. Cutoff frequency 20 Hz.

With the stage order shown in Figure 12.16 this is not an issue, as the first low-Q stage attenuates the signal before it reaches the second stage. If the third stage was first in the cascade there would be a serious loss of headroom in the region 1 kHz–2 kHz.

Component values for the other types of three-stage 4th-order highpass filter are given in Table 12.15. All capacitors have been selected as preferred values, giving nonpreferred values for the resistors. The capacitors are chosen to prevent any of the resistors from becoming too

small and so presenting excessive loading. The maximum Q of 1.932 occurs in the second 2nd-order stage of the filter; note the very large ratio between R3 and R4. The amount of gain peaking is different for each filter characteristic, increasing as the filter moves from Bessel through linear-phase to Butterworth.

So far as I can see it is not possible to make a 6th-order filter in two stages, because each stage will only implement one complex pole pair and one single pole, whereas what is needed is three complex pole pairs.

In Table 12.15 the component values were calculated using the tables of stage frequency and Q given earlier in this chapter and have been checked by simulation. The design process depends on what definition of cut-off attenuation is used; throughout this book I have employed the most common definition, set out in the rightmost column of the table. The 2xE24 resistor values are shown in Table 12.16.

Sixth-Order Butterworth Filter: Single Stage

A 6th-order highpass filter can be built in a single stage, as in Figure 12.17. The lowpass capacitor ratios were derived from Aikens and Kerwin[17] and transformed into highpass resistor values by taking the reciprocal and appropriate scaling.

Table 12.15 6th-order three-stage Sallen and Key highpass: component values for various filter types. Cutoff 1 kHz.

Type	C1 = C2 nF	R1 Ω	R2 Ω	C3 = C4 nF	R3 Ω	R4 Ω	C5 = C6 nF	R5	R6	Cutoff dB
Bessel	220	56.919k	59.288k	220	50.047k	74.783k	220	33.703k	141.19k	−3
Linear-phase 0.5°	220	23.662k	34.074k	220	22.389k	120.51k	220	17.708k	342.82k	−3
Butterworth	220	34.935k	37.452k	220	25.577k	51.154k	220	9.361k	139.77k	−3

Table 12.16 6th-order three-stage Sallen and Key highpass: resistor values in 2xE24 format

Type	R1 Ω	R1a	R1b	Error %	R2 Ω	R2a	R2b	Error %	R3 Ω	R3a	R3b	Error %
Bessel	56.919k	75k	240k	+0.39	59.288k	68k	470k	+0.20	50.047k	100k	100k	−0.09
Linear-phase 0.5°	23.662k	30k	110k	−0.38	34.074k	62k	75k	−0.39	22.389k	43k	47k	+0.30
Butterworth	34.935k	51k	110k	−0.26	37.452k	75k	75k	+0.13	25.577k	56k	47k	−0.09

Type	R4 Ω	R4a	R4b	Error %	R5 Ω	R5a	R5b	Error %	R6 Ω	R6a	R6b	Error %
Bessel	74.783k	150k	150k	+0.29	33.703k	51k	100k	+0.21	141.19k	220k	390k	−0.38
Linear-phase 0.5°	120.51k	150k	620k	+0.22	17.708k	30k	43k	−0.21	342.82k	360k	7.5M	+0.20
Butterworth	51.154k	75k	160k	−0.18	9.361k	13k	33k	−0.37	139.77k	200k	470k	+0.38

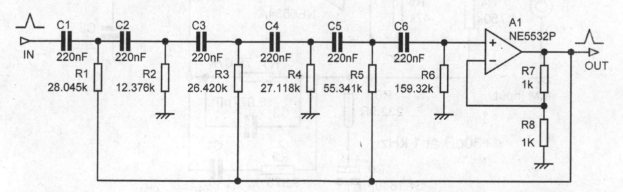

Figure 12.17 6th-order Sallen and Key Butterworth highpass filter, implemented as single stage, with gain = 2 and exact values. Cutoff frequency 20 Hz.

Table 12.17 6th-order one-stage Sallen and Key highpass: 2xE24 resistor values

Type	R1 Ω	R1a	R1b	Error %	R2 Ω	R2a	R2b	Error %	R3 Ω	R3a	R3b	Error %
Bessel	50.102k	100k	100k	−0.20	20.095k	33k	51k	−0.29	47.674k	91k	100k	−0.06
Butterworth	28.045k	51k	62k	−0.22	12.376k	13k	270k	+0.22	26.420k	39k	82k	+0.04

Type	R4 Ω	R4a	R4b	Error %	R5 Ω	R5a	R5b	Error %	R6 Ω	R6a	R6b	Error %
Bessel	44.500k	75k	110k	+0.21	92.819k	160k	220k	−0.20	239.70k	390k	620k	−0.12
Butterworth	27.118k	36k	110k	+0.02	55.341k	68k	300k	+0.17	159.32k	270k	390k	+0.14

The component sensitivity will be inferior to that of a three-stage 6th-order filter. I regret that a linear-phase design is not available for this filter configuration.

Subsonic Filters Integrated With the Phono Stage

We have noted that subsonic disturbances can be at worryingly high levels, and it has occurred to some people that it would be a good idea to get rid of them before they reach any active electronics, where they might cause intermodulation distortion, or in bad cases even erode the available headroom. There is no apparent need for anything more radical than putting an effective subsonic filter immediately after a phono stage having the relatively modest gain of +30 dB (1 kHz) and applying any further gain after that. A filter could be placed *before* the phono amp, but that would be sure to degrade the noise performance.

Tomlinson Holman presented in his "New Factors" paper[3] the interesting idea that the phono amp itself could also act as a 3rd-order subsonic filter. My version, using an opamp and the efficient RIAA Configuration-C (see Chapter 7) is designed to make C1 3 × 10 nF = 30 nF, and the other values fall where they may, as shown in Figure 12.18; the original Holman design used discrete transistors and RIAA Configuration-A. C4 and C5 together with R4 and R5 form a 2nd-order Sallen and Key highpass filter with a Q of 1.0 that acts as the first stage of a 3rd-order Butterworth filter. Since the feed to R4 is taken from the inverting input of the opamp, the whole phono stage has unity gain to this point and, as far as R4 is concerned, is working as a voltage-follower.

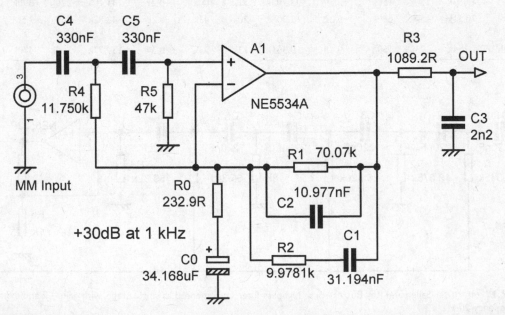

Figure 12.18 3rd-order Butterworth subsonic filter at the input of phono amplifier; C0 is critical as it forms the third pole with R0

The second stage of the filter is implemented by restricting the value of C0 to give a single pole which combines with the first stage to give an 18 dB/octave roll-off. Holman's paper floated the idea in a single paragraph that gave few details and illustrated it with a schematic that gave no values and did not even reveal which capacitors were electrolytics.

No problem, we'll just work it out for ourselves. I decided to go for cutoff at −3 dB at 20 Hz to allow direct comparison with the other subsonic filters in this chapter. The 2nd-order part is straightforward to design for fc = 20 Hz and Q =1.0 using the standard equations, but a complicating factor is that R5 must be fixed at 47 kΩ to give the correct cartridge loading. (R4 has no effect in the audio band because it is bootstrapped.) As it happens C4 and C5 came out as 338 nF, which is conveniently close to the E6 value of 330 nF, and using the preferred value makes very little difference. Calculating R0–C0 for 20 Hz is likewise simple, though an awkward value for C0 is inevitable.

In SPICE simulation this works very nicely, with the −3 dB point at 20 Hz as expected and an 18 dB/octave slope down to about 3 Hz, where it begins to lessen slightly. As noted earlier in this chapter, a 3rd-order Butterworth filter does intrude a bit on the audio band; if C4 and C5 are increased to 470 nF, the −3 dB point falls to 16 Hz.

While this is an ingenious circuit, its practicality is in doubt for three reasons:

1) C4 and C5 are much smaller than the usual input coupling capacitor, which would be 47 uF as a minimum. This implies the circuit will be noisy at low frequencies because current noise of the opamp is flowing through a much higher impedance than usual. For this reason I have not yet attempted to use this concept in a manufactured product.

2) The opamp end of the input circuit is at a much higher impedance than usual and will be very susceptible to hum pickup.

   Is there anything to be done? No. C4 and C5 could be made larger, say 470 nF, but then R5 would have to be reduced proportionally and the loading on the cartridge would be wrong.

3) C0 is an electrolytic, and its tolerance is unlikely to be better than ±20%. C0 at 20% high gives a broad peak of +0.22 dB around 45 Hz, and C0 at 20% low a corresponding droop, which will make life very hard if you are looking for quality RIAA accuracy.

   Is there anything to be done? Oh yes! If C0 is made much larger so it has no influence on the filtering action, then the third pole can be implemented as a passive C-R network after the amplifier stage. This arrangement is shown in Figure 12.19, where C0 has

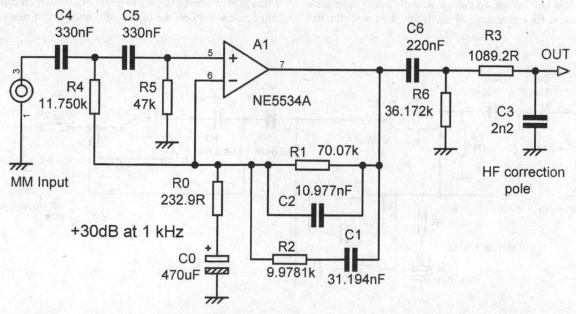

Figure 12.19 3rd-order Butterworth subsonic filter at the input of phono amplifier, C0 is now noncritical as C6–R6 make up the third pole

been increased to the point where it has no effect on the response and C6–R6 make up the third pole. This works very nicely giving the expected 18 dB/octave roll-off slope. Third-order Butterworth filters have a Q of 1.00 in the 2nd-order stage, and this leads to peaking of +1.2 dB at 30 Hz; given the likely levels from vinyl at this frequency there will be no headroom issues.

If you think like I do, the obvious next step is to try to exploit this idea to make a 4th-order subsonic filter; and yes, it works. The 4th-order filter is built in the usual way with a first 2nd-order stage with a Q of 0.541 and a second 2nd-order stage with a Q of 1.306. The first stage is the phono amplifier, with the value of R4 adjusted to reduce the Q, and the second stage is a conventional Sallen and Key configuration based on A2; see Figure 12.20. Since there is no gain peaking in the first stage, there are no extra headroom limitations.

This filtering method could be extended to 5th- and 6th-order versions, and other filter characteristics, in the usual way.

To put this early filtering scheme in perspective, opamps or other amplifiers have maximum loop gain and so maximum negative feedback at low frequencies, and subsonic disturbances are not likely to cause significant intermodulation distortion. The situation is quite different at ultrasonic frequencies, where amplifier gain will be low and falling, so there is actually much more justification for integrating an ultrasonic filter with the first

amplifier stage rather than a subsonic filter. This will be more difficult, as a lowpass filter requires series resistors, which will introduce extra noise, rather than series capacitors.

All-Pole Filters: Conclusions

At the start of this chapter we decided that a really good subsonic filter should be down −40 dB around 10 Hz, but we're clearly not going to get it with any of the all-pole filters examined so far; they are all useful filters but do not meet this rather severe criterion. A 7th-order Butterworth should do it, with a small margin of 2 dB in hand, but that means moving to a four-stage filter, which is a relatively complex design and may have problems with component sensitivities. Increasing the safety margin beyond 2 dB means an 8th-order Butterworth, and that is getting into difficult territory.

Notch Subsonic Filters

As we've seen earlier in this chapter, ideally we would like an attenuation of −40 dB in the band 8–12 Hz. All-pole filters are not going to give that without unduly encroaching on the passband and/or being unduly complex. An alternative approach is to use a notch filter that, in theory at least, gives infinite attenuation in the centre of the notch.

There is little point in trying to guess exactly which cart/arm combination on the market gives the worst results

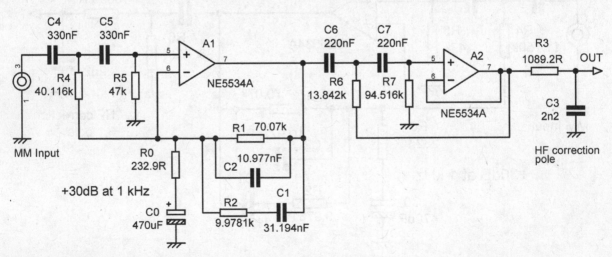

Figure 12.20 4th-order Butterworth subsonic filter with first 2nd-order stage built into phono amplifier and second 2nd-order stage A2

and focusing on that exact frequency, so I have decided to put the notch at 10 Hz, with due attention paid to how much it attenuates from 8–12 Hz. A very narrow notch would not be useful as it cannot deal with cart/arm variations.

You can always make a notch rumble filter by taking a 3rd-order Butterworth and cascading it with a standard symmetrical 10 Hz notch filter, but that is not efficient; it is no way to design filters. A true filter cunningly fits together the various peakings and roll-offs of its cascaded stages to make the overall turnover and roll-off as clean and steep as possible.

My initial plan was to use an Inverse Chebyshev highpass filter, which, as any filter textbook will tell you, has a faster roll-off than Butterworth but still has a maximally flat passband with no ripples. There are one or more notches in the stopband, and one of them can be plonked exactly on 10 Hz. The price of this useful behaviour is that the stopband gain keeps bouncing back up again between the notches, but it will not exceed a level chosen at the design stage.

I ran immediately into the problem of designing an Inverse Chebyshev highpass filter. There are many filter textbooks, and I must have read a good proportion of them; but I can tell you that the Inverse Chebyshev highpass is the red-headed stepchild of filters, and they are rarely mentioned. Daniels[18] is the only useful reference I found.

This forced me to widen my research, and it looked like elliptical filters were the only option left. You can design elliptical filters with no passband ripple, and this is (I think) exactly the same as an Inverse Chebyshev. The word "elliptical" comes from the underlying mathematics, which is pretty fearsome and has only the remotest connection with geometrical ellipses. Elliptically shaped PCBs are not required.

However, while elliptical filters are frequently discussed in the textbooks, procedures for designing them are rarely given. There are remarks like ". . . the design process is highly involved . . ." and ". . . reference to published tables is advised . . .". If you can find these tables, which tend to be in obscure journals, they are anything but easy to interpret. Having stalked the book stacks of both Imperial College and the Cambridge University libraries, I eventually worked out my own ways of designing what I believe are "official" elliptical filters and not mere approximations. That's three weeks I can never get back, though the techniques will be very useful in my consultancy work.

You'd better face up to the fact right away that you're not going to learn how to design elliptical filters from scratch here. The procedures would take up the whole volume. However, I will give you designs that have been built and proven and which can be easily scaled to change the cutoff frequency.

Elliptical Highpass Filters

There are various inputs to the elliptical filter design process:

1) The filter order. As with any filter, higher order means more stages and so more parts, and crucially, higher Q's that can cause high component sensitivities. The lower the order that meets the spec, the better. With elliptical filters as used here, odd-order designs have the disadvantage that the ultimate rate of roll-off is only 6 dB/octave, which is not helpful when warp frequencies go down to 0.55 Hz for a disc rotating at 33 1/3 rpm. Only 4th-order, 6th-order, etc. give an ultimate roll-off at 12 dB/octave; compare the Butterworth versions with 24 dB/oct and 36 dB/oct roll-offs.

2) Cutoff frequency. Elliptical filters do not have a natural cutoff level like the Butterworth does at −3dB. Here I have defined the cutoff as −3dB to make comparisons with the Butterworth easier, but the real driving factor was putting a notch exactly on 10 Hz, and the −3 dB frequency comes out as whatever it does.

3) Stopband attenuation. This is a very important specification because it has a strong effect on how sharply the filter turns over. The lower the stopband attenuation, the more rapid the turnover and the steeper the initial slope, but the lower the attenuation below the notch. For highpass filters this is usually called A0 as it is the gain at zero frequency, i.e. DC. The filters here all have A0 = −35 dB, as it seems the best compromise.

4) Use multiple critical poles? MCP = 2 means that two pole pairs have been coupled, and so on.

5) Finite or infinite zero version? Only the latter have zero response at DC (i.e. infinitely-low frequency) and will therefore head on down forever after the notches have happened.

Here is my take on the properties of highpass elliptical filters. I have used combinations of 2nd-order highpass notches implemented as Bainter filters (described in the next section) and 1st or 2nd order highpass stages. The Bainter filter has unity gain in the passband for highpass notches and component sensitivities of 0.5, which is about as low as it gets. The number of opamps quoted is for a single channel and assumes that the filter will be driven from a low impedance, and must be able to drive a reasonable load, so a final CR pole will need to be buffered. The schematics show 220 nF capacitors in all positions, this being in my view a good compromise between cost and keeping the circuit impedances down; arranging this outcome has required a lot of calculation. Polypropylene types are commonly used to eliminate capacitor distortion, but read the rest of the chapter before putting in a big order. They make little difference in this application, and since they should also be close tolerance, they are going to be distinctly expensive. Resistors are by comparison cheap.

First-order Elliptical Filters

Do not exist as you cannot have a 1st-order notch.

Second-order Elliptical Filters

Consist of a single highpass notch that gives very limited attenuation (A0) at low frequencies and a very slow roll-off. They are not useful to us here and, for the same reasons, are rarely if ever used anywhere. Two capacitors and three opamps are required.

Third-order Elliptical Filters

Must consist of one 2nd-order highpass notch stage and one highpass pole. Despite the steepness of the notch sides, the ultimate roll-off is only 6dB/octave, from the highpass pole. Third-order filters cannot have multiple critical poles (MCP) because they have only one pole pair. Three capacitors and four opamps are required. All the resistors except R2 are set at their nonstandard values by the notch frequency and the 220 nF capacitor value chosen. See Figure 12.24 for the response and Figure 12.25 for the schematic.

Fourth-order Elliptical Filters

Can consist of two 2nd-order highpass notch stages or one 2nd-order highpass notch stage and one 2nd-order highpass stage. In filterspeak that is the difference between finite zero or infinite zero designs. Only the second is useful here, giving an ultimate roll-off of a more satisfactory 12 dB/octave from the 2nd-order highpass stage. MCP = 2 is possible, giving lower Q's. Four capacitors and four opamps are required. See Figure 12.26 for the response and Figure 12.27 for the schematic.

Fifth-order Elliptical Filters

Must consist of two 2nd-order highpass notch stages and one highpass pole. There are now two notches in the stopband, at 10 Hz and 6 Hz. The upper notch has a higher Q than the 4th-order, for faster roll-off, so the extra lower notch is needed to keep the response down in the stopband. The ultimate roll-off is only 6 dB/octave from the single highpass pole, and this is not much return for the greater complexity of the filter. MCP = 2 is possible, giving lower Q's. Five capacitors and seven opamps are now required, which is a significant increase in complexity. See Figure 12.28 for the response and Figure 12.29 for the schematic.

Sixth-order Elliptical Filters

Can consist of three 2nd-order highpass notch stages (finite zero) or two 2nd-order highpass notch stages and one 2nd-order highpass stage (infinite zero). There are two notches in the stopband, at 10 Hz and 7.5 Hz. Only the infinite zero version is useful, giving an ultimate roll-off of 12dB/octave from the 2nd-order highpass stage. With the infinite zero option, only MCP = 2 is possible. Six capacitors and seven opamps are required; only one more capacitor and one more resistor are needed to go from 5th-order to 6th-order. See Figure 12.30 for the response and Figure 12.31 for the schematic.

Higher-order Elliptical Filters

Will require at least three highpass notch stages, using two capacitors and three opamps each. A 7th-order filter will use seven capacitors and 10 opamps, and I consider this jump in complexity objectionable. I have therefore stopped at 6th-order.

You can see from this list that a very large number of options are available, and I have explored and simulated a very large number of them. As a result, I decided that the best way to keep this chapter to a manageable length was to concentrate on A0 = −35 dB filters, because they seem to me to give the best compromise between rapid initial roll-off and good stopband attenuation for subsonic filtering. I used the MCP option in every case (except the 3rd-order, where it is impossible) to make the circuitry more buildable. Each is compared with the 20Hz 3rd-order Butterworth in the accompanying figures.

The Bainter Notch Filter

The Bainter filter is an extremely convenient building-block for highpass filters. For lowpass filters, not so much, for above the notch the gain is always 0 dB, and so making a lowpass notch must mean gain on the low-frequency side of the notch, and this will often be highly inconvenient in the signal path. The Bainter filter is intriguing, not least because it is a relatively recent invention, dating from 1975.[19] Remarkably, the notch depth depends only on the gain of the opamps and is very deep without any tweaking or matching. It is however a sad fact that in this application a really good notch depth is not of much value.

A Bainter filter consists of three opamps, the shunt-feedback stage A1, the integrator A2, and the output buffer A3. See Figure 12.21. The gain of A1 is here unity, unlike the filters in the rest of this chapter. As explained later, you can have any gain here so long as R3 is appropriate to inject the current amount of signal current into the summing-point of A2.

The Bainter filter is enigmatic in its operation. There is a notch at the output, so there must be the same notch at the junction of C2 and R6, and it is here that the cancellation occurs to create that notch; space forbids going any deeper. One thing is clear; at frequencies well above the notch the filter must be unity gain and noninverting, because there C2 is essentially a short circuit from the input to the unity-gain buffer, which is connected directly to the output. The signal at the A2 output is then irrelevant because of the presence of R5.

The input impedance below the notch is the value of R1. Above the notch it falls to R1 in parallel with R5 and R6, which in these filters is a very small reduction.

The Bainter filter type is set by the ratio of the notch frequency fz to the filter centre frequency f0. The square of this ratio gives the parameter Z, which is used to calculate R3, R4, and R5 (this has nothing to do with using Z to represent impedance). For Z > 1 you get a lowpass notch with R3 < R4. For Z = 1 you get a symmetrical notch with R3 = R4. For Z < 1 you get a highpass notch with R3 > R4. The gain of the shunt-feedback stage g can be set arbitrarily as it can be allowed for by the value of R3. The Bainter filters in this article all emerged from the design process with shunt-feedback stage gains of between 0.35 and 0.83. This seems to give good results for noise and distortion, but I am not claiming each value is optimal. It occurs to me that adjustments to R3 would allow both R1 and R2 to be preferred values.

The design procedure given here assumes the passband gain is unity. You can get gain by making A3 a series-feedback amplifying stage, but unity gain is usually what's wanted.

The design procedure:

1) Choose C = C1 = C2
2) Choose gain of shunt-feedback stage g
    (Different values can be used for C1 and C2, but it makes the calculation more complicated)
3) Choose R1
4) Calculate $R2 = gR1$
5) Choose the centre frequency of the filter f_0
6) Choose the notch frequency f_z
    (this is only the notch frequency for a symmetrical notch)

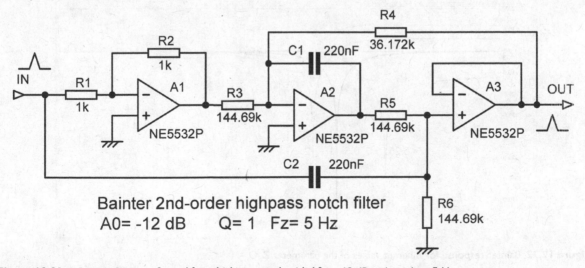

Bainter 2nd-order highpass notch filter
A0= -12 dB   Q= 1   Fz= 5 Hz

Figure 12.21 A Bainter filter configured for a highpass notch with A0 = −12 dB and notch at 5 Hz

7) Calculate: $Z = \left(\dfrac{f_z}{f_0}\right)^2$

   For $Z > 1$ you get a lowpass notch
   For $Z = 1$ you get a symmetrical notch
   For $Z < 1$ you get a highpass notch

8) Calculate $a = 2\pi f_0 C$

9) Choose Q

   The value of Q that gives maximal flatness in the passband depends on the value of Z

   For $Z = 1/2$, $Q = 1$ gives maximal flatness.

10) Calculate $R3 = \dfrac{g}{2ZQa}$ $R4 = \dfrac{1}{2Qa}$ $R5 = R6 = \dfrac{2Q}{a}$

   The effect of changing Z is illustrated in Figure 12.22. The gain is always unity above the notch frequency.

Figure 12.23 shows the effect of different values of Q with $Z = 1/2$ to give a highpass notch. The schematic for $Z = 0.25$ is shown in Figure 12.21.

Looking at Figure 12.21, we can attempt some deductions about the likely distortion performance. I never make predictions, and I never will, but I think we can say:

1) Opamp A1 is working in shunt-feedback mode, so there is no common-mode (CM) voltage on its inputs, and so no common-mode distortion.

2) A1 is also working at a low-noise gain, so there will be plenty of negative feedback (NFB) to reduce distortion.

3) A2 is working in shunt-feedback mode, so there is no CM voltage on its inputs.

4) A2 is working as an integrator, so there will be a lot of NFB at high frequencies where open-loop gain is falling.

5) A3 is a voltage-follower, so it has maximal CM voltage on its inputs, which may cause increased distortion.

6) A3 is a voltage-follower, so it has maximal NFB, which should reduce distortion.

7) Perhaps most importantly, in the audio passband the signal only goes through C2 and A3, so distortion from A1 and A2 does not reach the output.

All these statements except No 5 look promising. We need to bear in mind that a complete elliptical filter will require another voltage-follower to buffer the HP poles, and this will also generate CM distortion.

Some thoughts about the likely noise performance:

1) A1 has low-value resistors around it, set by R2, which loads its output; 1kΩ was chosen to give good THD results with 5532 opamps.

2) A1 is in shunt-feedback mode, so it is operating at a higher noise gain than its signal gain, increasing noise.

3) A2 is an integrator, and integrators are quiet because of their low gain at high frequencies.

4) A3 is working at unity signal gain and noise, gain so noise output should be low.

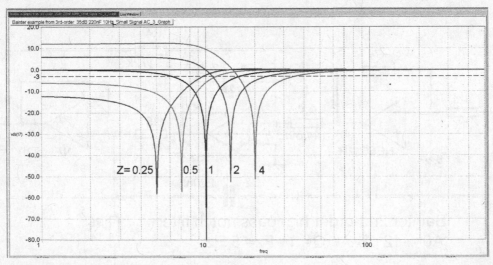

Figure 12.22 Bainter response for different values of the parameter Z. Q = 1

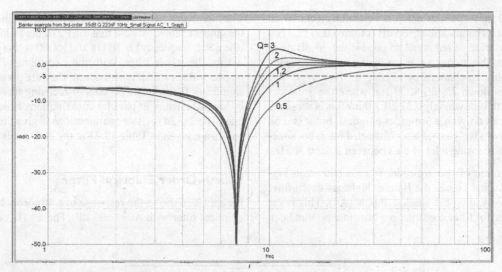

Figure 12.23 Bainter response for different values of Q. Z = 1/2 to give a highpass notch

5) Again, perhaps most importantly, over the audio band, A3 is directly connected to the input via C2, so only the input voltage noise of A3 will be seen at the output.

This also looks distinctly promising. Both sets of predictions were very largely confirmed by the measurements made, confirming that the Bainter is a very useful and well-behaved filter configuration. It may not be the simplest of filters, but its inherently deep notch and its good noise and distortion behaviour are a joy to work with. Mr Bainter, wherever you are, I salute you.

Third-Order Elliptical Filter

Figure 12.24 shows the response of a 3rd-order highpass elliptical filter with A0 = −35 dB; you will note that the stopband response comes back up to exactly −35 dB at 6 Hz and then goes down again, at a slow 6dB/octave,

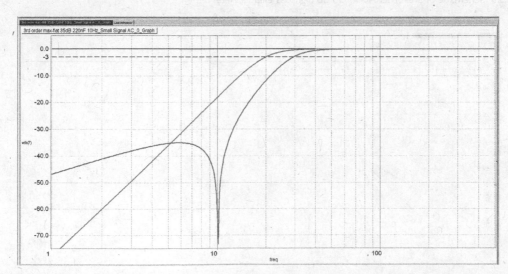

Figure 12.24 Frequency response of 3rd-order A0 = −35 dB highpass elliptical filter—3 dB at 29 Hz, compared with 3rd-order Butterworth that is −3 dB at 20 Hz

to −47 dB at 1 Hz. Our notch is beautifully centred on 10 Hz, and its bandwidth at −35 dB is from 6.0 to 11.7 Hz, i.e. 5.7 Hz wide. What is not so good is the −3 dB point at 29.5 Hz; this frequency is set by the notch frequency and can only be changed by altering the filter type. The response is down 2.8 dB at 30 Hz, where the 3rd-order Butterworth is down only 0.37 dB. I think this is about the best you can do with a 3rd-order elliptical, but it is a bit of a let-down; we've used a sophisticated filter, but nonetheless we're losing a lot of bass between 20 and 30 Hz.

The schematic of the 3rd-order filter is shown in Figure 12.25. First comes the Bainter highpass notch filter (A1, A2, A3), then the single HP pole (A4). This is the usual order for filter construction, but more on that later.

The filter parameters for each stage are shown in this and the other schematics; note that the Fz of the Bainter filter (the notch frequency) is 10 Hz and its A0 is less than the A0 for the whole filter. With the exception of R2, the resistor values are all nonstandard; they are given to five significant figures. The process of realising these values by single resistors or parallel combinations is dealt with later. The 2xE24 resistor combinations to give the desired values are given in Table 12.21 at the end of this section.

Fourth-Order Elliptical Filter

Figure 12.26 shows the response of a 4th-order highpass elliptical filter with A0 = −35 dB. The 10 Hz notch has

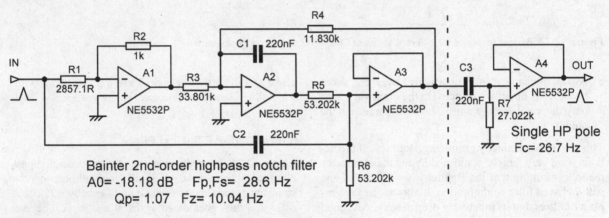

Figure 12.25 Schematic of 3rd-order A0 = −35 dB highpass elliptical filter

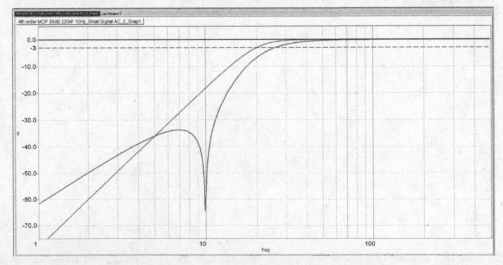

Figure 12.26 Frequency response of 4th-order A0 = −35 dB MCP highpass elliptical filter that is −3 dB at 26 Hz, compared with 3rd-order Butterworth that is −3 dB at 20 Hz

a higher Q and is narrower, giving a more rapid roll-off, moving the −3 dB point down from the 29.5 Hz of the 3rd-order filter to 26.2 Hz. A helpful improvement, but this is not the 20 Hz we are looking for. Below the notch the response again comes back up to −35 dB around 7 Hz, as expected, but then heads downwards forever at 12 dB/octave, giving much better rejection of VLF signals below 3 Hz than the 3rd-order filter. Our notch now has a bandwidth at −35 dB from 8.0 to 10.8 Hz, only 2.8 Hz wide, and a good deal narrower than the 5.7 Hz width of the 3rd-order notch.

Figure 12.27 shows the schematic of the 4th-order MCP filter, arranged in the usual way with the 2nd-order HP poles at the end. Note that all the resistor values in the Bainter filter have changed, with the exception of R2.

The resistor combinations to give the desired values are shown in Table 12.22.

Fifth-Order Elliptical Filter

Figure 12.28 shows the response of a 5th-order high-pass elliptical filter with A0 = −35 dB. There are now two Bainter highpass notch filters, at 10 Hz and 6 Hz, and a single highpass pole. The 10 Hz upper notch has a higher Q and is narrower, giving a more rapid roll-off, moving the −3 dB point down from 26.2 Hz (4th-order) to 16.9 Hz, giving at last an improvement on the 20 Hz of the 3rd-order Butterworth filter. The upper notch bandwidth is now from 8.3 Hz to 10.8 Hz, i.e. 2.5 Hz wide, at −35 dB. One begins to wonder of this is not a bit

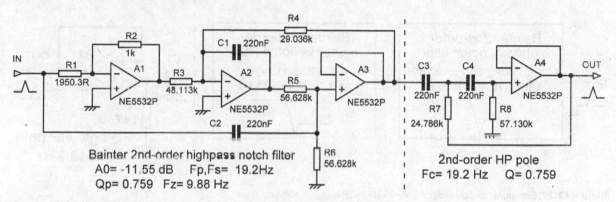

Figure 12.27 Schematic of 4th-order A0 = −35 dB MCP highpass elliptical filter

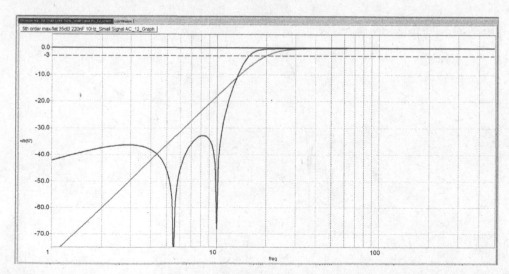

Figure 12.28 Frequency response of 5th-order A0 = −35 dB MCP highpass elliptical filter that is −3 dB at 17 Hz, compared with 3rd-order Butterworth that is −3 dB at 20 Hz

too narrow to give good attenuation with a wide range of cartridge/arm combinations, but the notch cannot be wider if we are to get the steep initial roll-off that gives a pleasingly low −3 dB frequency of 16.9 Hz. The lower notch is significantly wider to maximise attenuation in the 5–7 Hz region. Below it the response slope has gone back to 6 dB/octave, coming solely from the single output pole.

Figure 12.29 shows the schematic of the 5th-order MCP filter, using the conventional layout with the single pole at the end. Each Bainter block is identical to that in Figure 12.25, except that all but one (R2) of the resistor values are changed. The resistor values for the Bainter filters in Figure 12.29 are given in Table 12.23.

Sixth-order elliptical filter

Finally, Figure 12.30 shows the response of a 6th-order MCP highpass elliptical filter with A0 = −35 dB. The 10 Hz upper notch yet again has a higher Q and is narrower, for faster roll-off, moving the −3 dB point down from 16.9 Hz (5th-order) to 14.4 Hz. As a result, the lower notch has to move up to 7.4 Hz to stop the response from coming back up above the chosen −35 dB A0 level.

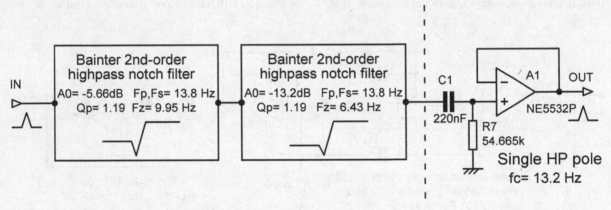

Figure 12.29 Schematic of 5th-order A0 = −35 dB MCP highpass elliptical filter

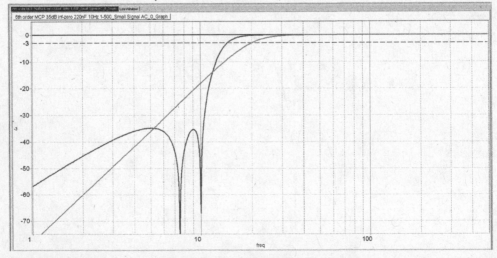

Figure 12.30 Frequency response of 6th-order A0 = −35 dB MCP highpass elliptical filter that is −3 dB at 14 Hz, compared with 3rd-order Butterworth that is −3 dB at 20 Hz

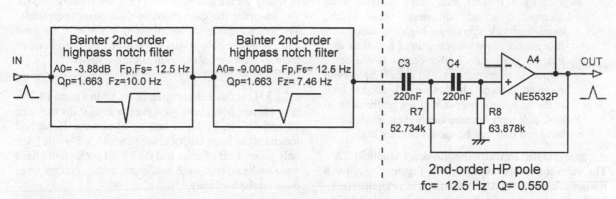

Figure 12.31 Schematic of 6th-order A0 = −35 dB MCP highpass elliptical filter

At the −35 dB level, the upper notch now only extends from 9.0 to 10.4 Hz, 1.4 Hz wide, which may be a bit too narrow. However, this filter still gives more attenuation than the 3rd-order Butterworth, up to 11.7 Hz.

Below the lower notch the response goes down at 12 dB/ octave because of the 2nd-order output poles.

Figure 12.31 shows the schematic of the 6th-order MCP filter, using the conventional layout with the 2nd-order poles at the end of the chain. The two highpass notches are now at 10 Hz and 7.5 Hz, followed by 2nd-order highpass poles. Each Bainter block is identical to that in Figure 12.25, except that most of the resistor values are changed. The resistor values for the Bainter filters in Figure 12.31 are given in Table 12.24.

Since the blessed advent of oversampling DACs we no longer need 9th-order elliptical reconstruction filters, so this 6th-order design is about as near to a brick-wall filter as you are likely to encounter in contemporary audio electronics.

The elliptical filter responses are summarised in Table 12.18; note the unhelpfully high −0.1 dB frequency

of 80.2 Hz for the 4th-order elliptical filter. The highest Q used is 1.663 in the 6th-order filter, which is really quite modest by active filter standards; I must admit this rather surprised me when it emerged. The highest Q used in the 4th-order Butterworth filter was 1.3065.

Elliptical Filter Internal Levels

You may think that after setting the various frequency responses, most of the design work is done. And yet not. In any multistage filter you must consider the level at the output of each stage, across the whole operating frequency range. There can be big internal gains at certain frequencies that are not revealed by looking just at the overall gain of the whole filter; these can cause unexpected clipping.

Things get a little more complicated when, as here, a single stage of the filter contains more than one opamp. In general each opamp output needs to be checked, thus:

Step 1) Check levels at the output of each Bainter stage and each 2nd-order pole stage (the single poles can always be ignored as they cannot give gain).

Table 12.18 Summary of elliptical filter attenuations + 3rd-order 20Hz Butterworth. The −70 dB entries at 10 Hz are nominal figures.

Freq	1 Hz	5 Hz	10 Hz	20 Hz	30 Hz	−3dB freq	−0.1dB freq
Elliptical 3rd-order	−47 dB	−35 dB	−70 dB	−12 dB	−2.8 dB	29.5 Hz	53.9 Hz
Elliptical 4th-order	−62 dB	−39 dB	−70 dB	−6.8 dB	−1.9 dB	26.2 Hz	80.2 Hz
Elliptical 5th-order	−47 dB	−45 dB	−70 dB	−0.5 dB	0.0 dB	16.9 Hz	22.4 Hz
Elliptical 6th-order	−57 dB	−35 dB	−70 dB	−0.2 dB	0.0 dB	14.4 Hz	22.1 Hz
3rd-order Butterworth	−78 dB	−36 dB	−18 dB	−3.0 dB	−0.37 dB	20 Hz	37 Hz

Step 2) Each Bainter filter stage contains three opamps. Overload will never occur in the shunt-feedback stage of a highpass notch version as this always has a gain of less than one, and the unity-gain buffer output is the stage output, so overload here will already have been assessed in Step 1. In fact we only need to look at the integrator output, but we really need to look at it hard because the highest levels occur here, and they can be inconveniently high.

As an example, consider the 4th-order elliptical filter. The relevant levels are shown in Figure 12.32, for a design calculated for A0 = −40 dB. It is a classical (non-MCP) filter, as this change increases the gain peaking and keeps the traces separate on the plot to make it comprehensible. The principles are the same for the A0 = −35 dB MCP filters we have actually examined.

In Step 1 we look at the outputs of the Bainter highpass notch and the 2nd-order pole stage. The gain of the Bainter stage peaks at +2.4 dB around 27 Hz, combining with the slow roll-off from the 2nd-order pole stage to give a maximally flat passband, then a rapid roll-off. Thus clipping will occur 2.4 dB prematurely at the Bainter output, in other words 2.4 dB before the final filter output. Since the 2nd-order pole stage has a Q of only 0.55 (less than 0.7071), it has no gain peak.

In Step 2, Figure 12.32 shows that the gain to the integrator output peaks at +10 dB at 23 Hz. This is a real problem. Clipping will occur here 7.6 dB before it occurs at the Bainter output.

In reality we are using A0 = −35 dB MCP filters, and for the 4th-order the gain problem disappears completely. The internal gains are summarised in Table 12.19. First and 2nd-order pole stages are omitted because none of them as used here show gain at their outputs.

In the 3rd-order filter we uncover gain peaking at +6 dB at 28.5 Hz at the Bainter integrator (A2 in Figure 12.25). It is a broad peak, with gain greater than 0 dB between 13 and 60 Hz, giving a loss of headroom. If we had looked at the stage output alone (Step 1), we would have only seen +1 dB of gain, and the +6 dB peak would have gone undiscovered until headroom measurements were done on the hardware.

In Figure 12.25 the single pole is placed at the end of the signal path, the order used in all the filter textbooks

Table 12.19 Summary of A0 = −35 dB MCP filter internal levels with conventional stage order

	Bainter 1 integrator	Bainter 1 output	Bainter 2 integrator	Bainter 2 output
Elliptical 3rd-order	+6.0 dB (28.5 Hz)	+1.04 dB (42.6 Hz)	not used	not used
Elliptical 4th-order	< 0 dB	< 0 dB	not used	not used
Elliptical 5th-order	+14.5 dB (16.6 Hz)	+3.3 dB (9.2 Hz)	+0.71 dB (18.3 Hz)	+1.2 dB (21.8 Hz)
Elliptical 6th-order	+9.7 dB (14.5 Hz)	+0.3 B(20.7 Hz)	+9.3 dB (11.6 Hz)	+2.2 dB (17 Hz)

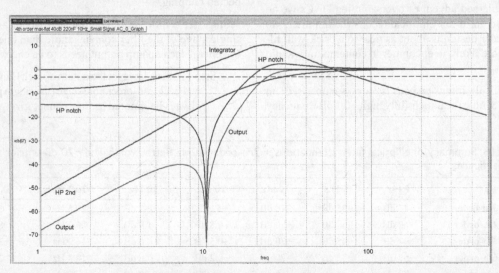

Figure 12.32 Internal levels of 4th-order A0 = −40dB classical (non-MCP) highpass elliptical filter. Note Y-axis change.

I have seen, and they are many. If instead we put it first, before the Bainter stage, in what I will call the New Order, things are improved, because the pole gives some low-frequency attenuation before the Bainter, reducing the effective gain peak to +3.4 dB. We have gained 2.6 dB of headroom for no cost at all. See the first row in Table 12.20.

The 4th-order MCP filter has no internal gain peaking (unless you count +0.08 dB in the 2nd-order HP poles), so no rearrangement is required.

There are however potentially serious problems with the 5th-order filter, because its first Bainter filter has an internal gain at the integrator of +14.5 dB at 16.6 Hz, which is more than five times. If we had looked only at the output of Bainter 1 then we would have remained in blissful ignorance of this. Looking at Figure 12.29, we see that the first Bainter highpass notch has a relatively high stage A0 = of −5.7 dB (the overall A0 for the whole elliptical filter is a different quantity), while the second has a stage A0 = of −13 dB. This suggests that if we swap the Bainter order, low-frequency signals will get more attenuation in the first stage. This works to an extent, as shown in the "Elliptical 5th-order A" row in Table 12.20; the gain peak is now in the second stage rather than the first and is reduced from +14.5 dB to +11.3 dB. We have acquired 3.2 dB more headroom at zero cost.

We then move the single HP pole to the input of the chain, as we did for the 3rd-order filter, and things improve again, with the gain peaking at only +9.3 dB (at 11.6 Hz) in the second Bainter. The effect is not dramatic

because the single pole only rolls off at 6 dB/octave; see the "Elliptical 5th-order B" row in Table 12.20. The gain is now above 0 dB only between 12 and 44 Hz, thanks to the New Order.

This would seem to be all we can do with the 5th-order filter; we are already using the MCP strategy, which gives lower Q's and so less peaking. The question is, will it do? With ±17 V rails the maximum signal level is about 9.5 Vrms, depending somewhat on opamp type. If the nominal audio signal level is 2 Vrms, the general headroom is only some 13 dB, and the +9.3dB peak will reduce this to 3.7 dB, which isn't exactly a stunning figure. But—this peak is at 11.6 Hz, and we have to ask what sort of signal amplitudes we are really going to see at this frequency. The literature suggests that even in bad situations the subsonic disturbance will be at least 20 dB below the audio signal. In many cases the rumble filter will be placed before the final amplification that gets signals up to nominal level, and then there are unlikely to be headroom issues.

Also, clipping in a Bainter filter is relatively benign, as described in the measurement section later. I therefore think we are good to go. If you don't agree, your only recourse is to reduce the nominal signal level in the filter and amplify afterwards to maintain unity gain. This will require extra circuitry and may degrade the noise performance.

In The Case of The 6th-order Filter (Quick, Watson, the game's afoot!) we have +9.7 dB and +9.3 dB peaking in the two filters. This suggests that swapping their order may not do any good, and so it proves; in fact it makes things slightly worse. See the "Elliptical 6th-order A" row in Table 12.20. We therefore leave the Bainter filters where they are in Figure 12.31. Fortunately, moving the 2nd-order pole stage to the front of the chain is now much more effective than it was for the 5th-order filter, as it rolls off at 12 dB/octave. The gain peaks are now at +4 dB and +5.9 dB, and it is of course the higher of the two that sets the headroom; see the "Elliptical 6th-order B" row in Table 12.20. Another triumph for the New Order! If the 5th-order filter is acceptable, then this one certainly is.

Putting the single or 2nd-order poles at the start has the further advantage that it acts as a "roofing filter" in that the worst subsonic disturbances are attenuated somewhat before they ever reach an opamp, which should soothe worries about intermodulation distortion.

Schematics of the four filters with optimised stage order are given in the next section.

Table 12.20 Summary of A0 = −35 dB MCP filter internal levels, stage orders optimised (NB: The A versions are *not* optimal)

	Bainter 1 integrator	Bainter 1 output	Bainter 2 integrator	Bainter 2 output
Elliptical 3rd-order	+3.4 dB (31.5 Hz)	< 0 dB	not used	not used
Elliptical 4th-order	< 0 dB	< 0 dB	not used	not used
Elliptical 5th-order A	< 0 dB	< 0 dB	+11.3 dB (17.6 Hz)	+1.2 dB (21.8 Hz)
Elliptical 5th-order B	< 0 dB	< 0 dB	+9.3 dB (17.7 Hz)	< 0 dB
Elliptical 6th-order A	+11.5 dB (12.7 Hz)	+2.2 dB (16.4 Hz)	+9.9 dB (13.4 Hz)	+2.2 dB (16.4 Hz)
Elliptical 6th-order B	+4.0 dB (12.8 Hz)	< 0 dB	+5.9 dB (15.4 Hz)	< 0 dB

The internal level problems are much milder in the Butterworth filters. The 3rd-order Butterworth (Figure 12.9) shows +1.2 dB at 29 Hz at the opamp output. The 4th-order Butterworth non-MCP filter does not show internal gain peaks if built with the low-Q stage first, as in Figure 12.9, but if the high-Q stage comes first there is a +3.0 dB peak at 24 Hz at its output. The 4th-order Butterworth MCP filter has no internal gain peaks.

Practical Elliptical Filter Designs

Having followed my philosophy of making all the capacitors one preferred value, you are lucky if you can choose even one resistor value arbitrarily; in this case R2 only. Thoroughly awkward values are the norm; as described in Chapter 2, there are three good ways to get these values:

1) Use the nearest E96 value and hope; this is simple; however, the way of least thought is rarely the best way in anything. The accuracy will simply be that of the resistor series chosen. Despite the close spacing of the values, at about 2%, E96 resistors are often available at 1% tolerance. I call this 1xE96 format.

2) Use two E24 1% resistors, Ra, Rb, in parallel, making them as equal as possible to get the best reduction in effective tolerance. Sometimes it is necessary to balance accuracy of nominal value against reduction of effective tolerance. A criterion that the nominal value should be accurate to better than half of the usual resistor tolerance of 1% was used here; with that, achieved reduction in effective tolerance was pursued. I call this 2xE24 format.

3) Using three E24 1% resistors in parallel not only allows us to get ten times nearer to a desired nominal value than 2xE24 but also gives a better chance of getting near-equal resistors giving most of the potential $1/\sqrt{3}$ (= 0.577) reduction in effective tolerance, as there are many more combinations. The design process is not obvious; I used a Willmann table; see Chapters 2 and 7. I call this 3xE24 format.

Tables 12.21, 12.22, 12.23, and 12.24 give the resistor values for the filters in 1xE96, 2xE24, and 3xE24 format, together with the error compared with the desired nominal value.

Table 12.21 Resistor values for 3rd-order A0 = −35 dB MCP elliptical filter. Error is % deviation from nominal 5 sig fig values

	Nominal Ω	1xE96	Error	2xE24	Error	3xE24	Error
R1a	2857.1	2870	+0.45%	3k3	+0.436%	7k5	+0.0015%
R1b		–		22k		7k5	
R1c		–		–		12k	
R2a	1k	1k	0.00%	2k	0.00%	3k	0.00%
R2b		–		2k		3k	
R2c		–		–		3k	
R3a	33.801k	34k	+0.59%	62k	+0.416%	75k	+0.0054%
R3b		–		75k		100k	
R3c		–		–		160k	
R4a	11.803k	11.8k	−0.02%	15k	+0.237%	20k	−0.0093%
R4b		–		56k		39k	
R4c		–		–		110k	
R5a=R6a	53.202k	53.6k	+0.75%	82k	−0.347%	120k	+0.0094%
R5b=R6b		–		150k		120k	
R5c=R6c		–		–		470k	
R7a	27.022k	26.7k	−1.19%	36k	+0.375%	39k	+0.0059%
R7b		–		110k		120k	
R7c		–		–		330k	
Average abs err			0.601%		0.362%		0.0065%

Table 12.22 Resistor values for 4th-order 35 dB MCP elliptical filter. Error is % deviation from nominal 5 sig fig values.

	Nominal Ω	1xE96	Error	2xE24	Error	3xE24	Error
R1a	1950.3	1960	+0.50%	3k9	−0.015%	3300	+0.029%
R1b		–		3k9		6800	
R1c		–		–		16k	
R2a	1k	1k	0.00%	2k	0.00%	3k	0.00%
R2b		–		2k		3k	
R2c		–		–		3k	
R3a	48.113k	48.7k	+1.22%	51k	−0.21%	91k	−0.041%
R3b		–		820k		120k	
R3c		–		–		680k	
R4a	29.036k	28.7k	−1.16%	51k	+0.37%	56k	−0.0021%
R4b				68k		62k	
R4c		–		–		2.2M	
R5a=R6a	56.628k	56.2k	−0.76%	100k	−0.19%	110k	+0.013%
R5b=R6b		–		130k		120k	
R5c=R6c		–		–		4.3M	
R7a	24.786k	24.9k	+0.46%	39k	−0.004%	47k	−0.020%
R7b		–		68k		56k	
R7c		–		–		820k	
R8a	57.130k	57.6k	+0.82%	110k	+0.46%	100k	+0.022%
R8b		–		120k		240k	
R8c		–		–		300k	
Average abs err			0.983%		0.206%		0.026%

Table 12.23 Resistor values for 5th-order 35 dB MCP elliptical filter. Error is % deviation from nominal 5 sig fig values.

	Nominal Ω	1xE96	Error	2xE24	Error	3xE24	Error
R1a	2369	2370	+0.042%	2.7k	+0.42%	5.6k	+0.077%
R1b		–		20k		7.5k	
R1c		–		–		9.1k	
R2a	1k	1k	0.00%	2k	0.00%	3k	0.00%
R2b		–		2k		3k	
R2c		–		–		3k	
R3a	87.200k	86.6k	−0.69%	110k	+0.45%	150k	+0.083%
R3b		–		430k		240k	
R3c		–		–		1.6M	
R4a	36.799k	36.5k	−0.81%	62k	+0.21%	62k	−0.022%
R4b		–		91k		110k	
R4c		–		–		510k	
R5a=R6a	68.888k	68.1k	−1.14%	120k	−0.46%	120k	+0.017%
R5b=R6b		–		160k		180k	

(*Continued*)

Table 12.23 (Continued)

	Nominal Ω	1xE96	Error	2xE24	Error	3xE24	Error
R5c=R6c		–		–		1.6M	
R1a	1655	1650	−0.30%	3300	−0.30%	4.7k	+0.021%
R1b		–		3300		4.7k	
R1c		–				5.6k	
R2a	1k		0.00%	2k	0.00%	3k	0.00%
R2b		–		2k		3k	
R2c		–		–		3k	
R3a	17.963k	17.8k	−0.91%	36k	+0.21%	39k	−0.042%
R3b		–		36k		56k	
R3c		–		–		82k	
R4a	10.850k	11.0k	+1.38%	18k	−0.46%	22k	+0.040%
R4b				27k		22k	
R4c		–		–		820k	
R5a=R6a	175.275k	174k	−0.73%	180k	+0.05%	510k	−0.034%
R5b=R6b		–		6.8M		510k	
R5c=R6c		–		–		560k	
R7a	54.665k	54.9k	+0.43%	100k	−0.22%	82k	+0.004%
R7b		–		120k		200k	
R7c		–		–		910k	
Average abs err			0.715%		0.308%		0.038%

Table 12.24 Resistor values for 6th-order 35 dB MCP elliptical filter. Error is % deviation from nominal 5 sig fig values.

	Nominal Ω	1xE96	Error	2xE24	Error	3xE24	Error
R1a	1236	1240	+0.29%	1800	−0.39%	2.4k	−0.003%
R1b		–		3900		5.1k	
R1c		–		–		5.1k	
R2a	1k	1k	0.00%	2k	0.00%	3k	0.00%
R2b		–		2k		3k	
R2c		–		–		3k	
R3a	21.821k	22.1k	+1.28%	36k	−0.42%	51k	+0.063%
R3b		–		56k		56k	
R3c		–		–		120k	
R4a	17.649k	17.8k	+0.85%	30k	+0.13%	27k	−0.034%
R4b		–		43k		56k	
R4c		–		–		560k	
R5a=R6a	193.566k	196k	+1.26%	330k	+0.16%	390k	−0.024%
R5b=R6b		–		470k		430k	
R5c=R6c		–		–		3.6M	
R1a	1689	1690	+0.77%	3000	+0.41%	3300	+0.063%
R1b		–		3900		3600	
R1c		–		–		91k	

Table 12.24 (Continued)

	Nominal Ω	1xE96	Error	2xE24	Error	3xE24	Error
R2a	1k	1k	0.00%	2k	0.00%	3k	0.00%
R2b	–	–		2k		3k	
R2c	–	–		–		3k	
R3a	29.080k	29.4k	+1.10%	51k	+0.22%	82k	−0.007%
R3b	–	–		68k		82k	
R3c	–	–		–		100k	
R4a	17.649k	17.8k	+0.86%	30k	+0.13%	27k	−0.034%
R4b	–	–		43k		56k	
R4c	–	–		–		560k	
R5a=R6a	189.082k	191k	+1.01%	300k	−0.10%	390k	−0.015%
R5b=R6b	–	–		510k		390k	
R5c=R6c	–	–		–		6.2M	
R7a	52.734k	52.3k	−0.82%	75k	+0.39%	100k	+0.025%
R7b	–	–		180k		120k	
R7c	–	–		–		1.6M	
R8a	63.878k	63.4k	−0.75%	75k	−0.026%	120k	−0.011%
R8b	–	–		430k		220k	
R8c	–	–		–		360k	
Average abs err			0.922%		0.264%		0.025%

Figures 12.33–12.36 show the practical 3rd, 4th, 5th, and 6th-order filter designs, using the New Order and the 3xE24 resistor format.

Every set of resistor values in Tables 12.21, 12.22, 12.23, and 12.24 has been checked against the original design spreadsheets and has been fed back into the simulator to check they are correct. The values for the 2xE24 format have been checked a third time by building the circuits

and measuring them. It's a lot of data to produce without a single mistake, but I am cautiously hopeful that I have achieved it. If you think otherwise, let me know.

Taking all four tables together, 36 nominal values were dealt with. The global average absolute error for 1xE96 is 0.805%, for 2xE24 is 0.285%, and for 3xE24 only 0.025%. Thus 2xE24 is three times better, and 3xE24 ten times better again. It is questionable if 3xE24 is

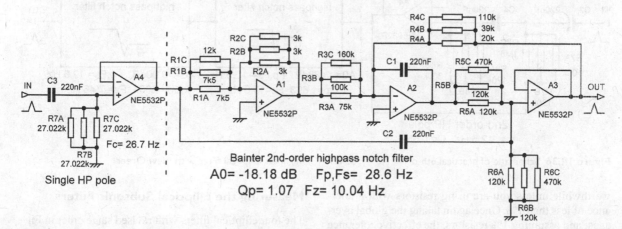

Figure 12.33 Schematic of practical 3rd-order A0 = −35 dB MCP highpass elliptical filter, with New Order

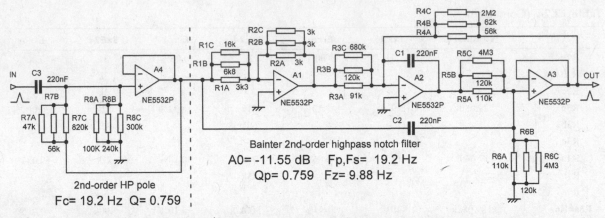

Figure 12.34 Schematic of practical 4th-order A0 = −35 dB MCP highpass elliptical filter, with New Order

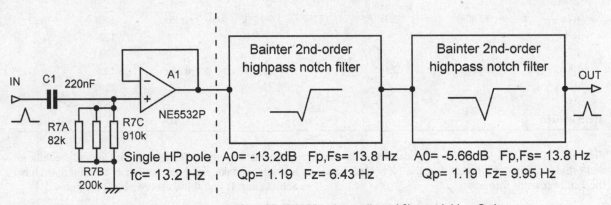

Figure 12.35 Schematic of practical 5th-order A0 = −35 dB MCP highpass elliptical filter, with New Order

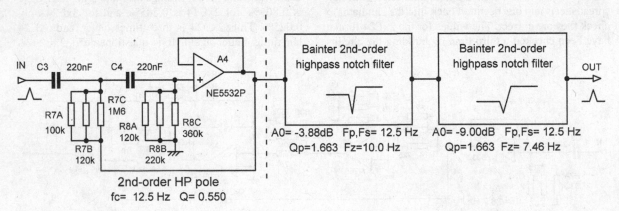

Figure 12.36 Schematic of practical 6th-order A0 = −35 dB MCP highpass elliptical filter, with New Order

worthwhile unless you are using resistors with a tolerance of less than 1%. Once again taking the global averages, and assuming 1% resistors, the effective tolerance for 2xE24 is 0.764% and for 3xE24 is 0.659%. This was quite an experiment in itself.

Measuring the Elliptical Subsonic Filters

The four elliptical filters with revised stage order in Figures 12.33–12.36 were built using 1% resistors in the 2xE24 format, 5% polyprop capacitors, and Fairchild

5532 opamps. The 5% capacitors did not cause any significant response deviations, underlining that these circuits have low component sensitivity and can be built without difficulty.

The only practical issue likely to be encountered is that in every Bainter filter the circuit nodes at C3-R7, C4-R8, and C2-R5-R6 are at a high impedance and are very susceptible to electric field pickup. A grounded screening plate under the prototype and a grounded biscuit-tin covering it solves that problem. Full screening is essential for meaningful noise and THD measurements. Bear this in mind if you plan to have the PSU in the same box as the subsonic filtering.

Audio measuring systems do not usually go below 10 Hz, allowing little exploration of subsonic filtering. Function generators can provide test signals as low as 0.01 Hz, but there are difficulties in measuring the filter output; a digital scope is one way, but the accuracy will not be great, and you will have to plot the response yourself. A better scheme is to change all the 220 nF capacitors to 22 nF parts of the same precision. This will shift the filter frequency up a decade without altering any other parameter, and you can use an audio measuring system down to the equivalent of 1 Hz. I confirm that this works well in practice, the only issue being that the circuit nodes mentioned are at an even higher impedance and more vulnerable to electric fields. Once again, thorough screening eliminates the problem, but you have to be careful to do it *really* thoroughly.

Measurements: 3rd-Order Elliptical Filter

THD+N is shown in Figure 12.37 at a hefty 7 Vrms. There is no detectable distortion below about 3 kHz, and above this second harmonic can be seen. Half of this comes from the HP pole buffer and half from the Bainter

filter buffer, as they are the only opamps with a significant common-mode voltage on their inputs.

Measurements confirm that the unmodified 3rd-order filter has an internal gain peak of +6 dB, as determined earlier by simulation, and this limits the signal level that can be handled across the 1 Hz—20kHz band to 5 Vrms. The smaller internal gain peak of +1 dB was also confirmed, but this has a negligible effect. Rearranging the stage order as in Figure 12.33, to put the 1st-order roll-off first, increases the whole-band maximum signal level to 7.3 Vrms, as simulated. THD+N results for the New Order version are shown in Figure 12.37; the rise below 50 Hz is wholly due to the relative increase in noise as the filter attenuates the signal.

The noise output was −111.4 dBu (22Hz–22kHz, rms sensing, corrected). The improvement in signal-handling capability achieved by changing to the New Order (Table 12.20) was confirmed.

Notch depth was measured as −93 dB, which is pretty stunning considering it was achieved with no adjustment of any kind. A great pity such a depth is not really useful here.

As always I started out with polyprop caps to eliminate possibly confusing capacitor distortion. As I have noted before,[20] often it is not necessary for all the capacitors in the filter to be polyprop. In this case making the integrator cap C1 polyester produces no detectable extra distortion down to 10 Hz, by which point the attenuation is so great that distortion is irrelevant. Replacing C2 and C3 does introduce detectable distortion between 20 and 60 Hz, but it is at a very low level. Changing from all polyprop to all polyester only increased the THD+N at 40 Hz and 7 Vrms from 0.00041% to 0.00050%. This is one case where the advantage from polyprop capacitors is very small indeed, and the cost difficult to justify.

Measurements: 4th-Order Elliptical Filter

There were no surprises in measuring the 4th-order filter. THD+N is shown in Figure 12.38; the step at about 300Hz is an artefact of the measuring system as it attempts to extract a meaningful distortion figure from random noise. The rise below 50 Hz is wholly due to the relative increase in noise as signal level falls rapidly.

The noise output was −113.7 dBu (22 Hz–22 kHz, rms sensing). As expected there were no problems with internal overload, and the filter easily handled 10 Vrms in/out without clipping or a disproportionate increase in distortion.

Notch depth was −88 dB.

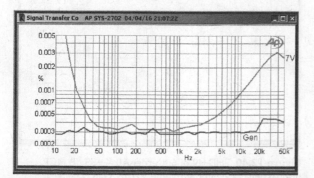

Figure 12.37 THD+N of 3rd-order A0 = −35 dB MCP highpass elliptical filter at 7 Vrms. Gen is testgear output.

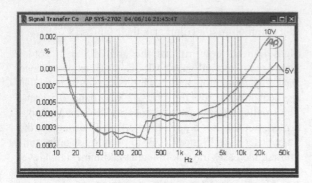

Figure 12.38 THD+N of 4th-order A0 = −35 dB MCP highpass elliptical filter at 5 and 10 Vrms

Measurements: 5th-Order Elliptical Filter

THD+N is shown in Figure 12.39. With the original stage order an input of 20 Hz at only 2.9 Vrms gave clipping at the 10 Hz Bainter integrator, as expected. The overload behaviour was however remarkably benign; the THD at the integrator was 6.0%, with clearly visible clipping. However, at the Bainter output the THD was only 0.96% and was not visible as a misshaped sine wave. This can cause confusion when measuring the frequency response; even with a 5 Vrms input the output sine wave still looks clean, but the notch is some 30 dB less deep than simulation predicts, and the passband flatness is poor because the clipping prevents this part of the filter from peaking properly.

Altering the stage order to the New Order shown in Figure 12.35 increases the clipping level to 4.4 Vrms at 20 Hz. Above this frequency the usual opamp limit of about 10 Vrms applies. As noted earlier, if you are

running at nominal signal levels of 1 or 2 Vrms, you will never see that level at that frequency from a vinyl source.

The noise output was −111 dBu (22Hz–22kHz, rms sensing). The depth of the first (10 Hz) notch was −61 dB.

Measurements: 6th-Order Elliptical Filter

THD+N is shown in Figure 12.40. As Tables 12.19 and 12.20 show, the 6th-order filter has fewer problems with its internal levels, particularly after changing to the New Order so the 2nd-order highpass filter is at the start of the signal path.

The noise output was −111.4 dBu (22Hz–22kHz, rms sensing). Depth of the first (10 Hz) notch was −66 dB.

This filter was also initially built with polyprop caps. Replacing all six of them with polyester showed only very marginal increases in THD+N. In light of this, polyester replacement was not tried for the 4th- and 5th-order filters; I am confident the results will be the same.

Elliptical Filter Modifications

The frequency of filter operation can be changed easily by scaling resistors R3, R4, R5, R6 in the Bainter stages and R7, R8 in the pole stages by the same ratio. However, the −3 dB frequency and the notch frequencies will all change together, so only small changes are recommended. Altering other parameters such as A0 requires the filter to be redesigned from scratch.

The second-harmonic distortion shown in the THD+N plots could be reduced by changing to more expensive opamps, such as perhaps the LM4562. Alternatively the

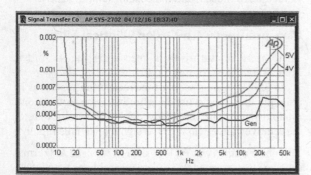

Figure 12.39 THD+N of 5th-order A0 = −35 dB MCP highpass elliptical filter at 4V and 5 Vrms. Gen is testgear output.

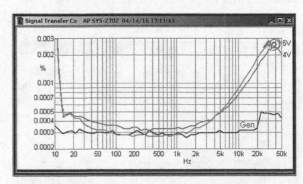

Figure 12.40 THD+N of 6th-order A0 = −35 dB MCP highpass elliptical filter at 4V and 5 Vrms. Gen is testgear output.

single and 2nd-order poles could be redesigned to be inverting stages with no common-mode voltage, but it is hard to see how this could be applied to the Bainter filter output buffer, so HF distortion would be at best halved. A phase inversion would be introduced that will probably be highly inconvenient.

It is sometimes possible to make single-opamp highpass notches with the Boctor filter circuit,[21] but my first attempt gave +12 dB of gain in the passband, which is definitely not wanted, so I did not pursue that further. There are however potential parts savings, so do feel free to have a go.

The Story So Far

While this examination of elliptical filters has to a large extent proved successful, there were some unexpected results.

- I had high hopes of the 3rd-order elliptical filter, but it turned out to have poor passband flatness and is not an improvement on the 3rd-order Butterworth unless you suffering truly awful cart/arm resonance problems; if so you need to fix the rude mechanicals rather than rely wholly on filtering.
- The 4th-order elliptical is flatter in the passband, with a minimal increase in complexity, and has the great advantage that its ultimate roll-off is 12 dB/octave rather than the 6 dB/octave of the 3rd-order elliptical.
- The 5th-order elliptical is much better than the 4th in the passband, but it is more complicated, and the ultimate roll-off falls back to 6 dB/octave.
- The 6th-order elliptical exhibits another small increase in parts count, and the ultimate roll-off is 12 dB/octave. Right at the moment, either the 4th- or the 6th-order elliptical filters look to be the best choices.
- Distortion produced by internal clipping in the filter was much more benign than expected.
- When I began, I was much concerned that component sensitivity would cause problems. It did not, not even in the 6th-order elliptical, and the measured responses were very close indeed to the calculations and simulations. With the famed clarity of hindsight, using the MCP strategy was over-cautious.
- It was unexpected that polyester capacitors were very nearly as good as low-distortion polyprop types *in this particular application*. It will save you a lot of money.
- It was also unforeseen that making up the nominal resistances required with two E24 values in parallel would be more than three times more accurate than

using a single E96 part. Three E24 values in parallel was ten times more accurate again, but that was not a surprise.

In the quest for optimal rumble filtering, I have so far provided you with an armoury of four elliptical filters that, despite their sophisticated operation, are eminently buildable. However, looking at the points raised just now, despite a lot of hard work and some quite sophisticated circuitry, none of them would be described as the ideal or even optimal subsonic filter.

The Optimal 4th-Order Elliptical Filter (Classical)

As noted earlier, I used the MCP elliptical filter strategy to be as sure as possible that there were not intractable issues with component sensitivity. Designing an elliptical filter is a lot of work, and you really don't want to do it more often than necessary, so I began with what I considered the safest option. However, experience with the actual construction and measurement of the MCP elliptical filters showed that they were so docile that there was every chance that the "classical" (non-MCP) versions would be no less buildable. The 4th-order elliptical filter looked like a promising candidate, as the 3rd-order elliptical does not have a classic/MCP option, so it is probably as good as it gets already. The parameter A0 was kept at −35 dB to allow comparison with the MCP versions; it seems unlikely that a different value would be better. A0 can only be changed by redesigning the whole filter from scratch.

Figure 12.41 recapitulates the response of the 4th-order MCP highpass elliptical filter with A0 = −35 dB. Recalculating the 4th-order elliptical filter to use the "classical" rather than the MCP strategy gives Figure 12.42, which shows an obviously faster roll-off. The −3 dB frequency is lowered from 26.2 Hz to 21.0 Hz, the new figure being very close to that of the 3rd-order Butterworth. The −0.1 dB frequency is significantly lowered from 80.2 Hz to a very useful 30.5 Hz. The notch bandwidth at −35 dB changes from 8.0–10.8 Hz to 8.5–10.9 Hz, so it is slightly narrower. Below the notch the response is the same for both versions. The attenuations at some important frequencies are given in Table 12.25.

It is notable that the resistor values that emerge are very different from those of the 4th-order MCP filter.

Note the unhelpfully high −0.1dB frequency of 80 Hz for the 4th-order MCP elliptical filter. If it is redesigned in the "classic" alignment, i.e. not using MCP, then this is reduced to 30.5 Hz.

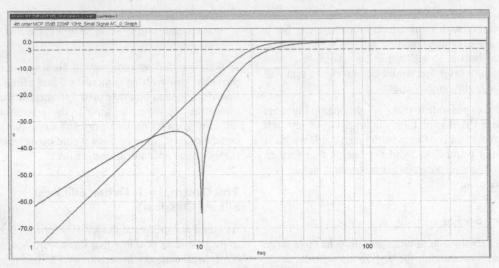

Figure 12.41 Frequency response of 4th-order A0 = −35 dB MCP highpass elliptical filter that is−3 dB at 26 Hz, compared with 3rd-order Butterworth that is −3 dB at 20 Hz

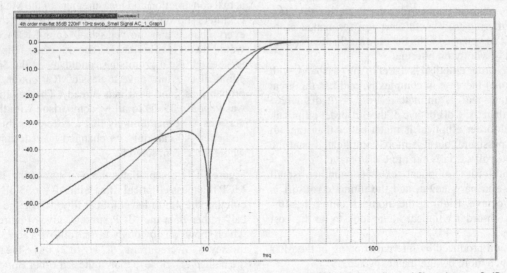

Figure 12.42 Frequency response of 4th-order A0 = −35 dB classic (non-MCP) highpass elliptical filter that is −3 dB at 21 Hz, almost the same as the 3rd-order Butterworth that is −3 dB at 20 Hz

The differences between the classic and MCP 4th-order filter are shown in close-up in Figure 12.43, where it is obvious that the classic version gives a steeper roll-off where it counts and so is much flatter in the pass-band, making it realistic to claim an RIAA accuracy of ±0.1 dB down to about 30 Hz.

Figure 12.44 shows the schematic of the 4th-order classic elliptical filter, arranged in the usual way with the 2nd-order HP poles at the end and with exact component values. The resistor combinations to give the desired values are shown in Table 12.26.

Optimal 4th-Order Elliptical Filter: Internal Levels

When we looked at the MCP elliptical filters, three out of four of them showed significant internal gain, especially at the Bainter filter integrators, limiting the headroom.

Table 12.25 Summary of elliptical filter attenuations + 3rd and 4th-order 20 Hz Butterworth
The −70 dB entries are nominal notch depth.

Freq	1 Hz	5 Hz	10 Hz	20 Hz	30 Hz	−3dB freq	−0.1dB freq
Elliptical 3rd-order	−47 dB	−35 dB	−70 dB	−12 dB	−2.8 dB	29.5 Hz	53.9 Hz
Elliptical 4th-order classic	−61 dB	−35.7 dB	−70 dB	−4.0 dB	−0.14 dB	21.0 Hz	30.5 Hz
Elliptical 4th-order MCP	−62 dB	−39 dB	−70 dB	−6.8 dB	−1.9 dB	26.2 Hz	80.2 Hz
Elliptical 5th-order MCP	−47 dB	−45 dB	−70 dB	−0.5 dB	0.0 dB	16.9 Hz	22.4 Hz
Elliptical 6th-order MCP	−57 dB	−35 dB	−70 dB	−0.2 dB	0.0 dB	14.4 Hz	22.1 Hz
3rd-order Butterworth	−78 dB	−36 dB	−18 dB	−3.0 dB	−0.37 dB	20 Hz	37 Hz
4th-order Butterworth	−104 dB	−48 dB	−24 dB	−3.0 dB	−0.16 dB	20 Hz	32 Hz

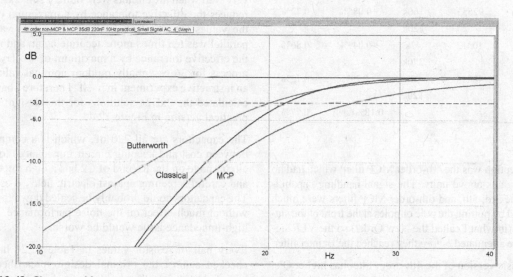

Figure 12.43 Close-up of frequency response of 4th-order A0 = −35 dB classical and MCP highpass elliptical filters. 4th-order classical is −3 dB at 21 Hz, 3rd-order Butterworth −3 dB at 20 Hz.

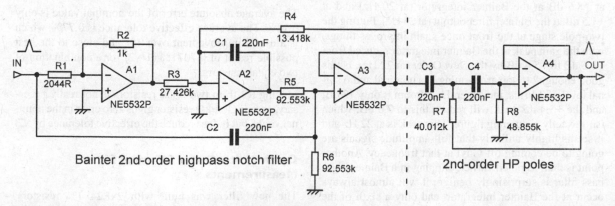

Figure 12.44 Schematic of 4th-order A0 = −35 dB classical highpass elliptical filter

Table 12.26 Resistor values for 4th-order −35 dB classical (non-MCP) elliptical filter
Error is % deviation from nominal 5 sig fig values.
Effective tolerance is for 1% resistors.

	Nominal Ω	2xE24	Error	Effective tolerance
R1a	2044.0	4k3	−0.055%	0.71%
R1b		3k9		
R2a	1k	2k	0.00%	0.71%
R2b		2k		
R3a	27426	43k	−0.35%	0.73%
R3b		75k		
R4a	13418	16k	−0.22%	0.85%
R4b		82k		
R5a=R6a	92553	160k	+0.085%	0.72%
R5b=R6b		220k		
R7a	40012	47k	+0.049%	0.86%
R7b		270k		
R8a	48855	82k	−0.29%	0.72%
R8b		120k		
Average			0.18% (abs)	0.77%

The exception was the 4th-order MCP filter, which had no internal gains above unity. The signal-handling capabilities of the 3rd, 5th, and 6th-order MCP filters were much improved by putting the pole or poles at the front of the signal path (in what I called the New Order) so the VLF signals were attenuated *before* they reached the Bainter filter.

The classic version of the 4th-order filter has higher Q's, and so, as expected, there are now points in the filter where the gain is greater than unity and so the headroom is eroded. With the two HP poles at the end, gain peaks at +8.6 dB at the Bainter integrator (at 20 Hz) and at +1.5 dB at the Bainter filter output (at 25 Hz). Putting the two-pole stage at the front once again improves things, with the gain peak at the Bainter integrator reduced from +8.6 dB to +5.4 dB by the New Order (at 22 Hz). Is this good enough? If you are running at a nominal audio signal level of 2 Vrms, the general headroom is only 13 dB, and the +5.4 dB peak will reduce this to 7.6 dB, which isn't exactly a stunning figure. This peak is at 22 Hz and it seems highly unlikely that full-amplitude signals are going to be coming off vinyl at that frequency. Another point is that, as noted earlier, clipping in a Bainter high-pass filter is surprisingly benign; it will almost always occur at the Bainter integrator, and only a sixth of the resulting distortion gets through to the Bainter output.

Detecting such clipping by simply looking at the output waveform is very difficult—a THD analyser is required.

Optimal 4th-Order Elliptical Filter: Practical Design

Three options for obtaining the nonstandard resistor values required; the nearest E96 value (1xE96), or two E24 resistors in parallel (2xE24), or three E24 resistors in parallel (3xE24). Putting resistors in series would work equally well but is less convenient for PCB layout. When all the 36 resistor value combinations were determined, I found that on average using two E24 values in parallel was more than three times more accurate in setting the nominal value than a single E96 part and worked very well when the circuits were built. Using 2xE24 also reduces the effective tolerance by a maximum of √2 if the values are near equal. Using three E24 values in parallel was ten times more accurate again and reduces the effective tolerance by a maximum of √3. Trying the process for 36 essentially random nominal values was an instructive experiment in itself. Therefore I have only calculated the 2xE24 solutions for this design; see the practical version in Figure 12.45.

The capacitors are all 220 nF, which is a compromise between cost and keeping circuit impedances low. The circuit node on the R6 end of C2 has a high impedance, and careful screening against electric fields is essential. The capacitors could probably be scaled down to 100 nF without much effect on the noise performance, but the high-impedance issue would be worse.

Every pair of resistor values in Table 12.26 has been checked against the original design spreadsheets and been fed back into the simulator to check they are correct. The values were checked again by building the circuit and measuring it.

The average absolute error of the nominal value is only 0.18%. The average effective tolerance is 0.77%, which is a useful improvement over 1% and close to the best possible result of 0.7071% (1/√2) for a combination of two equal resistors.

Making R2 from two 2 kΩ resistors is not really necessary, as a single 1 kΩ resistor gives zero error in the nominal value, but it does reduce the effective tolerance by √2.

Optimal 4th-Order Elliptical Filter: Measurements

The new filter was built with 2xE24 1% resistors, 5% polyprop caps, and Fairchild 5532s. Most audio

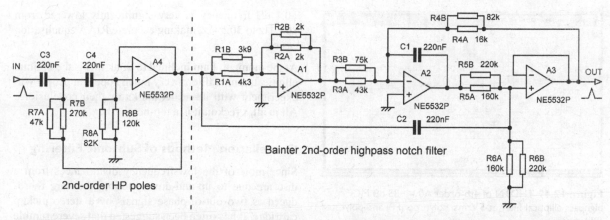

Figure 12.45 Schematic of practical 4th-order A0 = −35 dB classical highpass elliptical filter, using the New Order

measuring systems do not go below 10 Hz; a good way to check the response below this is to change all the 220 nF capacitors for 22 nF parts, raising all the frequencies by a factor of ten. As shown in Figure 12.46, this works well and corresponds closely with the simulated response in Figure 12.39; any deviations are due to the 5% caps. No less than 222 frequency steps were used to give a good representation of the notch depth.

THD+N is shown in Figure 12.47. The rise below 50 Hz is wholly due to the relative increase in noise as the signal is attenuated. The second-harmonic distortion visible above 2 kHz is due to the relatively high common-mode voltage on A4 and A3 inputs; this generates distortion in 5532s. A1 and A2 are shunt-feedback stages and have no common-mode voltage.

Figure 12.48 shows the significant improvement gained by replacing A3, A4 only with LM4562, which produces much less common-mode distortion. No other opamp types were tested.

After the results from the MCP elliptical filters, I expected that polyester capacitors would be very nearly as good as low-distortion polypropylene types. I was quite wrong,

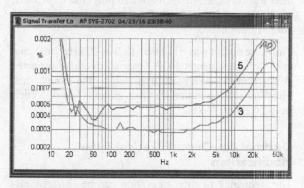

Figure 12.47 THD+N of 4th-order A0 = −35 dB MCP highpass elliptical filter at 3 and 5 Vrms. Polyprop caps.

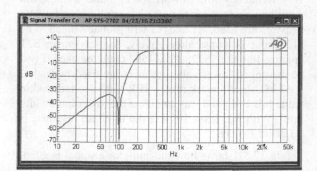

Figure 12.46 Frequency response of 4th-order A0 = −35 dB classical highpass elliptical filter with 22 nF caps

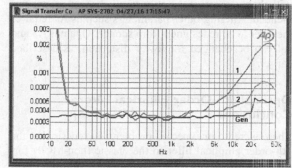

Figure 12.48 THD+N of 4th-order A0 = −35 dB MCP highpass elliptical filter at 5 Vrms. Trace 1 = all 5532, Trace 2 = A3, A4 are LM4562. Gen = testgear output. Polyprop caps.

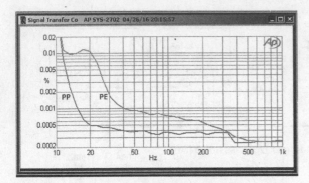

Figure 12.49 THD+N of 4th-order A0 = −35 dB MCP highpass elliptical filter at 5 Vrms. Polyprop (PP) and polyester (PE) caps.

and it just shows that you should always check things by measurement, no matter how plausible they are. The change from MCP to classical seems to have made the filter a good deal more sensitive to polyester capacitors, as shown in Figure 12.49. The extra distortion is fairly pure 3rd-harmonic.

The noise output was −113 dBu (22Hz–22kHz, rms sensing) with all opamps 5532. Internal overload occurred at 5.9 Vrms at 20 Hz, as predicted by the simulations. The filter easily handled 10 Vrms in/out at low distortion from 50 Hz to 50 kHz.

Optimal 4th-Order Elliptical Filter: Modifications

The frequency of filter operation can be changed easily by scaling resistors R3, R4, R5, R6 in the Bainter stages and R7, R8 in the pole stages by the same ratio. However, the −3 dB frequency and the notch frequency will change together, so only small changes are recommended. Altering other parameters such as A_0 requires the filter to be redesigned from scratch. Opamps A3 and A4 can be changed for lower distortion, as described earlier.

Optimal 4th-Order Elliptical Filter: Conclusions

There were no unexpected problems in redesigning the 4th-order elliptical filter to the classic (non-MCP) format. The result has a flatter passband, with the −3 dB frequency lowered from 26.2 Hz to 21.0 Hz. The latter figure is now very close to the −3 dB at 20 Hz of the 3rd-order Butterworth used for comparison. The

−0.1 dB frequency is very significantly lowered from 80.2 Hz to 30.5 Hz, making accurate RIAA equalisation easier.

At the moment, I am inclined to think this is the optimal elliptical rumble filter. I very much hope that people will experiment with it and let me know their conclusions. All in all, I reckon it is a top-notch filter.

Cancellation Methods of Subsonic Filtering

Since most of the low-frequency disturbances from a disc are due to up-and-down motion, they are reproduced as two out-of-phase signals by a stereo pickup cartridge. It has often been suggested that severe rumble overlapping the audio band can be best dealt with by reducing the stereo signal to mono at low frequencies, cancelling the disturbances but leaving the bass, which is usually panned towards the middle, relatively unaffected. This is usually done by cross-feeding the outputs of two lowpass filters between the channels.

Subsonic Filtering by Passive VLF Crossfeed

A simple circuit that gives crossfeed increasing at 6 dB/octave can be implemented with a series resistor in each channel joined by a capacitor.

If however you are happy with a roll-off of the anti-phase signal that does not exceed 6 dB/octave, then the delightfully simple circuit described by Renardsen[22] is recommended. See Figure 12.50; as in all the circuits in this book, I assume that the driving impedance is effectively zero, as from an opamp output. It needs to be connected to a relatively load high impedance to function properly, so voltage followers are shown, underlining that a way to bias them is required, and this is done by R2 and R3. These resistors cause a very frequency low roll-off for in-phase signals, but with the values given the effect is negligible. If 1 MΩ resistors are used, A1 and A2 will probably need to be FET-input opamps; alternatively, for low-noise BJT opamps like the 5532 or LM4562 could be used if two bootstrapped lower-value resistors are used instead; 2 × 47 kΩ should work nicely. If R2, R3 are reduced in value this is a good way to define the lower bandwidth limit. With R2, R3 = 100 kΩ the in-phase response is −3 dB at 1.6 Hz.

Figure 12.51 shows how the anti-phase signal is attenuated at 6dB/octave. The Y-axis calibration of Figure 12.51 is used for all plots in this section to make comparisons easier.

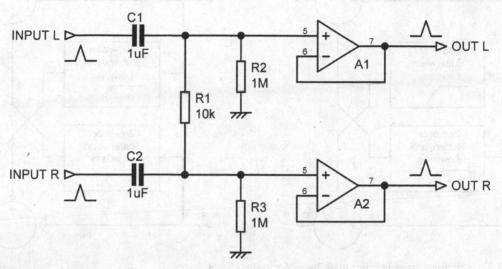

Figure 12.50 Passive crossfeed circuit

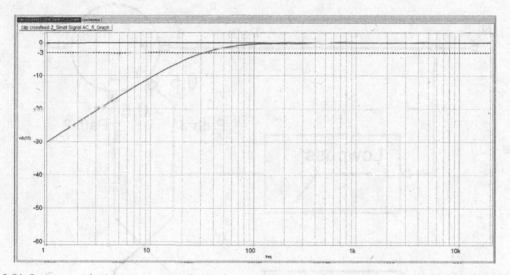

Figure 12.51 Passive crossfeed circuit response for in-phase and anti-phase signals. Anti-phase is −3 dB at 32 Hz and −10.4 dB at 10 Hz.

Subsonic Filtering by Active Filter VLF Crossfeed

You may be feeling that 6 dB/octave hardly counts as a "filter", and I am inclined to agree. To obtain a faster anti-phase roll-off, the apparently obvious solution is to use a lowpass active filter, of 2nd, 3rd or higher order, to control the crossfeed signal in each direction, as in Figure 12.52a. The direct path is Path 1 and the crossfeed path is Path 2. Note the gain is unity for each into the summer for each path.

The arrangement of Figure 12.52a is not usable. If the lowpass filters have a cutoff frequency of 34 Hz (explained later), their phase-shift causes in-phase signals to be rolled-off at 6dB/octave below 40 Hz, with a +2 dB peak centred on 40 Hz, while anti-phase signals are not only not reduced but are actually amplified by +6 dB below 40 Hz. This situation can be partly corrected by inverting the phase of the filter outputs, as in Figure 12.52b; the minus signs on the summers are changed to pluses. The 6 dB/octave roll-off is now on the

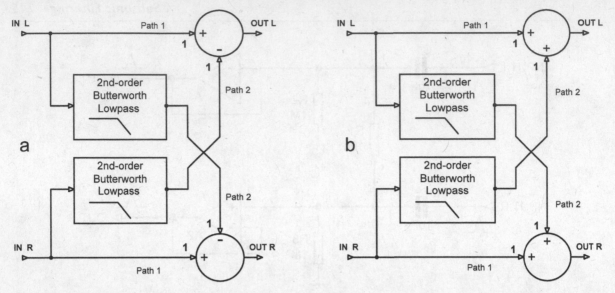

Figure 12.52 Notional crossfeed circuits using 2nd-order lowpass filters

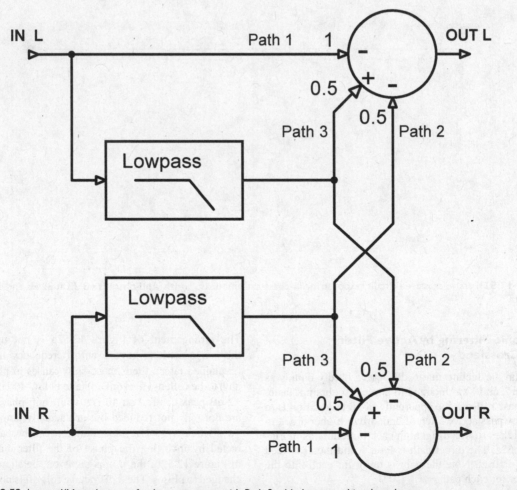

Figure 12.53 Langvad/Macaulay crossfeed arrangement, with Path 3 added to cancel in-phase boost

anti-phase signal, and the +6 dB boost is on the in-phase signal.

The latter is clearly still unacceptable, but it can be remedied by adding Path 3 from each filter output summing back into the same channel, as in Figure 12.53; note the gain changes for the summer inputs. The basic idea was put forward by Macaulay in 1979,[23] but Langvad in 1980[24] presented a simpler arrangement as Figure 12.53. I have not so far found earlier implementations.

The unwanted +6 dB boost is in theory eliminated, though the flatness of the resulting response depends on signal cancellation and so is critically affected by the accuracy of the gains involved.

Figure 12.54 shows my version of the Langvad arrangement; the 2nd-order lowpass Butterworth filters have been redesigned to use 220 nF capacitors; with 15 kΩ resistors the cutoff frequency is 34 Hz; this is not necessarily optimal, but it seems plausible as a starting point. The overall gain is set to unity for easy integration into signal paths.

Its simulated anti-phase response is shown in Figure 12.55. While the arrangement of Figure 12.54 is now usable, it still has an anti-phase slope of only 6 dB/octave, despite all the extra hardware, and still has an unwanted +2 dB peak just before roll-off. It is still inferior to a simple passive crossfeed circuit.

The Devinyliser: Phase Correction of the Lowpass Filters

Figure 12.55 was immediately identified as looking very similar to the results obtained from attempts to make subtractive crossovers for loudspeakers, where filters of any order give only a 6 dB/octave slope for the output derived by subtraction. This was addressed by Lipshitz and Vanderkooy,[25] who showed that adding a suitable group delay in the unfiltered path, as in Figure 12.56 (only left channel shown), compensates for the group delay in the lowpass filter; the same slope as that of the filter is obtained, and the +2 dB peak disappears. The delay required for this subsonic filter application is small and can be obtained with simple 1st-order allpass filters.[26]

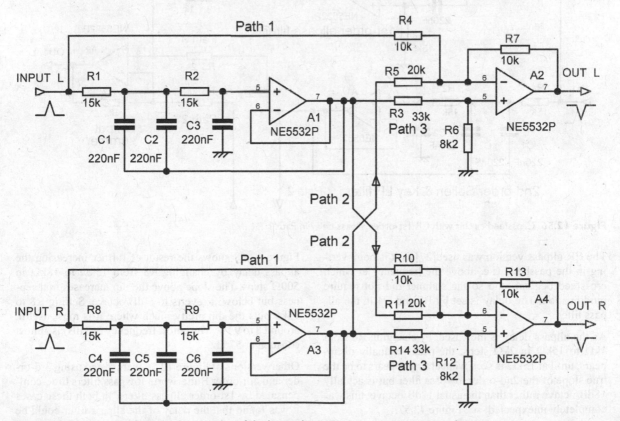

Figure 12.54 Low-impedance implementation of the Langvad crossfeed arrangement: not satisfactory

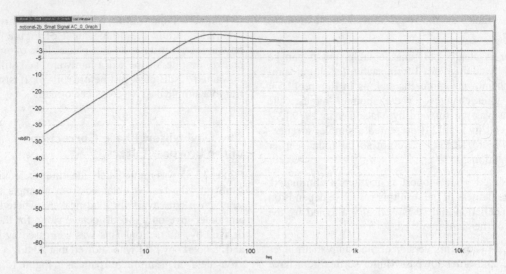

Figure 12.55 Simulated response for anti-phase signals of the arrangement in Figure 12.54

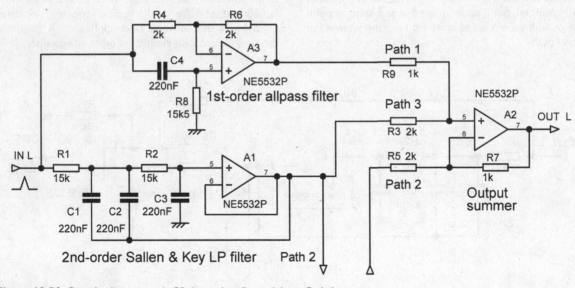

Figure 12.56 Crossfeed circuit with CR 1st-order allpass delay in Path 1

The CR allpass version was used[27] as it is noninverting in the passband (i.e. above the frequency at which crossfeed begins). and so the summer did not require modification. The delay is set by R8 and C4 in the allpass filter.

As the allpass delay is increased, by changing R8 from 1kΩ to 15kΩ in 2kΩ steps, the peak gradually disappears, until at 15kΩ is seen what at first appears to be the true slope of the 2nd-order lowpass filter but is actually 18dB/octave rather than the usual 12dB/octave; this was completely unexpected; see Figure 12.57.

Figure 12.58 shows the result of further increasing the allpass delay by changing R8 from 15kΩ to 18kΩ in 500Ω steps. The slope above the dip increases in steepness but below it reverts to 6 dB/octave. Setting R8 to 15k5 puts the dip pretty much where it is most useful, around 8 to 9 Hz. The cutoff frequency (−3 dB) is almost constant at 57 Hz.

Other versions of Figure 12.56 were tested using 3rd-order and 4th-order Butterworth lowpass filters time-compensated by 1st-order allpass filters. In both these cases it was found that the delay of the allpass filter could be

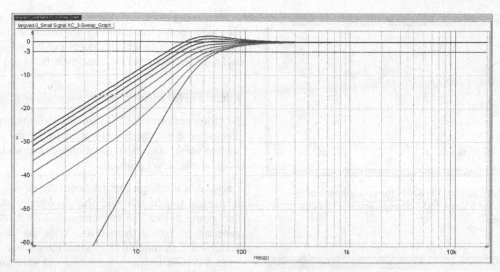

Figure 12.57 Simulated anti-phase freq response; R8 set from 1kΩ to 15kΩ in 2kΩ steps

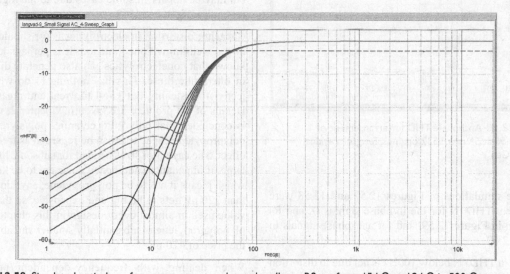

Figure 12.58 Simulated anti-phase frequency response: 1st-order allpass. R8 set from 15 kΩ to 18 kΩ in 500 Ω steps.

adjusted to give a linear-in-dB slope, but in both its slope was 18 dB/octave. This unexpected behaviour of the filtering slopes is under investigation. In both cases a slight increase in delay from this value gave a deep notch that could be placed between 8 and 9 Hz. Other anti-phase cutoff frequencies can be obtained by scaling R1, R2, and R8 in Figure 12.56.

I call this concept for the effective removal of anti-phase subsonic disturbances the Devinyliser. It was first published in Jan Didden's *Linear Audio* in April 2016;[28]

further developments were described in a paper at the Paris AES Convention in June 2016,[29] and a more detailed description at the London AES meeting in October 2016.

The Devinyliser: Measured Performance

The arrangement of Figure 12.56 was built with 0.1% resistors, 1% capacitors, and 5532 opamps and measured with an Audio Precision SYS-2702. The frequency

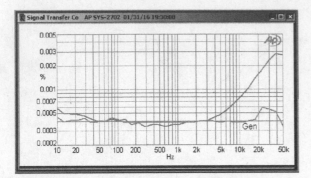

Figure 12.59 In-phase THD+N for arrangement of Figure 12.56 at 5 Vrms. 5532 opamps. Testgear residual marked "Gen".

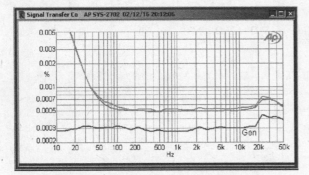

Figure 12.60 Anti-phase THD for arrangement of Figure 12.56 at 5 Vrms. 5532 opamps. Testgear residual marked "Gen".

response simulations of Figures 12.57 and 12.58 were confirmed. THD+N for the in-phase signals (L and R) is shown in Figure 12.59, and for anti-phase signals in Figure 12.60.

The rise in THD+N below 100Hz is not due to distortion but to relatively increasing noise as the anti-phase signal falls in amplitude.

These results are reassuring because they show that the circuitry behaves as expected and does not have any opamps with nonobvious heavy loading on their outputs. The noise output with the input terminated with 40Ω was measured at −105 dBu for both channels (22–22kHz, rms sensing).

The Devinyliser: Conclusions

The addition of simple 1st-order allpass filters to a known crossfeed arrangement allows effective highpass filtering of anti-phase signals with a slope of 18 dB/

octave, using only 2nd-order lowpass filters. Comparison with Butterworth (up to 4th-order) and elliptical filters (up to 6th-order) shows greater circuit simplicity and a flat passband for in-phase signals.

Filter Performance

By "filter performance" I mean the performance parameters apart from the finely-honed frequency response, such as noise, distortion, input impedance, and freedom from internal level problems.

Because of the large capacitances, the noise generated by the resistors in subsonic filters is usually well below the opamp noise. The capacitances do not, of course, generate any noise themselves. With the values used here, SPICE simulation shows that the resistors produce −125.0 dBu of noise at the output (22 kHz bandwidth, 25 °C). The use of the LM4562 will reduce voltage-follower CM distortion compared with the 5534/5532 but may be noisier in some cases due to the higher current noise of the LM4562.

Capacitor distortion in electrolytics is (or should be) by now a well-known phenomenon. It is perhaps less well known that nonelectrolytics can also generate distortion in filters like these. This has nothing to do with Subjectivist musicality but is all too real and measurable. Details of the problem are given in Chapter 2, where it is concluded that only NP0 ceramic, polystyrene, and polypropylene capacitors can be regarded as free of this effect. The capacitor sizes needed for subsonic filters are large, if impedances and hence noise are to be kept low, which means it has to be polypropylene; anything larger than 220 nF gets to be big and expensive, so that is the value used in almost every design in this chapter—220 nF polypropylene is substantially smaller and about half the price of 470 nF.

When dealing with frequency-dependent networks like filters you need to keep an eye on the input impedance, because it can drop to unexpectedly low values, putting excessive loading on the stage upstream and degrading its linearity. In a highpass Sallen and Key filter, the input impedance is high at low frequencies but falls with increasing frequency. In the 3rd-order version, it tends to the value of R1 in parallel with R3, which here is 10.6 kΩ. This should not worry the previous stage.

Measuring Subsonic Filters

Audio measuring systems do not usually go below 10 Hz. Function generators can provide test signals as low as 0.01 Hz, but there are difficulties in measuring the filter output; a digital scope is one way, but the accuracy will not be great, and you will have to plot the response yourself. A better scheme is to change all the 220 nF capacitors to 22 nF parts of the same precision. This will shift the filter frequency up a decade without altering any other parameter, and you can use an audio measuring system down to the equivalent of 1 Hz. The lowest frequency of interest is 0.55 Hz for a disc rotating at 33 and 1/3 rpm, and to measure down in these depths the 220 nF capacitors will have to be scaled more radically to 10 nF, which makes the 10 Hz bottom limit equivalent to 0.45 Hz.

I confirm that this works well in practice, the only issue being that some of the circuit nodes will be at a higher impedance than normal and ten times more vulnerable to electric fields; this is particularly true of the Bainter stages in the elliptical filters. Straightforward electrical screening eliminates the problem, but you have to be careful to do it thoroughly to get accurate noise and THD measurements. A grounded screening plate under the circuitry is essential (I have a big grounded plate under the wooden top of my workbench), and the top and sides must likewise be fully screened. Grounded biscuit-tins are very good for this.

References

1. Happ, L., and Karlov, F. "Record Warps and System Playback Performance" presented at the 46th Convention of the Audio Engineering Society, New York, 1973 Sept 10–13, preprint no. 926.

2. Ladegaard, P. "Audible Effects of Mechanical Resonances in Turntables" Bruel and Kjaer Application Note, 1977.

3. Holman, T. "New Factors in Phonograph Preamplifier Design" *Journal of Audio Engineering Society*, Volume 24, May 1975, p. 263.

4. Holman, T. "Phonograph Preamplifier Design Criteria: An Update" *Journal of Audio Engineering Society*, Volume 28, May 1980, p. 325.

5. Taylor, D. L. "Measurement of Spectral Content of Record Warps" *Journal of Audio Engineering Society*, Volume 28, Dec 1980, p. 263.

6. Allmaier, H. "The Ins and Outs of Turntable Dynamics" *Linear Audio*, Volume 10, Sept 2015, pp. 9–24.

7. Self, D. *The Design of Active Crossovers*. Focal Press, 2011, pp. 269–301. ISBN 978-0-240-81738-5.

8. Hickman, I. "Top Notches" (Notch Filters) *Electronics World*, Feb 2000, p. 120.

9. Thomson, W. E. "Delay Networks Having Maximally Flat Frequency Characteristics" *Proceedings of IEEE*, Part 3, Volume 96, Nov 1949, pp. 487–490.

10. Self, D. "An Advanced Preamplifier Design" *Wireless World*, Nov 1976.

11. Self, D. "High Performance Preamplifier" *Wireless World*, Feb 1979.

12. Self, D. "A Precision Preamplifier" *Wireless World*, Oct 1983.

13. Self, D. "Precision Preamplifier 96" *Electronics World*, July/Aug and Sept 1996.

14. Williams, A., and Taylor, F. J. *Electronic Filter Design Handbook*. 3rd edition. McGraw-Hill, 1995, p. 3.17. ISBN 0-07-070441-4.

15. Billam, P. "Harmonic Distortion in a Class of Linear Active Filter Networks" *Journal of the Audio Engineering Society*, Volume 26, No. 6, June 1978, p. 426.

16. Wanhammar, L. *Analog Filters Using MATLAB*. Springer Science & Business Media, 2009. ISBN: 0387927670, 9780387927671 (MCP).

17. Aikens, R. S., and Kerwin, W. J. "Single Amplifier, Minimal RC, Butterworth, Thomson, & Chebyshev Filters to Sixth Order" Proceedings of International Filter Symposium, 1972, pp. 81–82.

18. Daniels, R. *Approximation Methods for Electronic Filter Design*. New York: McGraw-Hill.

19. Bainter, J. "Active filter has stable notch, and response can be regulated" *Electronics*, Oct 2 1975, p. 115.

20. Self, D. *The Design of Active Crossovers*. Focal Press, 2010, pp. 232–236.

21. Zumbahlen, H. *Opamp Applications Handbook*. Chapter 5, p. 371, p. 389 (Boctor filter).

22. Renardson, M. www.renardson-audio.com/phono-1. html Accessed Jan 2016.

23. Macaulay, J. P. "Differential Rumble Filter" Circuit Ideas, *Wireless World*, Sept 1979, p. 75.

24. Langvad, J. "Rumble Cancellation Filter" Letters, *Wireless World*, Mar 1980, p. 61.

25. Lipshitz, Stanley P., and Vanderkooy, John. "A Family of Linear-Phase Crossover Networks of High Slope Derived by Time Delay" *JAES*, Volume 31, Jan/Feb 1983, p. 2.

26. Self, D. *The Design of Active Crossovers*. 2nd edition. (multiple allpass) Chapter 10, pp. 296–298.

27. Self, D. *The Design of Active Crossovers*. 2nd edition. (multiple allpass) Chapter 10, pp. 271–273 (CR and RC allpass filters).

28. Self, D. "The Devinyliser" *Linear Audio*, Volume 11, Apr 2016, pp. 77–103.

29. Self, D. www.douglas-self.com/ampins/ampins. htm/Paris 2016 devinyliser.ppt Accessed Nov 2016.

Chapter 13

Ultrasonic and Scratch Filtering

Ultrasonic Filters

Scratches and groove debris create clicks that have a large high-frequency content, some of it ultrasonic and liable to cause slew rate and intermodulation problems further down the audio chain. The transients from scratches can easily exceed the normal signal level. It is often considered desirable to filter this out as soon as possible (though of course some people are only satisfied with radio-transmitter frequency responses).

If an MM input stage is provided with an HF correction pole, in the form of an RC 1st-order roll-off after the opamp, this in itself provides some protection against ultrasonics because its attenuation continues to increase with frequency and it is inherently linear. The opamp ahead of it naturally does not benefit from this; while it might be desirable to put some ultrasonic filtering in front of the first active stage, it is going to be very hard to do this without degrading the noise performance.

An ultrasonic filter could be a passive LC design, but inductors are not much loved in audio. A more likely choice is a 2nd- or 3rd-order active filter, probably opamp-based, but if Sallen and Key filters are used then a discrete emitter-follower is an option, and this should be free from the bandwidth and slew rate limitations of opamps. If an ultrasonic filter is incorporated it is usually 2nd-order, very likely due to misplaced fears of perceptible phase effects at the top of the audio band. If a Sallen and Key filter with an opamp is used, be aware that the response does not keeping going down forever but comes back up due to the nonzero output impedance of the opamp at high frequencies; the multiple-feedback (MFB) filter configuration is free from this problem.

Butterworth lowpass filters are popular for this work because their maximal flatness in the passband and rapid roll-off causes minimal intrusion into the audio band,

though naturally this depends on the cutoff frequency chosen. The other important filter types are Bessel and linear-phase. As in the subsonic filter case, Bessel ultrasonic filters give a much slower roll-off as the price for keeping the group delay more constant. There seems to be a vague general feeling that phase issues are more audible at the top end of the audio band than the bottom. If you feel this is the case (and all the evidence is that it is not), you may want to use the Bessel filter characteristic. Linear-phase filters provide a compromise between the Butterworth and Bessel characteristics.

Chebyshev filters are not likely to be useful here as they introduce ripples into the passband frequency response. Elliptical filters are more complicated, and while they make excellent subsonic filters, they are not likely to be necessary for ultrasonic filtering because, unlike subsonic filters, there is usually no need for very high attenuation just outside the audio band.

Butterworth Filters from 2nd- to 6th-Order

Figure 13.1 shows the circuits for Butterworth lowpass filters from 2nd-order to 6th-order; their frequency responses are shown in Figure 13.2. In each case the cutoff (−3 dB) frequency is 50 kHz, which I think gives a good compromise between passband flatness, fast roll-off, and filter complexity. For audio band flatness, the worst case is the 2nd-order Butterworth, which when designed for −3 dB at 50 kHz is only 0.11 dB down at 20 kHz. This is not much, but if you are specifying an overall frequency response of ±0.1 dB, then all of that tolerance and more is used up without considering any other part of the system. On moving to the 3rd-order Butterworth, the response droop at 20 kHz is reduced to 0.02 dB, only a fifth of the specified tolerance. The 4th-order Butterworth droop is only 0.004 dB down at

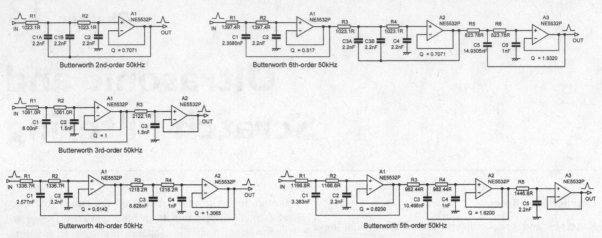

Figure 13.1 Ultrasonic Butterworth filters from 2nd to 6th order. Cutoff 50 kHz.

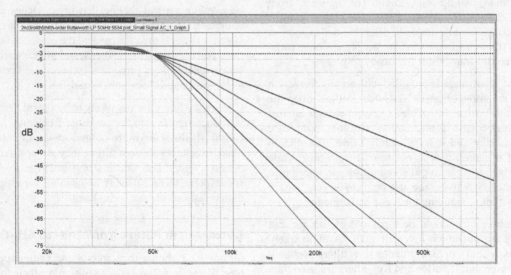

Figure 13.2 The response of Butterworth ultrasonic filters from 2nd to 6th order. Cutoff 50 kHz.

20 kHz, and for the 5th- and 6th-order Butterworths it is negligible. Considering this, it is doubtful if anything more complicated than 4th-order filter will be required, but 5th- and 6th-order filters are here if you need them. Other cutoff frequencies can be obtained by scaling the resistor values.

Figure 13.1 gives the component values for Butterworth filters; Tables 13.1 to 13.5 give the component values for Butterworth, linear-phase 0.5°, and Bessel ultrasonic filters, all with a cutoff frequency of 50 kHz.

A simple test that the 4th-order Bessel filter is working correctly is to check for −25 dB at three times the cutoff frequency.

The filters are made in the conventional way as a chain of second- and (sometimes) 1st-order stages. This process in covered in detail at the start of Chapter 12. The capacitor values have been chosen to keep the series resistors roughly equal to 1 kΩ, to minimise Johnson noise, the effect of current noise, and common-mode distortion without putting excessive loading on a preceding stage.

The opamps are shown as 5532/2. It is not advisable to use TL072 or similar opamps with poor gain-bandwidth product, as this will have two adverse effects:

1) There will be unwanted peaking of the response just before roll-off. In the case of the 6th-order filter, this will be as high as +0.5 dB around 35 kHz. If 5534

Table 13.1 Resistor values for 2nd-order 50 kHz ultrasonic filter in 2xE24 format
Error refers to deviation from nominal value and excludes component tolerances

Type	R1 = R2 Ω	R1a = R2a Ω	R1b = R2b Ω	Error	Effective tolerance	C1 nF	C2 nF
Bessel	1443.8	2400	3600	−0.26%	0.72%	2	1.5
Linear-phase 0.5°	1021.1	1300	4700	−0.27%	0.81%	3.6045	2.2
Butterworth	1023.1	1300	4700	−0.47%	0.81%	4.4000	2.2

The capacitor ratio is 2:1, so using three capacitors with two in parallel for C1 gives the exact result (apart from component tolerances).

Table 13.2 Resistor values for 3rd-order 50 kHz ultrasonic filter in 2xE24 format

Type	R1 = R2 Ω	R1a = R2a	R1b = R2b	Error	Effective tolerance	R3 Ω	R3a	R3b	Error	Effective tolerance	C1 nF	C2 nF	C3 nF
Bessel	1585.8	2400	4700	+0.18%	0.74%	1090.3	2000	2400	+0.06%	0.71%	1.910	1	2.2
Linear-phase 0.5°	1260.6	1300	39k	−0.20%	0.97%	1167.8	1800	3300	−0.26%	0.74%	3.6100	1	3.3
Butterworth	1061.0	1500	3600	−0.20%	0.76%	2122.1	3900	4700	+0.44%	0.71%	6	1.5	1.5

The capacitor ratio in the first stage is 4:1, so C1 can be made up exactly from four C2 capacitors in parallel; only one value is used and purchasing simplified.

Table 13.3 Resistor values for 4th-order 50 kHz ultrasonic filter in 2xE24 format

Type	R1 = R2 Ω	R1a = R2a	R1b = R2b	Error	Effective tolerance	R3 = R4 Ω	R3a = R4a	R3b = R4b	Error	Effective tolerance	C1 nF	C2 nF	C3 nF	C4 nF
Bessel	1432.2	2000	5100	+0.31%	0.77%	1240.9	2200	3300	+0.35%	0.72%	1.6349	1.5	2.5985	1
Linear-phase 0.5°	1310.4	1800	3300	−0.26%	0.74%	1081.5	1800	2700	−0.14%	0.72%	3.2745	2.2	4.8840	0.68
Butterworth	1336.7	2400	3000	−0.20%	0.71%	1218.2	2200	2700	−0.49%	0.71%	2.5770	2.2	6.828	1

Neither of the two capacitor ratios are integers for any of the three filter types, so C1 and C3 will have to be made up with parallel capacitors. How this is best done is determined very much by what capacitor values are available. An E6 capacitor series will require much more paralleling than an E12 series. The same considerations apply to the 5th-order and 6th-order filters.

Table 13.4 Resistor values for 5th-order 50 kHz ultrasonic filter in 2xE24 format

Type	R1 = R2 Ω	R1a = R2a	R1b = R2b	Error	Effective tolerance	R3 = R4 Ω	R3a = R4a	R3b = R4b	Error	Effective tolerance
Bessel	1206.2	1500	6200	+0.13%	0.83%	986.28	1800	2200	+0.21%	0.71%
Linear-phase 0.5°	1064.5	1200	9100	−0.40%	0.89%	1010.5	1100	12k	−0.28%	0.92%
Butterworth	1166.8	1800	3300	−0.18%	0.74%	982.44	1100	9100	−0.11%	0.90%

Type	R5 Ω	R5a	R5b	Error	Effective tolerance	C1 nF	C2 nF	C3 nF	C4 nF	C5 nF
Bessel	1408.2	2200	3900	−0.12%	0.73%	1.9052	1.5	3.3599	1	1.5
Linear-phase 0.5°	1366.3	2200	3600	−0.06%	0.73%	4.4376	1.5	5.8235	0.47	3.3
Butterworth	1446.8	2400	3600	−0.47%	0.72%	3.3830	2.2	10.498	1	2.2

Table 13.5 Resistor values for 6th-order 50 kHz ultrasonic filter in 2xE24 format

Type	R1 = R2 Ω	R1a = R2a	R1b = R2b	Error	Effective tolerance	R3 = R4 Ω	R3a = R4a	R3b = R4b	Error	Effective tolerance
Bessel	1297.4	1800	4700	+0.32%	0.77%	1026.4	1800	2400	+0.21%	0.71%
Linear-phase 0.5°	1024.0	1800	2400	+0.45%	0.71%	955.45	1600	2400	+0.48%	0.72%
Butterworth	1397.4	2000	4700	+0.40%	0.76%	1023.1	1300	4700	−0.47%	0.81%

Type	R5 = R6 Ω	R5a = R6a	R5b = R6b	Error	Effective tolerance	C1 nF	C2 nF	C3 nF	C4 nF	C5 nF	C6 nF
Bessel	815.46	1500	1800	+0.33%	0.71%	1.5624	1.5	2.2414	1.5	4.1894	1
Linear-phase 0.5°	714.58	1300	1600	+0.37%	0.71%	4.752	3.3	5.382	1	9.099	0.47
Butterworth	823.78	1100	3300	+0.15%	0.79%	2.3585	2.2	4.4	2.2	14.930	1

The 6th-order Butterworth filter has a convenient 2:1 ratio between C3 and C4.

opamp models are used, this peaking is entirely absent and the proper Butterworth response is obtained.

2) The response will cease falling and start to come up again, above approx 300 kHz. In the 2nd-order filter it has risen back up to −30 dB at 1 mHz. If 5534 opamp models are used, the response never comes back up again.

There are no equivalent problems with subsonic filters.

Figure 13.3 compares the responses of 2nd-order Butterworth, linear-phase 0.5°, and Bessel ultrasonic filters, showing how the linear-phase characteristic chosen (0.5°) offers a good compromise between the Butterworth and Bessel characteristics. The opamp models

used were 5534A; use of TL072 models gives responses that only fall to about −30 dB before coming back up again.

The component values for the three filter characteristics are shown in Table 13.6. Different cutoff frequencies can be obtained by scaling the component values, keeping the two resistors the same value and the ratio between the capacitors the same.

Combining Subsonic and Ultrasonic Filters in One Stage

An obstacle to the inclusion of an ultrasonic filter is the extra cost and power consumption of another filter stage.

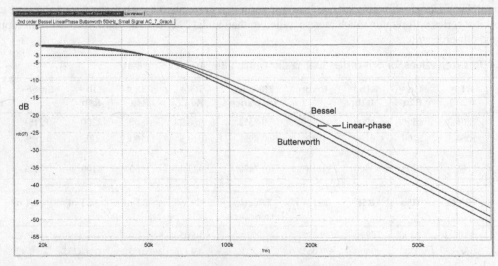

Figure 13.3 Response of 2nd-order Butterworth, linear-phase 0.5°, and Bessel ultrasonic filters. Cutoff 50 kHz.

Table 13.6 Component values for 2nd-order filters with 50 kHz cutoff and resistors in 2xE24 format. Error refers to deviation from nominal value and excludes component tolerances.

Type	R1 = R2 Ω	R1a = R2a Ω	R1b = R2b Ω	Error %	C1 nF	C2 nF
Bessel	1448.3	1800	7500	+0.23	1.997	1.5
Linear-phase 0.5°	1021.1	1300	4700	−0.27	3.604	2.2
Butterworth	1023.1	1200	6800	−0.30	4.400	2.2

This difficulty can be resolved by combining it with a subsonic filter in the same stage. Combined filters also have the advantage that the signal now passes through one opamp rather than two and can be extremely useful if you only have one opamp section left. The combination of a subsonic filter and an ultrasonic filter is sometimes called a bandwidth definition filter.

This cunning plan is workable only because the highpass and lowpass turnover frequencies are widely different. Figure 13.4 shows the 3rd-order Butterworth subsonic filter combined with a 2nd-order 50 kHz Butterworth lowpass filter; the response of the combination is exactly the same as expected for each separately. The lowpass filter is cautiously designed to prevent significant loss in the audio band and has a −3 dB point at 50 kHz, giving very close to 0.0 dB at 20 kHz. The response is −12.6 dB down at 100 kHz and −24.9 dB at 200 kHz. C4 is made up of two 2n2 capacitors in parallel.

Note that the passband gain of the combined filter is −0.15 dB rather than exactly unity. The loss occurs because the series combination of C1, C2, and C3, together with C5, forms a capacitive potential divider

with this attenuation, and this is one reason why the turnover frequencies need to be widely separated for filter combining to work. If they were closer together then C1, C2, C3 would be smaller, C5 would be bigger, and the capacitive divider loss would be greater. That is why this filter is one of the few in this book that uses 470 nF capacitors rather than 220 nF.

While this is an ingenious circuit, if I do say so myself, it occurred to me that it could be improved if designed as two combined 3rd-order filters, which could be implemented by just one amplifier so long as it can be assumed that the loading on the output is sufficiently light for R6–C6 to not be significantly affected. There is some flexibility here because an ultrasonic filter does not need to be so accurate as, say, an RIAA network, because almost everything it does is above the range of audibility. The 2nd-order part of the lowpass filter is set by C4 and C5 to a Q of 1.00, as in conventional two-stage 3rd-order filters. This as usual causes a gain peak of +0.87 dB at 35 kHz; it seems unlikely this is going to cause any headroom problems. Note that a 5534 model was used as the amplifier; a TL072 model gave the usual oh-no-it's-coming-back-up-again behaviour above 100 kHz.

The circuit is shown in Figure 13.5. In this case I used 220 nF capacitors, and as predicted the passband attenuation was a bit greater at −0.28 dB. I think this is still small enough to be ignored and definitely saves significant money on capacitors. The frequency response with its two 18 dB/octave slopes is shown in Figure 13.6.

I was going to leave it there, but the temptation to explore a topic just a little further is irresistible. To me, anyway. The result of a bit more night thought was a 4th-order highpass combined with a 4th-order lowpass in just two stages, with each stage implementing both a 2nd-order

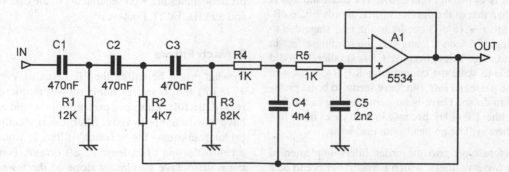

Figure 13.4 A 3rd-order Butterworth 20 Hz subsonic filter combined with a 2nd-order Butterworth 50 kHz ultrasonic filter

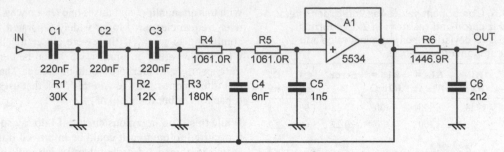

Figure 13.5 A 3rd-order Butterworth 20 Hz subsonic filter combined with a 3rd-order Butterworth 50 kHz ultrasonic filter

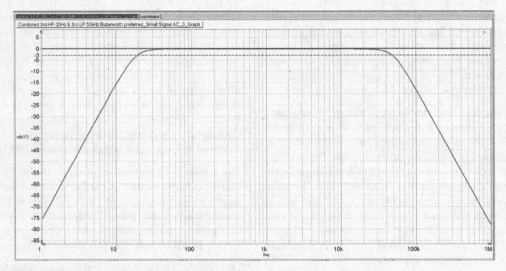

Figure 13.6 Frequency response of the combined 3rd-order/3rd-order filter

highpass and a 2nd-order lowpass, with different Q's in the two stages, as required for Butterworth or other filter types. All the combined filters in this section are Butterworths.

The result is shown in Figure 13.7. The passband loss is smaller than that of the previous filter, at only −0.26 dB. Most of this (−0.18 dB) occurs in the first stage due to the loading of C4 on C1 and C2; the loading effect in the second stage is less because C8 is smaller. It would be possible to scale the components R3, R4, C3, C4 to reduce the passband loss, but there seems to be no pressing need to do so. There is no gain peaking in the first stage at either LF or HF because in both cases it has low Q's, so there will be no headroom problems.

We therefore have two 4th-order filters implemented with just two amplifiers, which I fondly believe to be a new idea. The frequency response with its two 24 dB/octave slopes is shown in Figure 13.8.

It might be worth repeating at this point that combined filters only work because the highpass and lowpass cut-off frequencies are well separated—in the case of 20 Hz and 50 kHz, by 11.3 octaves.

Scratch Filters

In what might be called the First Age of Vinyl (say, 1948 to 1983, if we restrict ourselves to microgroove records), a fully equipped preamplifier would certainly have a switchable lowpass filter, usually called, with brutal frankness, the "scratch" filter. It would have a roll-off slope of at least 12 dB/octave, faster than the 6 dB/octave maximum slope of the tone-control

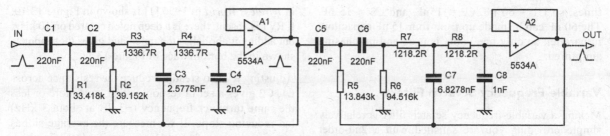

Figure 13.7 A 4th-order Butterworth 20 Hz subsonic filter combined with a 4th-order Butterworth 50 kHz ultrasonic filter.

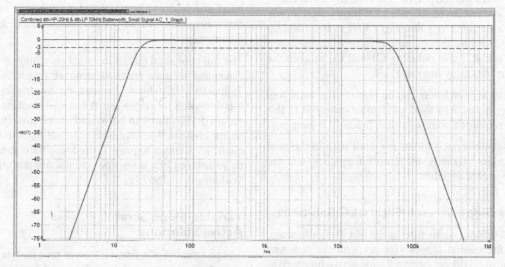

Figure 13.8 Frequency response of the combined 4th-order/4th-order filter

stage and beginning at a rather higher frequency. It was aimed at suppressing, or at any rate dulling and hopefully rendering acceptable, not just record surface noise and the inevitable ticks and clicks but also HF distortion. This function is quite separate from that of an ultrasonic filter, and the turnover frequency is very much in the audio band, usually in the range 3–10 kHz.

The more highly specified preamps would have two, three, or more alternative filter frequencies, and the really posh models had variable filter slope as well. The fact that the need was felt for some really quite sophisticated filtering to smooth the listening experience does rather emphasise the inherent vulnerability of a mechanical groove for delivering music. Now we seem to be in the Second Age of Vinyl, a reassessment of this once-abandoned bit of technology would seem to be timely.

Historically, scratch filters were often passive LC configurations, as it was much cheaper to design in a wound component than to add an extra valve to make an active filter. A modern scratch filter will almost certainly be an active filter based on one of the designs for ultrasonic filters given earlier, the only difference being that the cutoff frequency will be much lower, in the range 3 to 10 kHz, depending on just how badly the vinyl is scratched.

Fixed-frequency scratch filters from 2nd- to 6th-order can be quickly designed by modifying the ultrasonic filter designs given at the start of this chapter. The simplest way is to scale the capacitor values, keeping them in the same ratio. As an example, Figure 13.1 shows a 3rd-order Butterworth lowpass with a cutoff frequency of 50 kHz, with C1 = 6 nF, C2 = 1.5 nF, and C3 = 1.5 nF. To turn this into a 5 kHz cutoff scratch filter, the capacitors are increased by a factor of 10 times, so C1 = 60 nF, C2 = 15nF, and C3 = 15 nF. The 60 nF can be made up from four 15 nF capacitors in parallel, so only one value is used and purchasing simplified.

Variable-Frequency Scratch Filters

Making a variable-frequency scratch filter is relatively simple providing you are satisfied with a 2nd-order response. A lowpass 2nd-order Sallen and Key filter has two equal resistors, and these can be easily changed together with a ganged pot, though naturally this requires a 4-gang pot for stereo. Figure 13.9 shows such a filter with the wide frequency range of 1 kHz to 10 kHz. This can be reduced by increasing the value of the end-stop resistors R1, R2, and reducing C1 and C2 to keep the maximum frequency the same.

Third-order scratch filters can be made in the same way by adding a variable third pole after opamp A1; this however requires a somewhat less practical 6-gang pot.

Variable-slope scratch filters are dealt with in the next three sections. Usually the cutoff frequency and the slope are altered together.

Variable-Slope Scratch Filters: LC Solutions

Making a variable-slope filter is not that straightforward, because the natural slopes you get with resistors and capacitors are 6 dB per octave. Using active filters gives access to final slopes with 12, 18, 24, or more dB per octave, but intermediate slopes are usually only found in the transitions between flat and the ultimate roll-off slope.

A popular way of achieving variable slope back in the valve era involved an LC filter bypassed by a variable resistance. A classic example of this approach, published

in *Wireless World* in 1956,[1] is shown in Figure 13.10. If RV1 is absent, there is a deep notch centred on 10 kHz, but adding it abolishes the notch and gives the response shown in the bottom trace of Figure 13.11.

Adjusting RV1 to steadily reduce the resistance across L1, C2 gives a set of responses that have more or less the same turnover frequency (−3 dB at about 3 kHz) but reducing slope. If we measure the average slopes across the octave 5–10 kHz, we get Table 13.7, which shows a handy variation in slope from 4.5 to almost 18 dB/octave. As revealed in Figure 13.11, the filter has 6 dB loss in the passband, which is a long way from ideal. Noise or headroom will be seriously compromised.

The full published circuit included switching of all three capacitors to give nominal turnover frequencies of 5, 7, and 10 kHz, which of course means a rather clumsy 6-pole switch for stereo use. It had a 1 H inductor, which required 1100 turns to be wound on a Ferroxcube core, so its construction required some degree of commitment. RV1 was a log-law component. This form of filter resurfaced in a transistor preamplifier design by Reg Williamson in 1967.[2]

Variable-Slope Scratch Filters: Active Solutions

The prospect of winding 1100 turns on a former to make an inductor is not at all appealing, and would turn anybody's mind to RC active filters. This is quite apart from the well-known inductor drawbacks of weight, cost, nonlinearity, and susceptibility to hum fields. The well-known Leak Varislope preamplifiers of the 1950s used purely RC filtering, boasting specifically in their advertising material that there were no chokes to pick up hum. The earliest (mono) version had three switched filter

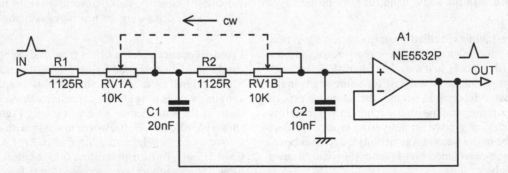

Figure 13.9 Variable-frequency 2nd-order Sallen and Key scratch filter, cutoff variable from 1 kHz to 10 kHz

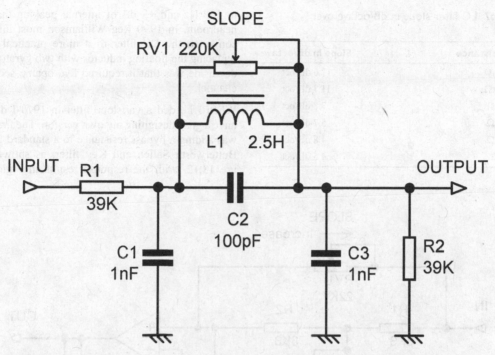

Figure 13.10 Historical variable-slope LC lowpass filter based on notch filter with bypass resistance RV1: 1956

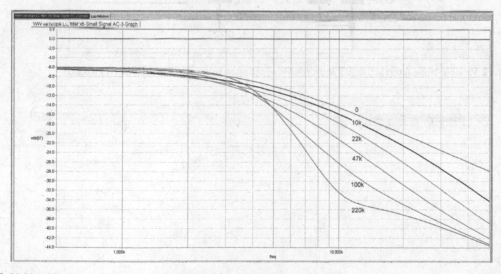

Figure 13.11 Variable-slope LC lowpass filter response for varying bypass resistance

frequencies plus "Off" and only offered two switched slope settings, "steep" and "gradual". The later Varislope II lived up to its name better by having switched filter frequencies but fully variable slopes using a two-gang pot. Still later, the stereo versions (Gold & Grey) reverted to switched slopes, perhaps because a four-gang

pot was not at the time a viable proposition. The circuitry can be found on the Web.[3]

In 1970 John Linsley-Hood published an RC variable-slope filter design,[4] but its operation was quite unacceptable, having irregularities in the passband down

Table 13.7 LC filter slope in dB/octave over 5–10 kHz

Bypass resistance	Slope in dB/octave
220 kΩ	17.6 dB/oct
100 kΩ	11.7 dB/oct
47 kΩ	7.8 dB/oct
22 kΩ	5.7 dB/oct
10 kΩ	4.8 dB/oct
0 Ω	4.5 dB/oct

to 100 Hz and +3 dB of internal peaking that eroded headroom. In 1990 Reg Williamson most ingeniously converted the LC filter to a more practical form by replacing the floating inductor with two gyrators;[5] the downside was that it required five opamp sections per channel.

When I needed a varislope filter in 1978 I decided to have a go at designing my own version. The first attempt was adding a bypass resistance to a standard 2nd-order Butterworth Sallen and Key filter, as shown in Figure 13.12, with the response results in Figure 13.13,

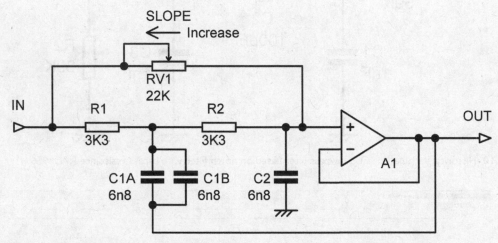

Figure 13.12 Variable-slope filter: active RC 2nd-order lowpass filter with bypass resistance RV1 added

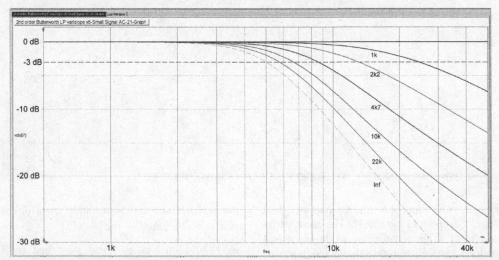

Figure 13.13 Variable-slope filter: active RC 2nd-order lowpass filter responses on varying bypass resistance RV1

where the dashed bottom trace shows the pure 2nd-order Butterworth characteristic obtained when the bypass resistance is entirely absent. The slopes are indeed varying smoothly and are summarised in Table 13.8. Naturally the maximum slope is somewhat less than 12 dB/octave, since we started out with a 2nd-order filter. The turnover frequency varies with the slope, though this is not necessarily a disadvantage. We are trying to make tolerable the reproduction from a very imperfect medium, not design a laboratory instrument. Here increasing the bypass resistance gives "more filtering" in two ways because the turnover frequency is reduced as the slope is increased.

In practice, adding a 1 kΩ end-stop resistor in series with the slope pot would probably be a good idea, as this will prevent C2 being connected directly to the output of the

previous stage, which may object by going unstable. It is assumed there is a way of switching out the filter stage completely.

While the second-order lowpass filter with bypass resistance delivers variable slopes quite well, it is doubtful if a maximum slope of 12 dB/octave is really enough; the historical LC filter gave a maximum of almost 18 dB/octave.

I therefore took another swing at the problem by starting with a 3rd-order Butterworth Sallen and Key filter, as in Figure 13.14. I have not converted the exact capacitor values to combinations of preferred values. Connecting a bypass resistance between the input and C3 gives no useful result, but wiring it between C1 and C3 gives the response in Figure 13.15. The operation is not the same as for the 2nd-order filter; there is something like a constant turnover frequency, but this does not align with the dashed line at −3 dB; rather it occurs around −7 dB. A variable-slope characteristic is obtained with slopes of between 5 and 15 dB/octave, as summarised in Table 13.9.

There is now no need to add an end-stop resistor in series with the slope pot, as C3 is never connected directly to the input. I am not claiming that these bypass filters are the last word on the subject, but they are certainly more economical of parts than the Williamson gyrator filter. To the best of my knowledge this is the first time this technique has been published.

Table 13.8 RC 2nd-order filter slope in dB/octave over 5–10 kHz

Bypass resistance	Slope in dB/octave
None	11.8 dB/oct
22 kΩ	10.5 dB/oct
10 kΩ	8.9 dB/oct
4k7	6.9 dB/oct
2k2	4.2 dB/oct
1 kΩ	1.7 dB/oct

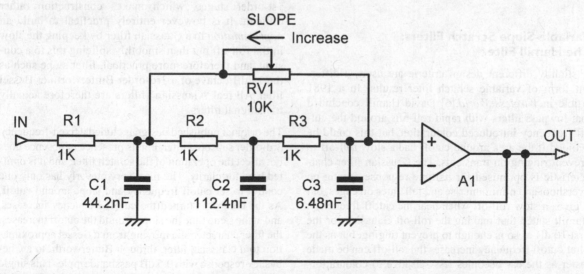

Figure 13.14 Variable-slope filter: active RC 3rd-order lowpass filter with bypass resistance RV1 added

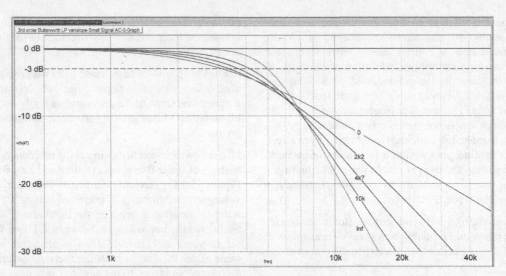

Figure 13.15 Variable-slope active RC 3rd-order lowpass filter response for varying bypass resistance

Table 13.9 RC 3rd-order filter slope in dB/octave over 5–10 kHz

Bypass resistance	Slope in dB/octave
None	15.2 dB/oct
10 kΩ	12.0 dB/oct
4k7	9.4 dB/oct
2k2	7.1 dB/oct
0 Ω	5.0 dB/oct

Variable-Slope Scratch Filters: The Hamill Filter

If slightly different design criteria are used, a different form of variable scratch filter results. In a 1981 article in *Wireless World*,[6] David Hamill concluded that lowpass filters with rapid roll-offs around the cutoff frequency introduced colouration, but this could be avoided if the area around cutoff had a slow roll-off to prevent ringing on transients. The Gaussian filter characteristic is optimised for its time response, giving no overshoot and minimum rise and fall times on edges, but it gives a slow roll-off when near the cutoff frequency. Hamill stated that making the roll-off Gaussian for the first 10 dB or so is enough to prevent ringing, but as the filter cutoff frequency increases the roll-off can be made faster as the ear becomes less sensitive to colouration. His design therefore had a variable slope around the cutoff frequency, transitioning to a fixed 18 dB/octave ultimate slope at higher frequencies, unlike the filters described earlier, which have variable ultimate slopes.

One slight problem with Gaussian filters is that they are impossible to construct. The slow roll-off that gives the good time response is obtained by cascading a series of 1st-order stages of differing and carefully chosen frequencies; the Gaussian roll-off slope steadily increases with frequency and is infinite at infinite frequency. You therefore need an infinite number of 1st-order stages, which makes construction rather difficult. It is however entirely practical to build an *approximation* to a Gaussian filter by keeping the slow initial roll-off but then smoothly splicing this to a constant, and therefore more practical, filter slope such as the 18 dB/octave of a 3rd-order Butterworth or Bessel filter. All real "Gaussian" filters are therefore actually transitional filters.

The original published design included a fixed-frequency 3rd-order subsonic filter, but its presence or absence does not affect the operation of the scratch filter, and it is omitted here for clarity. The filter very cleverly has only one control to set cutoff frequency and slope around cutoff. As the control is turned the cutoff frequency increases, and at the same time the slope around the cutoff increases, the filter characteristic moving from a Bessel approximation to a Gaussian filter, through Butterworth, to a Chebyshev response with 0.5 dB passband ripple. This single "ripple" is a very shallow peak around 15 kHz and is highly unlikely to be perceptible. The control is a single pot making stereo operation with a dual-gang pot simple. I think we should call this a Hamill filter.

My interpretation of this filter is shown in Figure 13.16. The basic configuration is that of a two-stage 3rd-order Sallen and Key filter, with 2nd-order stage around A2, but the innovation is that capacitors C2 and C4 are driven by a scaled version of the output voltage. The scaling

factor is set by RV1, with R6 bending the control law so it approximates to logarithmic, i.e. linear in octaves. The circuit impedances have been reduced by a factor of 4.25 to improve the noise performance with modern opamps (the original design used discrete transistors), the precise factor being chosen to make the largest capacitor C3 exactly 20 nF, so it can be made from two 10 nF polystyrene capacitors C3A, C3B in parallel. This luckily gave

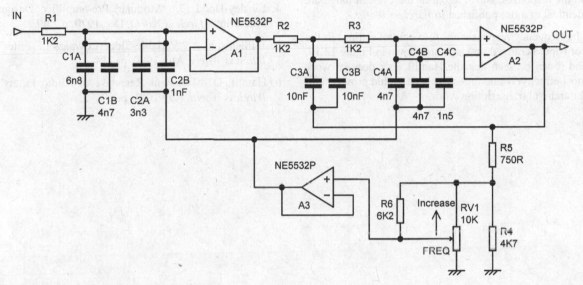

Figure 13.16 Hamill variable-slope filter circuit

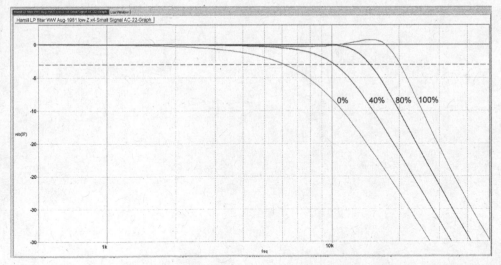

Figure 13.17 Hamill filter with variable slope around cutoff, merging into a 3rd-order 18 dB/octave roll-off. Four control settings shown.

a value of almost exactly 1.2 kΩ for R1, R2, and R3. The other capacitors, C1, C2, and C4, inevitably have awkward values; they are made up assuming only E6 series components are available. As luck would have it, C1 and C4 come out very nicely, their combined nominal values being within 0.1% of the target, though three parallel capacitors, C4A, C4B, and C4C, are required to achieve this for C4. C2 is a bit less favourable, coming out 1% high, but this seems to have very little effect on the responses, which as far as the eye can judge are identical to those published in *Wireless World*.

The frequency responses for four control settings as the pot wiper is moved upwards are shown in Figure 13.17, and it can be seen that the Hamill filter does its work most effectively and ingeniously. It should prove useful for archival transcription work.

References

1. Leakey, D. "Inexpensive Variable-Slope Filter" *Wireless World*, Nov 1956, p. 563.

2. Williamson, R. "All-Silicon Transistor Stereo Control Unit" *Hi-Fi News*, May 1967.

3. www.44bx.com/leak/leak_ccts.html Accessed Nov 2016.

4. Linsley-Hood, J. "Modular Pre-amplifier Design Postscript" *Wireless World*, Dec 1970, p. 609.

5. Williamson, R. "Variable-Slope Lowpass LF Filter" *Wireless World*, Aug 1990, p. 714.

6. Hamill, D. "Transient Response of Audio Filters" *Wireless World*, Aug 1981, p. 59.

Chapter 14

Line Outputs

Unbalanced Outputs

There are only two electrical output terminals for an unbalanced output—signal and ground. However, the unbalanced output stage in Figure 14.1a is fitted with a three-pin XLR connector to emphasise that it is always possible to connect the cold wire in a balanced cable to the ground at the output end and still get all the benefits of common-mode rejection if you have a balanced input. If a two-terminal connector is fitted, the link between the cold wire and ground has to be made inside the connector.

The output amplifier in Figure 14.1a is configured as a unity-gain buffer, though in some cases it will be connected as a series-feedback amplifier to give gain. A non-polarised DC-blocking capacitor C1 is included; 100 uF gives a –3 dB point of 2.6 Hz with one of those notional 600 Ω loads. The opamp is isolated from the line shunt capacitance by a resistor R2, in the range 47–100 Ω, to ensure HF stability, and this unbalances the hot and cold line impedances. A drain resistor R1 ensures that no charge can be left on the output side of C1; it is placed *before* R2, so it causes no attenuation. In this case the loss would only be 0.03 dB, but such errors can build up to an irritating level in a large system, and it costs nothing to avoid them.

If the cold line is simply grounded as in Figure 14.1a, then the presence of R2 degrades the CMRR of the interconnection to an uninspiring –43 dB even if the balanced input at the other end of the cable has infinite CMRR in itself and perfectly matched 10 kΩ input impedances.

To fix this problem, Figure 14.1b shows what is called an impedance-balanced output. There are now three physical terminals, hot, cold, and ground. The cold terminal is neither an input nor an output but a resistive termination R3 with the same resistance as the hot terminal output impedance R2. If an unbalanced input is being driven, this cold terminal can be either shorted to ground locally or left open circuit. The use of the word "balanced" is perhaps unfortunate, as when taken together with an XLR output connector it implies a true balanced output with anti-phase outputs, which is *not* what you are getting. The impedance-balanced approach is not particularly cost-effective, as it requires significant extra money to be spent on an XLR connector. Adding an opamp inverter to make it a proper balanced output costs little more, especially if there happens to be a spare opamp half available, and it sounds much better in the specification.

Zero-impedance Outputs

Both the unbalanced outputs shown in Figure 14.1 have series output resistors to ensure stability when driving cable capacitance. This increases the output impedance, which can impair the CMRR of a balanced interconnection and can also lead to increased crosstalk via stray capacitance in some situations. One such scenario that was authoritatively fixed by the use of a so-called "zero-impedance" output is described in my book *Small Signal Audio Design* in Chapter 22 in the section on mixer insert points.[1] The output impedance is of course not exactly zero, but it is very low compared with the average series output resistor.

Figure 14.2a shows how the zero-impedance technique is applied to an unbalanced output stage with 10 dB of gain. Feedback at audio frequencies is taken from outside isolating resistor R3 via R2, while the HF feedback is taken from inside R3 via C2, so it is not affected by load capacitance and stability is unimpaired. Using a 5532 opamp, the output impedance is reduced from 68 ohms to 0.24 ohms at 1 kHz—a dramatic reduction that would reduce capacitive crosstalk by 49 dB. Output impedance increases to 2.4 Ohms at 10 kHz and 4.8 Ohms at 20 kHz, as opamp open-loop gain falls with frequency. The impedance-balancing resistor on the cold pin has been replaced by a link to match the near-zero output impedance at the hot pin. More details

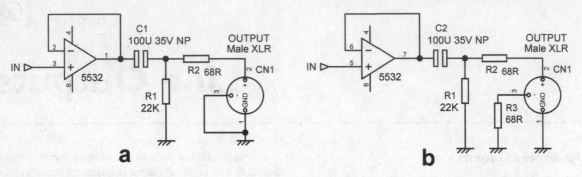

Figure 14.1 Unbalanced outputs; a) simple output and b) impedance-balanced output for improved CMRR when driving balanced inputs

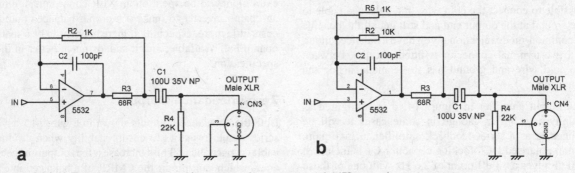

Figure 14.2 a) Zero-impedance output; b) zero-impedance output with NFB around output capacitor

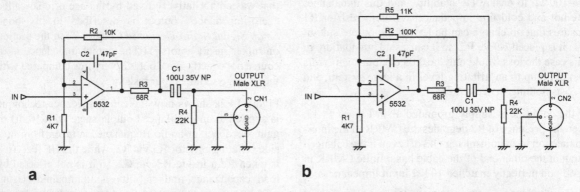

Figure 14.3 Voltage followers with zero-impedance output; a) simple, b) with negative feedback around output capacitor

on zero-impedance outputs are given later in this chapter in the section on balanced outputs.

There is no need for the output stage to have voltage gain for this to work, just a way to transfer the negative feedback point from outside to inside the series output resistor. Figure 14.3a shows a version using a unity-gain voltage-follower.

The quickest way to measure normal output impedances is to load each output as heavily as is practical (say with 560 Ω) and measure the voltage drop compared with the unloaded state. From this the output impedance is simply calculated. However, in the case of zero-impedance outputs the voltage drop is very small, and so measurement accuracy is poor.

A more sophisticated technique is the injection of a test signal current into the output and measuring the voltage that results. This is much more informative; the results for the hot output in Figure 14.3a are given in Figure 14.4. A suitable current to inject is 1 mA rms, defined by applying 1 Vrms to a 1 kΩ injection resistor. A 1 Ω output impedance will therefore give an output signal voltage of 1 mV rms (−60 dBV), which can be easily measured, as in Figure 14.4. Likewise, −40 dBV represents 10 Ω, and −80 dBV represents 0.1 Ω; a log scale is useful because it allows a wide range of impedance to be displayed and gives convenient straight lines on the plot. The opamp sections were both LM4562.

Below 3 kHz the impedance increases steadily as frequency falls, doubling with each octave due to the reactance of the output capacitor. This effect can be reduced simply by increasing the size of the capacitor. In Figure 14.4 the results are shown for 220uF, 470uF, and 1000uF output capacitors. For 1000uF a broad null occurs, reaching down to 0.06 Ω, and the output impedance is below 1 Ω between 150 Hz and 20 kHz. Bear in mind that output capacitors should be nonpolarised types, as they may face external DC voltages of either polarity and should be rated at no less than 35 V; this means that a 1000 uF capacitor can be quite a large component.

In Figure 14.4 the modulus of the output impedance falls to below 0.1 Ω between 1 and 5 kHz. It is easy to assume that the steady rise above the latter frequency is due to the open-loop gain of the opamp falling as frequency increases, and indeed I did, until recently. However, an apparent anomaly in some measured results (it was Isaac Asimov who said that most scientific advances start with "hmmm . . . that's funny . . ." rather than "Eureka!"), where a two-times change in noise gain (see the balanced output section later in this chapter) gave identical output impedance plots, led me to dig a little deeper.

Figure 14.5 shows the output impedance with the output blocking capacitor removed; R2 is 1 kΩ.

I'm not entirely sure that the line at the bottom for no C at all should be dead flat, but at any rate the point is illustrated.

Figure 14.6 shows the effect of adding a 10 pF capacitor across the external compensation pins of the 5534. Now the 0 pF and 1 pF curves are close to that for 10 pF, but the 100 pF curve stays where it was and is still ten times higher. When this simulation is repeated using a TL072 (which has less open-loop gain in the HF region) the curves are much closer together and the 100 pF passes through −60 dB at 10 kHz instead of −68 dB, so here the HF open-loop gain of the opamp really does define the situation. Historically many output amplifiers were built with TL072s, due to the higher cost and power consumption of the 5534/2, but the values used were frequently R2 = 10 kΩ and C2 = 100 pF, so then the

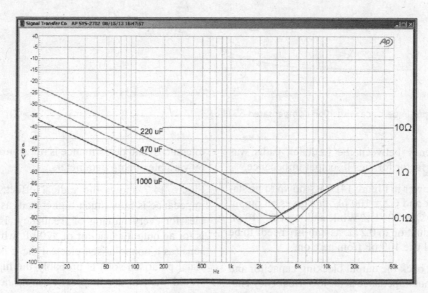

Figure 14.4 The signal voltage at the output of a zero-impedance voltage-follower for 1 mA rms injection current, and the output impedance modulus. Three values of output capacitor 5532 R2 = 1 kΩ, C2= 100 pF.

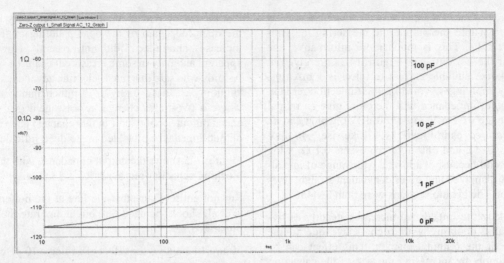

Figure 14.5 Output impedance with uncompensated 5534 varying C2

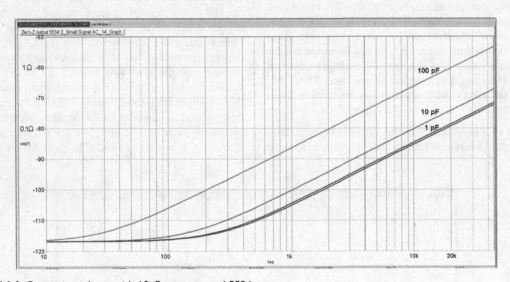

Figure 14.6 Output impedance with 10pF-compensated 5534

transition frequency between the feedback paths was still the defining factor for the HF output impedance.

The voltage-follower in Figure 14.3a is known to be stable with R2 = 1k and C2 = 100 pF; this gives an output impedance of 10 Ω at 20 kHz. While making C2 = 10 pF gives a useful reduction to 0.2 Ω, its stability needs to be checked. We will use 100 pF from here on.

We will now put back the output blocking capacitor and examine the output impedance at the low-frequency end. The situation is simpler at LF because the components are of known value (apart from the usual manufacturing

tolerances) and performance is not significantly affected by variable opamp parameters.

Figure 14.7 shows the radical effect; the LF output impedance is now much higher and, as expected, doubles for every halving of frequency. The middle trace is for an output capacitor of 470 uF, which for size reasons (as a nonpolarised electrolytic it is larger than the usual polarised sort) is probably the highest value you would want to use in practice; note how accurately it matches the measured impedance result for 470 uF in Figure 14.4. The 5534 is now uncompensated, as it gives

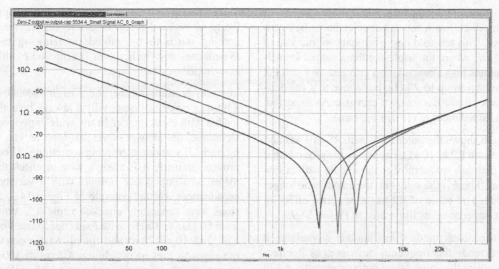

Figure 14.7 Output impedance with output cap 220, 470, and 1000 uF. 5534 R2 = 1 kΩ, C2 = 100 pF

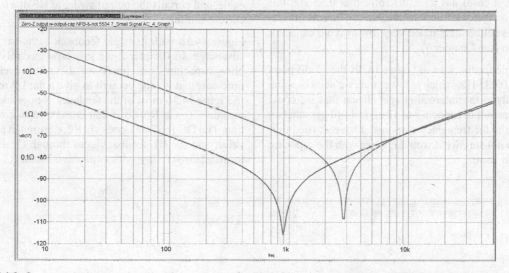

Figure 14.8 Output impedance with and without negative feedback around an output cap of 470 uF. 5534 R5 = 1k, R2 = 10k, C2 = 100 pF

a better match of HF open-loop gain with the 5532s or LM4562s that will be used in practice

The voltage at 50 Hz, the lowest frequency of interest, is −44 dB, equivalent to an output impedance of 6.3 Ω, which is much higher than the impedance set by the opamp and some way distant from "zero impedance". We can use a 1000 uF output capacitor instead, which reduces the output impedance at 50 Hz to 3.1 Ω, but the capacitor is now distinctly bulky. From here on we will stick with 470 uF.

What do we do now if we want a lower LF output impedance? As is so often the case, in electronics as in life, you can use either brawn (big capacitors) or brains. The latter, once again as is so often true, is the cunning application of negative feedback. The essence of the zero impedance technique is to have two NFB paths, with one enclosing the output resistor, which could be called double feedback. If we add a third NFB path, as in Figure 14.3b, then we can enclose the output capacitor as well and reduce the effects of its reactance; we'll call this triple feedback. Figure 14.8 shows the dramatic results. The output impedance at 50 Hz drops from 6.3 Ω to 0.63 Ω, which has a much better claim to be "zero impedance". This ten-fold improvement continues up to about 400 Hz, where the triple-feedback output impedance plummets into a cancellation crevasse. We still get a useful reduction up to 2 kHz, but above that the double-feedback circuit has its own crevasse. As expected, the output impedances at HF are the same.

Figure 14.3b shows the outer NFB path R5 as 1 kΩ and the middle path R2 as 10 kΩ, chosen on the grounds that the outer path has to overpower the middle path at low frequencies. There is the snag that the output is no longer DC blocked. The value of R2 has a powerful effect on the output impedance, as illustrated in Figure 14.9.

The triple-feedback method has another advantage; it radically extends the frequency response, as it effectively multiplies the value of the output capacitor. A 470 uF output capacitor with a 600 Ω load (a largely fictitious worst case) has a −3 dB roll-off at 0.58 Hz; this is what we get with double feedback. The roll-off frequency is deliberately made very low to avoid capacitor distortion (see Chapter 3).

With triple feedback as in Figure 14.3b, the −3 dB point drops to 0.035 Hz, which is lower than we need or want. This leads to the interesting speculation that we could reduce the output capacitor to 47 uF and still get satisfactory results for frequency response. I haven't tried this, and I am not sure what would happen with the capacitor distortion issue.

Since wiring resistances internal to the equipment are likely to be in the region 0.1 to 0.5 Ω, there seems nothing much to be gained by reducing the output impedances any further than is achieved by this simple zero-impedance technique.

You may be thinking that the zero-impedance output is a bit of a risky business; will it always be stable when loaded with capacitance? In my wide experience of this technique, the answer is yes, so long as you design it properly. If you are attempting something different from proven circuitry, it is always wise to check the stability of zero-impedance outputs with a variety of load capacitances. In the example of Figure 14.3, both versions were separately checked using a 5 Vrms sweep from 50 kHz to 10 Hz, with load capacitances of 470 pF, 1 nF, 2n2, 10 nF, 22 nF and 100 nF. At no point was there the slightest hint of instability. A load of 100 nF is of course grossly excessive compared with real use, being equivalent to about 1000 metres of average screened cable, and curtails the output swing at HF due to opamp current-limiting.

Figure 14.3b with its triple feedback paths has another advantage. In Chapter 2 we saw how electrolytic coupling capacitors can introduce distortion even if the time-constant is long enough to give a flat LF response. In Figure 14.3b most of the feedback is now taken from outside C1, via R5, so it can correct capacitor distortion. The DC feedback goes via R2, now much higher in value, and the HF feedback goes through C2 as before to maintain stability with capacitive loads. R2 and R5

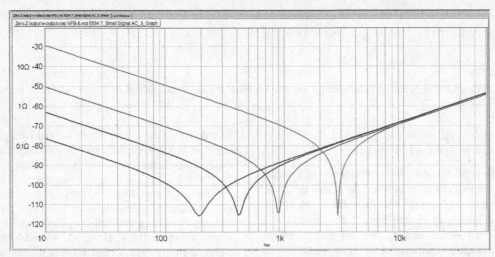

Figure 14.9 Output impedance with and without negative feedback around an output cap of 470 uF. 5534 R5 = 1k; R2=10k, 47k, 220k; C2 = 100 pF

in parallel come to 10 kΩ, so the gain is the same. Any circuit with separate DC and AC feedback paths must be checked carefully for frequency response irregularities, which may happen well below 10 Hz. A function generator is useful for this.

Ground-Cancelling Outputs: Basics

This technique, also called a ground-compensated output, appeared in the early 1980s in mixing consoles. It allows ground voltages to be cancelled out even if the receiving equipment has an unbalanced input; it prevents any possibility of creating a phase error by miswiring; and it costs virtually nothing except for the provision of a three-pin output connector.

Ground-cancelling (GC) separates the wanted signal from the unwanted ground voltage by addition at the output end of the link rather than by subtraction at the input end. If the receiving equipment ground differs in voltage from the sending ground, then this difference is added to the output signal so that the signal reaching the receiving equipment has the same ground voltage superimposed upon it. Input and ground therefore move together, and the ground voltage has no effect, subject to the usual effects of component tolerances. The connecting lead is differently wired from the more common unbalanced-out balanced-in situation, as now the cold line is joined to ground at the *input* or receiving end. This is illustrated in Figure 14.10, which compares a conventional balanced link with a GC link.

An inverting unity-gain ground-cancel output stage is shown in Figure 14.11a. The cold pin of the output socket is now an input and has a unity-gain path summing into the main signal going to the hot output pin to add the ground voltage. This path R3, R4 has a very low input impedance equal to the hot terminal output impedance, so if it *is* used with a balanced input, the line impedances will be balanced and the combination will still work effectively. The 6dB of attenuation in the R3-R4 divider is undone by the gain of two set by R5, R6.

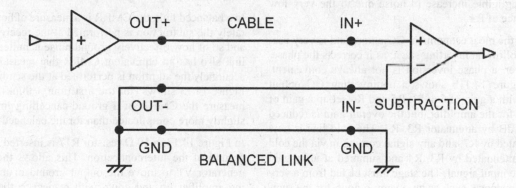

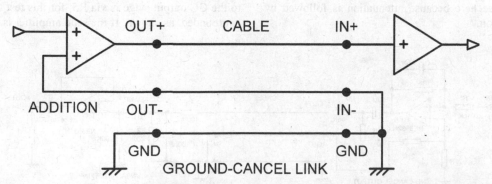

Figure 14.10 A balanced link uses subtraction at the receiving end to null ground noise, while a ground-cancel link uses addition at the sending end

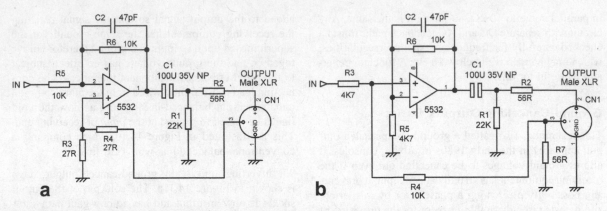

Figure 14.11 a) Inverting ground-cancelling output; b) non-inverting ground-cancelling output

It is unfamiliar to most people to have the cold pin of an output socket as a low-impedance input, and its very low input impedance minimises the problems caused by miswiring. Shorting it locally to ground merely converts the output to a standard unbalanced type. On the other hand, if the cold input is left unconnected then there will be a negligible increase in noise due to the very low resistance of R3.

This is the most economical GC output and is very useful to follow an inverting stage, as it corrects the phase. However, a phase inversion is not always convenient, and Figure 14.11b shows a noninverting GC output stage with a gain of 6.6 dB. R5 and R6 set up a gain of 9.9 dB for the amplifier, but the overall gain is reduced by 3.3 dB by attenuator R3, R4. The cold line is now terminated by R7, and any signal coming in via the cold pin is attenuated by R3, R4 and summed at unity gain with the input signal. The stage must be fed from a very low impedance, such as an opamp output, for the summation to be accurate and thus the ground-cancelling to work properly. There is a slight compromise on noise performance here because attenuation is followed by amplification.

Figure 14.12 shows the complete circuit of a GC link using an inverting ground-cancel output stage. EMC filtering and DC blocking are included for the unbalanced input stage.

Ground-Cancelling Outputs: CMRR

In a balanced link, the CMRR is a measure of how accurately the subtraction is performed at the receiving end and so of how effectively ground noise is nulled. A GC link also has an equivalent CMRR that measures how accurately the addition is performed at the sending end. Figure 14.13 shows (for the first time, I think) how to measure the CMRR of a ground-cancelling link. It is slightly more complicated than for the balanced case.

In Figure 14.13, a 10 Ω resistor R17 is inserted into the ground of the interconnection. This allows the signal generator V1 to move the output ground of the sending amplifier up and down with respect to the global ground, via R18. Quite a lot of power has to be supplied so that R17 can be kept low in value. Normally the input to the GC output stage is via R5; for this test the input is grounded, as shown. If the send amplifier is working

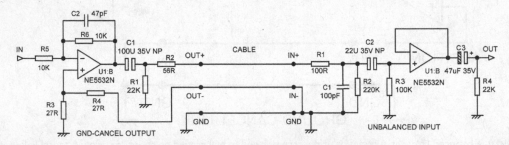

Figure 14.12 The complete circuit of a GC link using an inverting ground-cancel output

properly, the signal applied to OUT will cancel the signal on the output ground, so that as far as the input of the receiving amplifier is concerned, it does not exist. Be clear that here we are measuring the CMRR of the sending amplifier, not the receiving amplifier.

Just as the CMRR of a balanced link depends on the accuracy of the resistors and the open-loop gain of the

opamp in the receiving amplifier, the same parameters determine the CMRR of a ground-cancel send amplifier. The measured results from the arrangement in Figure 14.13 are given in Figure 14.14; the CMRR as built (with 1% resistors) was −50 dB, flat up to 10 kHz. The two lower traces were obtained by progressively trimming the value of R6 to minimise the output. If high

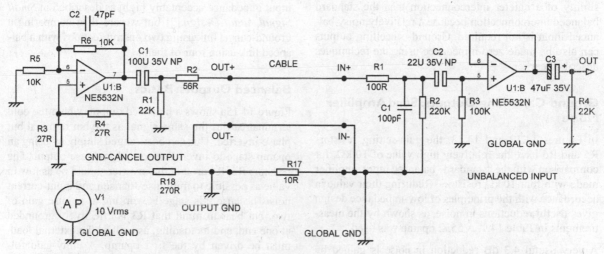

Figure 14.13 Measuring the CMRR of a GC link by inserting resistor R17

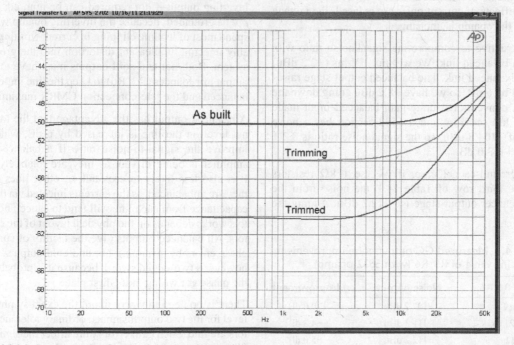

Figure 14.14 Optimising the CMRR of the ground-cancel output amplifier

CMRR is required a preset adjustment can be used, just as in balanced line input amplifiers.

Ground-cancelling outputs are an economical way of making ground loops innocuous when there is no balanced input, and it is rather surprising they are not more popular; perhaps it is because people find the notion of an input pin on an output connector unsettling. In particular GC outputs would appear to offer the possibility of a quieter interconnection than the standard balanced interconnection because a relatively noisy balanced input is not required. Ground-cancelling outputs can also be made zero impedance using the techniques described earlier.

Ground-Cancelling Outputs: Send Amplifier Noise

In Figures 14.12 and 14.13, the gain-setting resistors R5 and R6 have the relatively high value of 10 kΩ, for comparison with the "standard" balanced input amplifier made with four 10 kΩ resistors. Reducing their value in accordance with the principles of low-impedance design gives useful reductions in noise, as shown by the measurements in Table 14.1. A 5532 opamp was used.

A very useful 4.3 dB reduction in noise is gained by reducing R5 and R6 to 2k2, at zero cost, but after that the improvements become smaller, for while Johnson noise and the effects of current noise are reduced, the voltage noise of the opamp is unchanged.

It is instructive to compare the signal/noise ratio with that of a balanced link. We will put 1 Vrms (+2.2 dBu) down a balanced link. The balanced output stage raises the level by 6 dB, so we have +8.2 dBu going down the cable. A conventional unity-gain balanced input made with 10 kΩ resistors and a 5532 section has a noise output of −104.8 dBu, so the signal/noise ratio is 8.2 + 104.8 = −113.0 dB.

In the ground-cancel case, if we use 1 kΩ resistors as in the third row of Table 14.1, the noise from the ground-cancel output stage is −112.2 dBu. Adding the

Table 14.1 Measured GC output noise improvement by reducing value of R5, R6 (with 5532 opamp)

Value of R5, R6	Noise output	Improvement
10 kΩ	−107.2 dBu	0.0 dB
2k2	−111.3 dBu	−4.3 dB
1 kΩ	−112.2 dBu	−5.0 dB
560 Ω	−112.6 dBu	−5.4 dB

noise of the 5532 buffer at the receiving end in Figure 14.12, which is −125.7 dBu, we get −112.0 dBu as the noise floor. The signal/noise ratio is therefore 2.2 + 112.0 = −114.2 dB. This is 1.2 dB quieter than the conventional balanced link, and it only uses two opamp sections instead of three.

The noise situation could easily be reversed by using a low-impedance balanced input with buffers to make the input impedance acceptably high, as described in *Small Signal Audio Design*,[2] but we are then comparing a ground-cancel link using two opamp sections with a balanced link using four of them.

Balanced Outputs: Basics

Figure 14.15a shows a balanced output, where the cold terminal carries the same signal as the hot terminal but phase-inverted. This can be arranged simply by using an opamp stage to invert the normal in-phase output. The resistors R3, R4 around the inverter should be as low in value as possible to minimise Johnson and input-current noise, because this stage is working at a noise gain of two, but bear in mind that R3 is effectively grounded at one end, and its loading, as well as the external load, must be driven by the first opamp. A unity-gain follower is shown for the first amplifier, but this can be any other shunt or series-feedback stage, as convenient. The inverting output if not required can be ignored; it must *not* be grounded, because the inverting opamp will then spend most of its time clipping in current-limiting, probably injecting unpleasant distortion into the grounding system. Both hot and cold outputs must have the same output impedances (R2, R6) to keep the line impedances balanced and the interconnection CMRR maximised.

A balanced output has the advantage that the total signal level on the line is increased by 6 dB, which will improve the signal-to-noise ratio if a balanced input amplifier is being driven, as they are relatively noisy. It is also less likely to crosstalk to other lines even if they are unbalanced, as the currents injected via the stray capacitance from each line will tend to cancel; how well this works depends on the physical layout of the conductors. All balanced outputs give the facility of correcting phase errors by swapping hot and cold outputs. This is however a two-edged sword, because it is probably how the phase got wrong in the first place.

There is no need to worry about the exact symmetry of level for the two output signals; ordinary tolerance resistors are fine. Such gain errors only affect the signal-handling capacity of the interconnection by a small amount. This simple form of balanced output is the norm in hi-fi

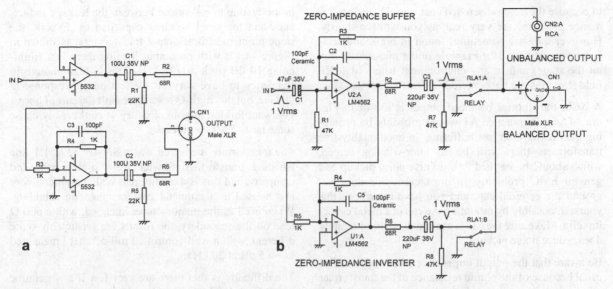

Figure 14.15 a) A conventional balanced output; b) a zero-impedance balanced output with muting relay

balanced interconnection but is less common in professional audio, where the quasi-floating output described here gives more flexibility.

Balanced Outputs: Output Impedance

The balanced output stage of Figure 14.15a has an output impedance of 68 Ω on both legs because this is the value of the series output resistors; however, as noted earlier, the lower the output impedance the better, so long as stability is maintained. This balanced output configuration can be easily adapted to have two zero-impedance outputs, as shown in Figure 14.15b. The unity-gain buffer that drives the hot output is as described earlier. The zero-impedance inverter that drives the cold output works similarly, but with shunt negative feedback via R4.

The output impedance plot for the cold output was identical to Figure 14.4. This came as rather a surprise because the inverter works at a noise gain of two times, as opposed to the buffer, which works at a noise gain of unity, and so I expected it to show twice the output impedance above 5 kHz. This observation prompted me to investigate the effect of the feedback capacitor value, as described earlier.

Balanced Outputs: Noise

The noise output of the zero-impedance balanced output of Figure 14.15b was measured with 0 Ω source

resistance, rms response, unweighted and measured at two bandwidths to demonstrate the absence of hum; the opamp sections were both LM4562. See Table 14.2.

Table 14.2 shows that reading the noise between the hot and cold outputs gives results 3 dB higher. This does not mean balanced operation is inferior; the total signal level is twice as high, and so the signal-to-noise ratio is in fact 3 dB better. This calculation does not take account of the noise added at the receiving amplifier end of a balanced link.

Transformer Balanced Outputs

If true galvanic isolation between equipment grounds is required, this can only be achieved with a line transformer, sometimes called a line isolating transformer. You don't use line transformers unless you really have

Table 14.2 Measured noise output of the zero-impedance balanced output of Figure 14.15b

Output	Noise out	Bandwidth
Hot output only	−113.5 dBu	(22–22 kHz)
Hot output only	−113.8 dBu	(400–22kHz)
Balanced output (hot & cold)	−110.2 dBu	(22–22 kHz)
Balanced output (hot & cold)	−110.5 dBu	(400–22kHz)

to because the much-discussed cost, weight, and performance problems are very real, as you will see shortly. However they are sometimes found in big sound reinforcement systems (for example in the mic-splitter box on the stage) and in any environment where high RF field strengths are encountered.

A basic transformer balanced output is shown in Figure 14.16a; in practice A1 would probably be providing gain rather than just buffering. In good-quality line transformers there will be an inter-winding screen, which should be earthed to minimise noise pickup and general EMC problems. In most cases this does *not* ground the external can, and you have to arrange this yourself, possibly by mounting the can in a metal capacitor clip. Make sure the can is grounded, as this definitely does reduce noise pickup.

Be aware that the output impedance will be higher than usual because of the ohmic resistance of the transformer windings. With a 1:1 transformer, as normally used, both the primary and secondary winding resistances are effectively in series with the output. A small line transformer can easily have 60 Ω per winding, so the output impedance is 120 Ω plus the value of the series resistance R1 added to the primary circuit to prevent HF instability due to transformer winding capacitances and line capacitances. The total can easily be 160 Ω or more, compared with, say, 47 Ω for nontransformer output stages. This will mean a higher output impedance and greater voltage losses when driving heavy loads.

DC flowing through the primary winding of a transformer is bad for linearity, and if your opamp output has anything more than the usual small offset voltages on it, DC current flow should be stopped by a blocking capacitor.

Output Transformer Frequency Response

Line input transformers give a nastily peaking high-frequency response if the secondary is not loaded properly, due to resonance between the leakage inductance and the stray winding capacitances. Exactly the same problem afflicts output transformers, as shown in Figure 14.17; with no output loading there is a frightening 14 dB peak at 127 kHz. This is high enough in frequency to have very little effect on the response at 20 kHz, but this high-Q resonance isn't the sort of horror you want lurking in your circuitry. It could easily cause some nasty EMC problems.

The transformer measured was a Sowter 3292 1:1 line isolating transformer. Sowter are a highly respected company, and this is a quality part with a mumetal core and housed in a mumetal can for magnetic shielding. When used as the manufacturer intended, with a 600 Ω load on the secondary, the results are predictably quite different, with a well-controlled roll-off that I measured as −0.5 dB at 20 kHz.

The difficulty is that there are very few if any genuine 600 Ω loads left in the world, and most output transformers are going to be driving much higher impedances. If we are driving a 10 kΩ load, the secondary resonance is not much damped, and we still get a thoroughly unwelcome 7 dB peak above 100 kHz, as shown in Figure 14.17. We could of course put a permanent 600 Ω load across the secondary, but that will heavily load the output opamp, impairing its linearity, and will give us unwelcome signal loss due in the winding resistances. It is also a profoundly inelegant way of carrying on.

A better answer, as in the case of the line input transformer, is to put a Zobel network, i.e. a series combination of resistor and capacitor, across the secondary, as in Figure 14.16b. The capacitor required is quite small and will cause very little loading except at high frequencies, where signal amplitudes are low. A little experimentation yielded the values of 1 kΩ in series with 15 nF, which gives the much improved response shown in Figure 14.17. The response is almost exactly 0.0 dB at 20 kHz, at the cost of a very gentle 0.1 dB rise around

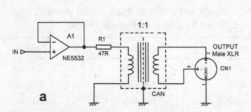

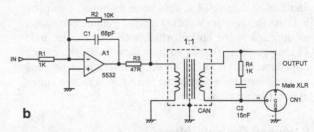

a   b

Figure 14.16 Transformer balanced outputs; a) standard circuit; b) zero-impedance drive to reduce LF distortion, also with Zobel network across secondary

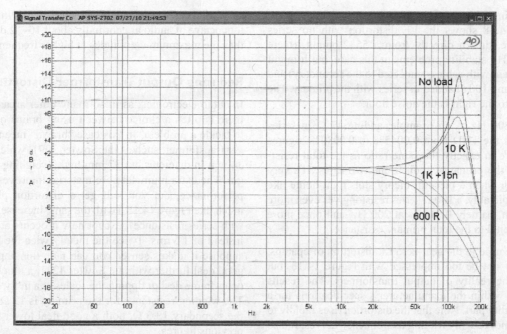

Figure 14.17 Frequency response of a Sowter 3292 output transformer with various loads on the secondary. Zero-impedance drive as in Figure 14.16b.

10 kHz; this could probably be improved by a little more tweaking of the Zobel values. Be aware that a different transformer type will require different values.

Output Transformer Distortion

Transformers have well-known problems with linearity at low frequencies. This is because the voltage induced into the secondary winding depends on the rate of change of the magnetic field in the core, and so the lower the frequency, the greater the change in magnetic flux must be for transformer action.[3] The current drawn by the primary winding to establish this field is nonlinear because of the well-known nonlinearity of iron cores. If the primary had zero resistance and was fed from a zero source impedance, as much distorted current as was needed would be drawn, and no one would ever know there was a problem. But . . . there is always some primary resistance, and this alters the primary current drawn so that third-harmonic distortion is introduced into the magnetic field established and so into the secondary output voltage. Very often there is a series resistance R1 deliberately inserted into the primary circuit, with the intention of avoiding HF instability; this makes the LF distortion problem worse, and a better means of

isolation is a low-value inductor of say 4 uH in parallel with a low-value damping resistor of around 47 Ω. This is more expensive and is only used on high-end consoles.

An important point is that this distortion does not appear only with heavy loading—it is there all the time, even with no load at all on the secondary; it is not analogous to loading the output of a solid-state power amplifier, which invariably increases the distortion. In fact, in my experience transformer LF distortion is slightly better when the secondary is connected to its rated load resistance. With no secondary load, the transformer appears as a big inductance, so as frequency falls the current drawn increases, until, with circuits like Figure 14.11a, there is a sudden steep increase in distortion around 10–20 Hz as the opamp hits its output current limits. Before this happens, the distortion from the transformer itself will be gross.

To demonstrate this I did some distortion tests on the same Sowter 3292 transformer that was examined for frequency response. The winding resistance for both primary and secondary is about 59 Ω. It is quite a small component, 34 mm in diameter and 24 mm high and weighing 45 gm, and is obviously not intended for transferring large amounts of power at low frequencies. Figure 14.18 shows the LF distortion with no series resistance, driven directly from a 5532 output, (there were no HF stability problems in this case, but it might be different with cables connected to the secondary) and with 47 and 100 Ω added in series with the primary. The flat part to the right is the noise floor.

Taking 200 Hz as an example, adding 47 Ω in series increases the THD from 0.0045% to 0.0080%, figures which are in exactly the same ratio as the total resistances in the primary circuit in the two cases. It's very satisfying when a piece of theory slots right home like that. Predictably, a 100 Ω series resistor gives even more distortion, namely 0.013 % at 200 Hz, and once more proportional to the total primary resistance.

If you're used to the near-zero LF distortion of opamps, you may not be too impressed with Figure 14.18, but this is the reality of output transformers. The results are well within the manufacturer's specifications for a high-quality part. Note that the distortion rises rapidly to the LF end, roughly tripling as frequency halves. It also increases fast with level, roughly quadrupling as level doubles. Having gone to some pains to make electronics with very low distortion, this nonlinearity at the very end of the signal chain is distinctly irritating. The situation is somewhat eased in actual use, as signal levels in the bottom octave of audio are normally about 10–12 dB lower than the maximum amplitudes at higher frequencies.

Reducing Output Transformer Distortion

In audio electronics, as in so many other areas of life, there is often a choice between using brains or brawn to tackle a problem. In this case "brawn" means a bigger transformer, such as the Sowter 3991, which is still 34 mm in diameter but 37 mm high, weighing in at 80 gm. The extra mumetal core material improves the LF performance, but you still get a distortion plot very much like Figure 14.18, (with the same increase of THD with series resistance), except now it occurs at 2 Vrms instead of 1 Vrms. Twice the metal, twice the level—I suppose it makes sense. You can take this approach a good deal further with the Sowter 4231, a much bigger open-frame design tipping the scales at a hefty 350 gm. The winding resistance for the primary is 12 Ω and for the secondary 13.3 Ω, both a good deal lower than the previous figures.

Figure 14.19 shows the LF distortion for the Sowter 4231 with no series resistance and with 47 and 100 Ω added in series with the primary. The flat part to the right

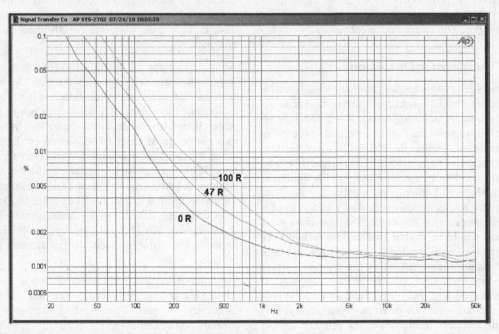

Figure 14.18 The LF distortion rise for a 3292 Sowter transformer, without (0R) and with (47 Ω and 100 Ω) extra series resistance. Signal level 1 Vrms.

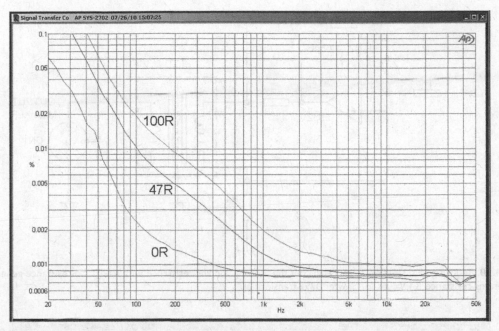

Figure 14.19 The LF distortion rise for the much larger 4231 Sowter transformer, without and with extra series resistance. Signal level 2 Vrms.

is the noise floor. Comparing it with Figure 14.13 the basic distortion at 30 Hz is now 0.015%, compared with about 0.10% for the 3292 transformer. While this is a useful improvement it is gained at considerable expense. Now adding 47 Ω of series resistance has dreadful results—distortion increases by about five times. This is because the lower winding resistances of the 4231 mean that the added 47 Ω has increased the total resistance in the primary circuit to five times what it was. Predictably, adding a 100 Ω series resistance approximately doubles the distortion again. In general bigger transformers have thicker wire in the windings, and this in itself reduces the effect of the basic core nonlinearity, quite apart from the improvement due to more core material. A lower winding resistance also means a lower output impedance.

The LF nonlinearity in Figure 14.19 is still most unsatisfactory compared with that of the electronics. Since the "My policy is copper and iron!"[4] approach does not really solve the problem, we'd better put brawn to one side and try what brains we can muster.

We have seen that adding series resistance to ensure HF stability makes things definitely worse, and a better means of isolation is a low-value inductor of say 4 uH paralleled with a low-value damping resistor of around 47 Ω. However, inductors cost money, and a more economic solution is to use a zero impedance output as shown in Figure 14.15b. This gives the same results as no series resistance at all but with wholly dependable HF stability. However, the basic transformer distortion remains because the primary winding resistance is still there, and its level is still too high. What can be done?

The LF distortion can be reduced by applying negative feedback via a tertiary transformer winding, but this usually means an expensive custom transformer, and there may be some interesting HF stability problems because of the extra phase-shift introduced into the feedback by the tertiary winding; this approach is discussed in *Interfacing Electronics and Transformers*.[5] However, what we really want is a technique that will work with off-the-shelf transformers.

A better way is to cancel out the transformer primary resistance by putting in series an electronically-generated negative resistance; the principle is shown in Figure 14.20, where a zero-impedance output is used to eliminate the effect of the series stability resistor. The 56 Ω resistor R4 senses the current through the primary and provides positive feedback to A1, proportioned so that a negative output resistance of twice the value of R4 is produced, which will cancel out both R4 itself and most of the primary winding resistance. As we saw

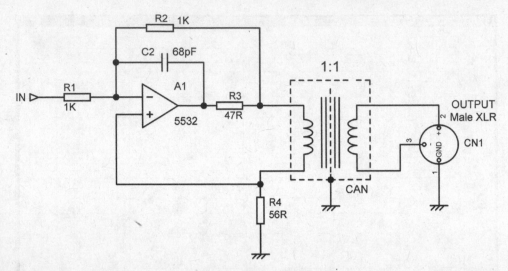

Figure 14.20 Reducing LF distortion by cancelling out the primary winding resistance with a negative resistance generated by current-sensing resistance R4. Values for Sowter 3292 transformer.

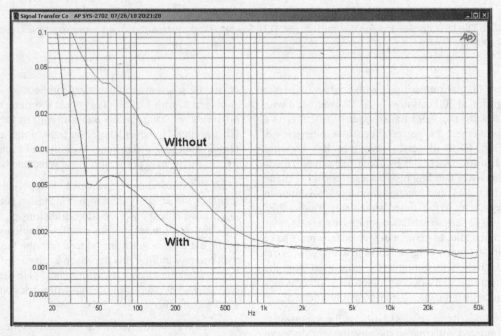

Figure 14.21 The LF distortion rise for a 3292 Sowter transformer, without and with winding resistance cancellation as in Figure 14.20. Signal level 1 Vrms.

earlier, the primary winding resistance of the 3292 transformer is approx 59 Ω, so if R4 was 59 Ω we should get complete cancellation. But . . .

It is always necessary to use positive feedback with caution. Typically it works, as here, in conjunction with good old-fashioned negative feedback, but if the positive exceeds the negative (this is one time you do *not* want to accentuate the positive) then the circuit will typically latch up solid, with the output jammed up against one of the supply rails. R4 = 56 Ω in Figure 14.20 worked reliably in all my tests, but increasing it to 68 Ω caused immediate problems, which is precisely what you would expect. No input DC-blocking capacitor is shown in Figure 14.20, but it can be added ahead of R1 without increasing the potential latch-up problems. The small Sowter 3292 transformer was used.

This circuit is only a basic demonstration of the principle of cancelling primary resistance, but as Figure 14.21 shows it is still highly effective. The distortion at 100 Hz is reduced by a factor of five, and at 200 Hz by a factor of four. Since this is achieved by adding one resistor, I think this counts as a definite triumph of brains over brawn, and indeed confirmation of the old adage that size is less important than technique.

The method is sometimes called "mixed feedback", as it can be looked at as a mixture of voltage and current feedback. The principle can also be applied when a balanced drive to the output transformer is used. Since the primary resistance is cancelled, there is a second advantage as the output impedance of the stage is reduced. The secondary winding resistance is however still in circuit, and so the output impedance is usually only halved.

If you want better performance than this—and it is possible to make transformer nonlinearity effectively invisible down to 15 Vrms at 10 Hz—there are several deeper issues to consider. The definitive reference is Bruce Hofer's patent, which covers the transformer output of the Audio Precision measurement systems.[6] There is also more information in *the Analog Devices Opamp Applications Handbook*.[7]

References

1. Self, D. *Small Signal Audio Design*. 2nd edition. Focal Press, 2015, pp. 607–608. ISBN: 978-0-415-70974-3 (hbk) ISBN: 978-0-415-70973-6 (pbk) ISBN: 978-1-315-88537-7 (ebk).

2. Self, D. *Small Signal Audio Design*. 2nd edition. Focal Press, 2015, pp. 527–535.

3. Sowter, G. A. V. "Soft Magnetic Materials for Audio Transformers: History, Production, and Applications" *Journal of Audio Engineering Society*, Volume 35, #10, Oct 1987, P. 769.

4. Otto von Bismarck Speech, 1862. (Actually, he said *blood* and iron).

5. Finnern, T. *Interfacing Electronics and Transformers*. Hamburg: AES preprint #2194 77th AES Convention, Mar 1985.

6. Hofer, B. *Low-Distortion Transformer-Coupled Circuit*. US Patent 4,614,914, 1986.

7. Jung, W., ed. *Opamp Applications Handbook*. Newnes, 2004. Chapter 6, pp. 484–491. ISBN 13: 978-0-7506-7844-5.

Chapter **15**

Level Indication

The Need for Level Indication

As noted in earlier chapters, moving-magnet (MM) cartridges do not have a wide range of output levels—about 7 dB covers almost all on the market—and so in general no gain adjustment is required or provided. There is therefore no need for an indication that the gain is wrongly set. On the other hand, the moving-coil (MC) cartridges available have a range of output levels that extends over more than 30 dB, and so gain adjustment is very much needed if a phono amplifier is to be able to cope with all of them. This makes some form of level indicator which allows that gain to be correctly set very desirable, because the alternative is to take cartridge sensitivities and estimated recorded velocities and start calculating. This could get very tedious if you change cartridges often.

This chapter therefore examines the various options for level indication, from a single LED indicating a single level, through the Log-Law Level LED which gives much more information but still uses a single LED to a complete bargraph meter. It is some decades since flashing lights were automatically regarded as enhancing the musical experience, and so once the gain has been correctly set for a given cartridge, it is nice to be able to switch the metering off, for example by means of a small slide-switch on the rear panel.

The brightness of the display is also an issue. A bargraph that can be easily read in a sunlit room is going to be over-bright in semi-darkness. It is of course possible to have manual control of display brightness, but few people are going to want to turn it up and down on a daily basis. A practical solution, and one which makes a nice feature on high-end equipment, is to have automatic brightness control using a light sensor such as a phototransistor. A simple way to do this is given at the end of the chapter.

Signal-Present Indication

Some amplifiers are fitted with a "signal-present" indicator that illuminates to give reassurance that a channel is receiving a signal and doing something with it. The level at which it triggers must be well above the noise floor but also well below the peak indication or clipping levels. Signal-present indicators are usually provided for each channel and are commonly set up to illuminate when the channel output level exceeds a threshold something like 20 or 30 dB below the nominal signal level, though there is a wide variation in this.

A simple signal-present detector is shown in Figure 15.1, based on an opamp rather than a comparator. The threshold is −32 dBu, which, combined with the −2 dBu nominal level which it was designed for, gives an indication at 30 dB below nominal. Since an opamp is used which is internally compensated for unity-gain stability, there is no need to add hysteresis to prevent oscillation when the signal lingers around the triggering point. U1 is configured as an inverting stage, with the inverting input biased slightly negative of the noninverting input by R4 in the bias chain R3, R4, R5. The opamp output is therefore high with no signal but is clamped by negative feedback through D1 to prevent excessive voltage excursions at the output, which might crosstalk into other circuitry; C2 is kept charged via R6 and R7, so Q1 is turned on and LED1 is off. When an input signal exceeds the threshold the opamp output goes low and C2 is rapidly discharged through R6 and D2. R6 limits the discharge current to a safe value; the overload protection of the opamp would probably do this by itself, but I have always been a bit of a belt-and-braces man in this sort of case. When C2 is discharged, Q1 turns off and LED1 illuminates as the 8 mA of LED chain current flows through it. When the input signal falls below the threshold, the opamp output goes high again, and C2 charges slowly through R7,

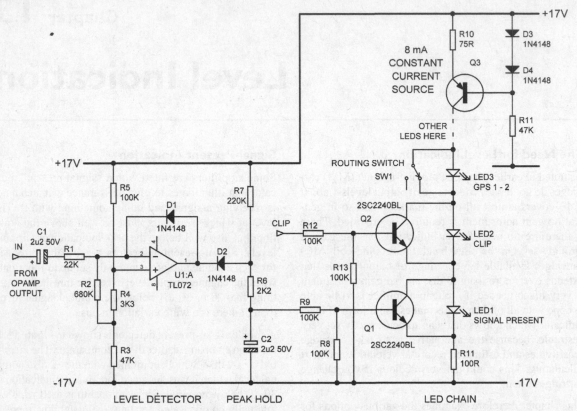

Figure 15.1 A simple unipolar signal-present detector, using an opamp. The threshold is −32 dBu. Other LEDs are run in the same constant-current chain.

giving a peak-hold action. This is a unipolar detector; only one polarity of signal activates it.

The LED chain current is provided by a wholly conventional constant-current source Q3, which allows any number of LEDs to be turned on and off without affecting the brightness of other LEDs. In this case the clip-detect LED is shown just above the signal-present LED; since it is only illuminated when the signal-present LED is already on, the drive requirements for Q2 are simple, and it can be driven from exactly the same sort of capacitor-hold circuitry. The other LEDs in the chain (here assumed to be routing switch indicators) simply "float" above the LEDs controlled by transistors. The LED chain is connected between the two supply rails, so there is no possibility of current being injected into the ground. This gives a span of 34 V, allowing a large number of LEDs to be driven economically from the same current. The exact number depends on their colour, which affects their voltage drop. It is of course necessary

to allow enough voltage for the constant-current source to operate correctly, plus a suitable safety margin.

I have used this circuit many times, and it can be regarded as well-proven.

A vital design consideration for signal-present indicators is that since they are likely to be active most of the time, the operation of the circuitry must not introduce distortion into the signal being monitored; this could easily occur by electrostatic coupling or imperfect grounding if there is a comparator switching on and off at signal frequency. Avoid this.

Peak Indication

A peak indicator is driven by fast-attack, slow-decay circuitry so that even brief peak excursions give a positive display. It is important that the circuitry should be bipolar, i.e. it will react to both positive and negative peaks.

The peak values of a waveform can show asymmetry up to 8 dB or more, being greatest for unaccompanied voice or a single instrument. This level of uncertainty in peak detection is not a good thing, so only the simplest implementations use unipolar peak detection. Composite waveforms, produced by mixing several voices or instruments together, do not usually show significant asymmetries in peak level.

Figure 15.2 shows a simple unipolar Peak LED driving circuit. This only responds to positive peaks, but it does have the advantage of using but two transistors and is very simple and cheap to implement. When a sufficient signal level is applied to C1, Q1 is turned on via the divider R1, R2; this turns off Q2, which is normally held

on by R4, and Q2 then ceases to shunt current away from peak LED D1. C2 acts as a Miller integrator to stretch the peak-hold time; when Q1 turns off again, R4 must charge C2 before Q2 can turn on again. Note that this circuit is integrated into an LED chain, with R5 setting the current through it; a mode indicator LED can be illuminated by removing the short placed across it by SW1. R5 is of high enough value, because it is connected between the two supply rails, for there to be no significant variation in the brightness of one LED when the other turns off. If for some reason this was a critical issue, R5 could be replaced by a floating constant-current source. Other LEDs switched in the same way, or nonswitched for a power indicator, can be included in the LED chain.

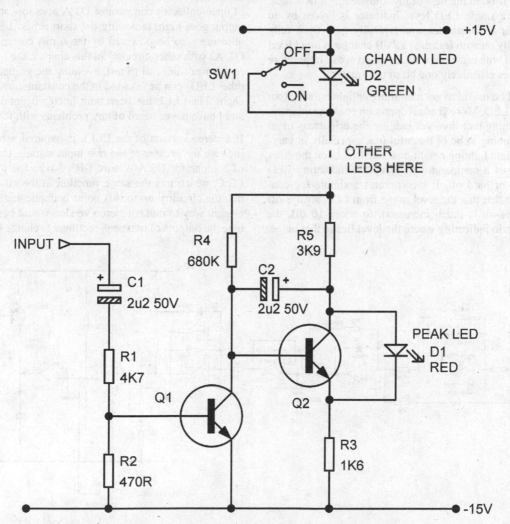

Figure 15.2 A simple unipolar peak detector, including powering for a mode indicator LED

This peak-detect circuit has a nonlinear input impedance and must only be driven from a low-impedance point, preferably direct from the output of an opamp. The peak LED illuminates at an input of 6.6 Vpeak, which corresponds to 4.7 Vrms (for a sine wave) and +16 dBu. For typical opamp circuitry running off the usual supply rails this corresponds to having only 3 or 4 dB of headroom left. The detect threshold can be altered by changing the values of the divider R1, R2.

The Log-Law Level LED (LLLL)

The Log-Law Level LED or LLLL was evolved for the Elektor preamplifier project of 2012 to aid in the adjustment of a phono input with several different gain options. It is, to the best of my knowledge, a new idea. Usually a single LED level indicator is driven by an opamp or a comparator, and typically goes from fully off to fully on with less than a 2 dB change in input level when fed with music (*not* steady sinewaves). It therefore only gives effectively one bit of information.

It would be useful to get a bit more enlightenment from a single LED. More gradual operation could be adopted, but anything that involves judging the brightness of an LED is going to be of doubtful use, especially in varying ambient lighting conditions. The LLLL, on the other hand, uses a comparator to drive the indicating LED hard on or hard off. It incorporates a simple log-converter so that that the level range from LED always-off to always-on is much increased, to about 10 dB, the on-off ratio indicating where the level lies in that range.

In some applications, such as the Elektor preamplifier, it may be appropriate to set it up so that the level is correct when the LED is on about 50% of the time. This gives a much better indication.

The circuitry of the LLLL is shown in Figure 15.3. The U1:A stage is a precision rectifier circuit that in conjunction with R3 provides a full-wave rectified signal to U1:B; this is another precision rectifier circuit that establishes the peak level of the signal on C1. This is buffered by U2:A and applied to the approximately log-law network around U2:B. As the signal level increases, first D6 conducts, reducing the gain of the stage, and then at a higher voltage set by R9, R10, D7 conducts and reduces the gain further. If sufficient signal is present to exceed the threshold set by R11, R12, the output of open-collector comparator U3:A goes low and U3:B output goes high, removing the short across LED1 and allowing it to be powered by the 6 mA current-source Q1. As with other circuitry in this chapter, the LED current is run from rail to rail, avoiding the ground. Many other LEDs can be inserted in the constant-current LED chain. The LLLL has been built in significant numbers, and I have never heard of any problems with it.

If a stereo version of the LLLL is required, which will indicate the greater of the two input signals, the output of comparator U3:A is wire-OR'ed with the output of U3:C, which has the same function in the other channel; the circuitry up to this point is duplicated. A more elegant way to make a stereo version would be to combine the outputs of two peak rectifiers to charge C1. This

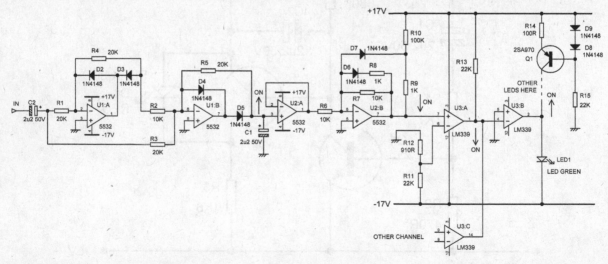

Figure 15.3 The Log-Law Level LED or LLLL

would save a handy number of components, but I have not yet actually tried it out.

I have spent some time testing the operation of this scheme, using various musical genres controlled by a high-quality slide fader. I believe it is a significant advance in signalling level when there is only one LED available, but in the words of Mandy Rice-Davies, "Well, he would say that, wouldn't he".[1] More opinions on the value of the LLLL would be most welcome; I haven't had a negative one yet.

Distributed Peak Detection

When an audio signal path consists of a series of circuit blocks, each of which may give either gain or attenuation, it is something of a challenge to make sure that excessive levels do not occur anywhere along the chain. Simply monitoring the level at the end of the chain is no use because a circuit block that gives gain, leading to clipping, may be followed by one that attenuates the clipped signal back to a lower level that does not trip a final peak-detect circuit. The only way to be absolutely sure that no clipping is happening anywhere along the path is to implement bipolar peak detection at the output of every opamp stage. This is however normally regarded as a bit excessive, and the usual practice in high-end equipment is to just monitor the output of each circuit block, even though each such block (for example an elliptical rumble filter) may actually contain several opamps. It could be argued that a well-designed circuit block should not clip anywhere except at its output, no matter what the control setting, but this is not always possible to arrange.

A multipoint or distributed peak detection circuit that I have made extensive use of is shown in Figure 15.4. It can detect when either a positive or negative threshold is

exceeded at any number of points desired; to add another stage to its responsibilities you need only add another pair of diodes, so it is very economical. However, if one peak detector monitors too many points in the signal path, it can be hard to determine which of them is causing the problem.

The operation is as follows. Because R5 is greater than R1, normally the noninverting input of the opamp is held below the inverting input and the opamp output is low. If any of the inputs to the peak system exceed the positive threshold set at the junction of R4, R3, one of D1, D3, D5 conducts and pulls up the noninverting input, causing the output to go high. Similarly, if any of the inputs to the peak system exceed the negative threshold set at the junction of R2, R6, one of D2, D4, D6 conducts and pulls down the inverting input, once more causing the opamp output to go high. When this occurs C1 is rapidly charged via D7. The output-current limiting of the opamp discriminates against very narrow noise pulses. When C1 charges, Q1 turns on and illuminates D8 with a current set by the value of R7. R8 ensures that the LED stays off when U4 output is low, as it does not get close enough to the negative supply rail for Q1 to be completely turned off.

Each input to this circuit has a nonlinear input impedance, and so for this system to work without introducing distortion into the signal path, it is essential that the diodes D1–D6 are driven directly from the output of an opamp or an equivalently low impedance. Do not try to drive them through a coupling capacitor, as asymmetrical conduction of the diodes can create unwanted DC-shifts on the capacitor.

The peak-detect opamp U4 must be a FET-input type to avoid errors due to bias currents flowing in the relatively

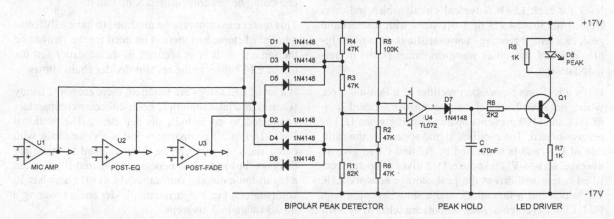

Figure 15.4 A multipoint bipolar peak detector, monitoring three circuit blocks

high-value resistors R1–R6, and a cheap TL072 works very nicely here; in fact the resistor values could probably be raised significantly without any problems.

As with other nonlinear circuits in this book, everything operates between the two supply rails, so unwanted currents cannot find their way into the ground system.

Combined LED Indicators

In the professional audio industry, there has for many years been a tendency towards very crowded channel front panels, driven by a need to keep the overall size of equipment within reasonable limits. One apparently ingenious way to gain a few more square millimetres of panel space is to combine the signal-present and peak indicators into one by using a bi-colour LED. Green shows signal present, and red indicates peak. One might even consider using orange (both LED colours on) for an intermediate level, giving three possible indications.

Unfortunately, such indicators are hard to read, even if with normal colour vision, because a light coming on is much more obvious than a light that is already on changing colour. If you have red-green colour-blindness, the most common kind (6% of males, 0.4% of females), they are useless. Space is unlikely to be in desperately short supply on the front of a phono amplifier, and combining indicators like this is really not a good idea.

LED Bargraph Metering

Bargraph meters are commonly made up of an array of LEDs. An LED bargraph meter can be made effectively with an active-rectifier circuit and a resistive divider chain that sets up the trip voltage of an array of comparators; this allows complete freedom in setting the trip level for each LED. A typical circuit which indicates from 0 dB to −14 dB in 2 dB steps with a selectable peak or average-reading characteristic is shown in Figure 15.5 and illustrates some important points in bargraph design.

U3 is a half-wave precision rectifier of a familiar type, where negative feedback servos out the forward drop of D11 and D10 prevents opamp clipping when D11 is reverse-biased. The rectified signal appears at the cathode of D11 and is smoothed by R7 and C1 to give an average, sort-of-VU response. D12 gives a separate rectified output and drives the peak-storage network R10, C9, which has a fast attack and a slow decay through R21. Either average or peak outputs are selected by SW1

and applied to the non-inverting inputs of an array of comparators. The LM2901 quad voltage comparator is very handy in this application; it has low input offsets and the essential open-collector outputs.

The inverting comparator inputs are connected to a resistor divider chain that sets the trip level for each LED. With no signal input, the comparator outputs are all low, and their open-collector outputs shunt the LED chain current from Q1 to −15V, so all LEDs are off. As the input signal rises in level, the first comparator U2:D switches its output off, and LED D8 illuminates. With more signal, U2:C also switches off and D7 comes on, and so on, until U1:A switches off and D1 illuminates. The important points about the LED chain are that the highest level LED is at the bottom of the chain, as it comes on last, and that the LED current flows from one supply rail down to the other and is not passed into a ground. This prevents noise from getting into the audio path. The LED chain is driven with a constant-current source to keep LED brightness constant despite varying numbers of them being in circuit; this uses much less current than giving each LED its own resistor to the supply rail and is universally used in mixing console metering. Make sure you have enough voltage headroom in the LED chain, not forgetting that yellow and green LEDs have a larger forward drop than red ones. The circuit shown has plenty of spare voltage for its LED chain, and so it is possible to put other indicator LEDs in the same constant-current path; for example D9 can be switched on and off completely independently of the bargraph LEDs and can be used to indicate Channel-on status or whatever. An important point is that in use the voltage at the top of the LED chain is continually changing in 2-volt steps, and this part of the circuit must be kept away from the audio path to prevent horrible crunching noises from crosstalking into it.

This meter can of course be modified to have a different number of steps, and there is no need for the steps to be the same size. It is as accurate in its indications as the use of E24 values in the resistor divider chain allows.

If a lot of LED steps are required, there are some handy ICs which contain multiple open-collector comparators connected to an in-built divider chain. The National LM3914 has 10 comparators and a divider chain with equal steps, so they can be daisy-chained to make big displays, but some law-bending is required if you want a logarithmic output. The National LM3915 also has 10 comparators, but a logarithmic divider chain covering a 30 dB range in 3 dB steps.

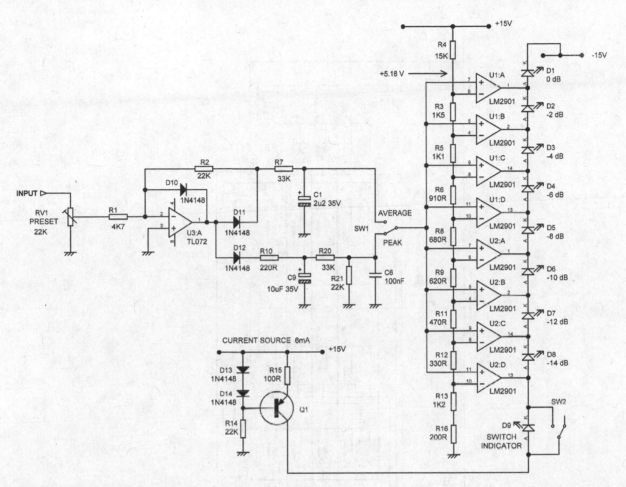

Figure 15.5 LED bargraph meter with selectable peak/average response

A More Efficient LED Bargraph

The bargraph meter shown in Figure 15.5 draws 6 mA from the two supply rails at all times, even if all the level LEDs are off for long periods, which is often the case for equipment owned by people who think that a warm-up period of a couple of months is reasonable (I am not one of them). This can be actually desirable in a simple system where the meter current is taken from the ±15V or ±17V rails used to power the audio circuitry and step changes in current taken by the meter could get into the ground system via decoupling capacitors and suchlike, causing highly unwelcome crunching noises.

More sophisticated phono amplifiers are likely to have a separate meter supply that is provided to prevent this problem, and this allows more freedom in the design of the meter circuitry. In the example here the meter supply available is assumed to be a single rail of +24V, which is the highest voltage that can conveniently be generated with standard IC regulators. It is further assumed that we wish to make a classy meter with 20 LEDs. An immediate problem is that you cannot power 20 LEDs of assorted colours in one chain running from +24V as there just is not enough voltage available; two LED chains are required, and the power consumption of the meter, even when completely dormant, becomes twice as great. I therefore devised a more efficient system, which not only saves a considerable amount of power but also actually economises on components.

The meter circuit is shown in Figure 15.6, and I must admit it is not one of those circuit diagrams where the

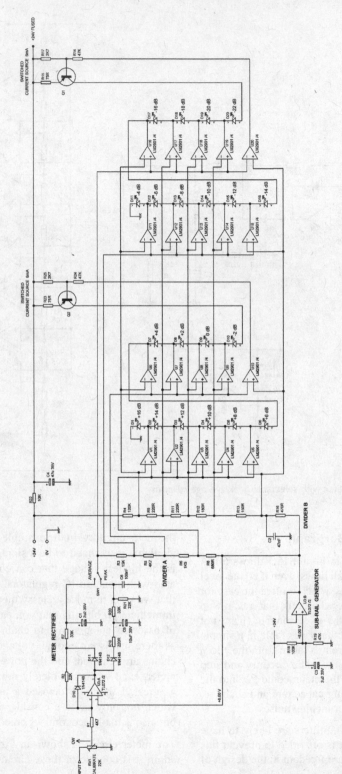

Figure 15.6 A more efficient LED bargraph meter

modus operandi exactly leaps from the page. However, stick with me.

There are two LED chains, each powered by its own constant-current source Q1, Q2. The relevant current source is only turned when it is needed. With no signal input, all LEDs are off; the outputs of comparators U10 and U20 are high (open-collector output off) and both Q1 and Q2 are off. The outputs of all other comparators are low. When a steadily increasing signal arrives, U20 is the first comparator to switch, and LED D20 turns on. With increasing signal, the output of U19 goes high, and the next LED, D19, turns on. This continues, in exactly the same way as the conventional bargraph circuit described earlier, until all the LEDs in the chain D11–D20 are illuminated. As the signal increases further, comparator U10 switches and turns on the second current source Q2, illuminating D10; the rest of the LEDs in the second chain are then turned on in sequence as before. This arrangement saves a considerable amount of power, as no supply current at all is drawn when the meter is inactive, and only half the maximum is drawn so long as the indication is below −2 dB.

There are 10 comparators for each LED chain, 20 in all, so a long potential divider with 21 resistors would be required to provide the reference voltage for each comparator if it was done in the conventional way, as shown in Figure 15.5. However, looking at all those comparator inputs tied together, it struck me there might be a better way to generate all the reference voltages required, and there is.

The new method, which I call a "matrix divider" system, uses only 10 resistors. This is more significant than it might at first appear, because the LEDs are on the edge of the PCB, the comparators are in compact quad packages, and so the divider resistors actually take up quite a large proportion of the PCB area. Reducing their number by half made fitting the meter into a pre-existing and rather cramped meter bridge design possible without recourse to surface-mount techniques. There are now two potential dividers. Divider A is driven by the output of the rectifier circuit, while Divider B produces a series of fixed voltages with respect to the +8.0V subrail. As the input signal increases, the output of the meter rectifier goes straight to comparators U16–U20, which take their reference voltages from Divider B and turn on in sequence as described earlier. Comparators U11–U15 are fed with the same reference voltages from Divider B, but their signal from the meter rectifier is attenuated by Divider A, coming from the tap between R3 and R5, and

so these comparators require more input signal to turn on. This process is repeated for the third bank of comparators U6–U10, whose input signal is further attenuated, and finally for the fourth bank of comparators U1–U5, whose input is still further attenuated. The result is that all the comparators switch in the correct order.

Since in this application there was only a single supply rail, a bias generator is required to generate an intermediate subrail to bias the opamps. This subrail is set at +8.0V rather than V/2 to allow enough headroom for the rectifier circuit, which produces only positive outputs; it is generated by R18, R19 and C3 and buffered by opamp section U3:B. The +24V supply is protected by a 10 Ω fusible resistor R22, so if a short circuit occurs the resistor will fail to open without flame. A small but vital point is that the supply for Divider B is taken from outside this fusing resistor; if it was not, the divider voltages would vary with the number of LED chains powered, upsetting meter accuracy.

The LM2901 quad voltage comparator is used again here, as it has low input offset voltages and the requisite open-collector outputs. Transistors Q1, Q2 can be any TO-92 devices with reasonable beta; their maximum power dissipation, which occurs with only one LED on in the chain, is a modest 128 mW. This meter system has been used by me in commercial products with great success, and its only downside is that it is a bit harder to understand than conventional meter circuitry.

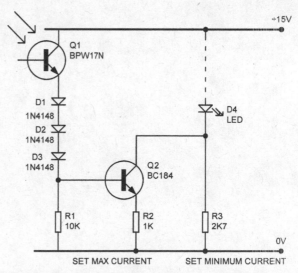

Figure 15.7 Automatic brightness control of LEDs with a phototransistor sensor on the front panel

Automatic Brightness Control

Level meters or other indicators have to be bright to be visible in daylight. This often means that they are excessively bright and distracting in a semi-darkened listening room. While manual brightness control is a possibility, it is clumsy, and a better way is automatic brightness control with a photo-sensor mounted on the front panel. Figure 15.7 shows a simple circuit that I have used several times. In darkness the phototransistor Q1 does not conduct and the LED current is set at its minimum by R3. When light hits the front panel Q1 conducts and turns on Q2, increasing the current through the LED chain to a maximum set by R2. D1–D3 modify the control law to give smooth operation. Take care that light from the LEDs cannot reach the phototransistor, or you may get optical oscillation.

Reference

1. https://en.wikipedia.org/wiki/Mandy_Rice-Davies Accessed Nov 2016.

Chapter 16

Power Supplies

"We thought, because we had power, we had wisdom." *Stephen Vincent Benet:* Litany for Dictatorships, *1935*

Power for Phono Preamplifiers

Moving-magnet cartridges generate small signals, at about 5 mV rms, while moving-coil cartridges give even smaller ones, down to 50 uV in some cases. This might lead you to suppose that very clean power supplies are essential to get the best noise and hum figures, but as with so many things that seem obvious in audio, it is not the case. I have designed many phono preamps that were powered directly by 78/79 or LM317/337 type IC regulators, with no additional filtering. In no case did rail noise or ripple make any detectable contribution to the output. To give this some perspective, professional microphone preamplifiers work with signal down to −80 dBu, which is 77 uV rms, and I have powered thousands of them directly from supply rails provide by quite conventional power supplies, with no hint of a problem. They were of course balanced preamplifiers, and this undoubtedly helps.

This immunity rail noise or ripple is very largely due to the excellent power supply rejection ratio (PSRR) of opamps such as the 5532 or the LM4562, but there are some important points to be made.

1) The decoupling capacitor grounds must be taken back to the point where power enters the PCB to prevent them injecting noise or ripple into the signal ground. In particular the input ground and the ground at the bottom of the negative-feedback network must be at the same potential, and allowing decoupling currents to flow through this sensitive part of the circuit will be disastrous.
2) It is assumed earlier that the phono preamp consists only of opamps with high PSRR. Some configurations however use a discrete transistor or transistors as a front end to an opamp, to reduce noise, and these

must be powered from one or both supply rails. Single-transistor configurations in particular are likely to have no PSRR at all, and heavy filtering of the supply will be necessary; simple R-C filters are usually all that is required, but the capacitances can get quite large (say 1000 uF) if the voltage drop across the associated resistor is to be kept low. Configurations with two discrete devices in a differential pair are likely to show some PSRR, but it will not be of the same order as an opamp, and supply filtering will almost certainly still be necessary.

3) It is assumed that the power supply is properly designed so that the IC regulators give the performance they are capable of. Specifically, ripple currents on the unregulated side must be kept out of the reference and regulated output circuitry. This is not hard to do.
4) The most difficult cause of power-supply-related hum problems is not so much the supply rails as the magnetic field from the mains transformer. If you have this inside the same box as the phono preamp, then in my experience it will define the hum performance of the circuitry. A separate external supply, even if it is only a humble wall-wart, i.e. plug-type AC adapter, gives physical distance and is the only real solution to this issue; there is more on this at the end of the chapter.

Opamp Supply Rail Voltages

It has been mentioned several times in the earlier chapters of this book that running opamps at the slightly higher voltage of ±17V rather than ±15V gives an increase in headroom and dynamic range of 1.1 dB for virtually no cost and with no reliability penalty. Soundcraft ran all the opamps in their mixing consoles at ±17V for at least two decades, and opamp failures were almost unknown. This recommendation assumes that the opamps concerned have a maximum supply voltage rating of ±18V, which is the case for the Texas TL072, the new LM4562, and many other types.

The 5532 is (as usual) in a class of its own. Both the Texas and Fairchild versions of the NE5532 have an absolute maximum power supply voltage rating of ±22V, (though Texas also gives a "recommended supply voltage" of ±15V), but I have never met any attempt to make use of this capability. The 5532 runs pretty warm on ±17V when it is simply quiescent, and my view (and that of almost all the designers I have spoken to) is that running it at any higher voltage is simply asking for trouble. This is a particular concern in the design of mixing consoles, which may contain thousands of opamps—anything that that impairs their reliability is going to cause a *lot* of trouble. In any case, moving from ±17V rails to ±18V rails only gives 0.5 dB more headroom. Stretching things to ±20V would give 1.4 dB more than ±17V, and running on the ragged edge at ±22V would yield a more significant 2.2 dB more than ±17V, but you really wouldn't want to do it. Pushing the envelope like this is also going to cause difficulties if you want to run opamps with maximum supply ratings of ±18V from the same power supply.

We will therefore concentrate here on ±17V supplies for opamps, dealing first with what might be called "small power supplies" i.e. those that can be conveniently built with TO-220 regulators. This usually means an output current capability that does not exceed 1.5 amps, which is plenty for even complicated phono amplifiers, preamplifiers etc.

An important question is: how low does the noise and ripple on the supply output rails need to be? Opamps in general have very good power supply rejection ratios (PSRR), and some manufacturer's specs are given in Table 16.1.

The PSRR performance is actually rather more complex than the bare figures given in the table imply; PSRR is typically frequency-dependent (deteriorating as frequency rises) and different for the +V and −V supply pins. It is however rarely necessary to get involved in this degree of detail. Fortunately even the cheapest IC regulators (such as the venerable 78xx/79xx series) have low enough noise and ripple outputs that opamp PSRR performance is rarely an issue.

There is however another point to ponder; if you have a number of electrolytic-sized decoupling capacitors between rail and ground, enough noise and ripple can be coupled into the nonzero ground resistance to degrade the noise floor. Intelligent placing of the decouplers can help—putting them near where the ground and supply rails come onto the PCB means that ripple will go straight back to the power supply without flowing through the ground tracks on the rest of the PCB. This is of limited effectiveness if you have a number of PCBs connected to the same IDC cable, as in many small mixing desks, and in such cases low ripple power supplies may be essential.

Apart from the opamp supply rails, audio electronics may require additional supplies, as shown in Table 16.2.

It is often convenient to power relays from a +9V unregulated supply that also feeds the +5V microcontroller regulator—see later in this chapter. The use of +24V to power LED metering systems is dealt with in Chapter 15 on metering.

Designing a ±15V Supply

Making a straightforward ±15V 1 amp supply for an opamp-based system is very simple, and has been ever since the LM7815/7915 IC regulators were introduced (which was a long time ago). They are robust and inexpensive parts with both overcurrent and overtemperature protection and give low enough output noise for most purposes. We will look quickly at the basic circuit because it brings out a few design points which apply equally to more complex variations on the theme. Figure 16.1 shows the schematic, with typical component values; a centre-tapped transformer, a bridge rectifier, and two reservoir capacitors C1, C2 provide the unregulated rails that feed the IC regulators. The secondary fuses must be of the slow-blow type. The small capacitors C7–C9 across the input to the bridge reduce RF emissions from the rectifier diodes; they are shown as X-cap types not because they have to withstand 230 Vrms but to underline the need for them to be rated to

Table 16.1 PSRR specs for common opamps

Opamp type	PSRR minimum dB	PSRR typical dB
5532	80	100
LM4562	110	120
TL072	70	100

Table 16.2 Typical additional supply rails for opamp-based systems

Supply voltage	Function
+5V	Housekeeping microcontroller
+9V	Relays
+24V	LED bargraph metering systems, discrete audio circuitry, relays

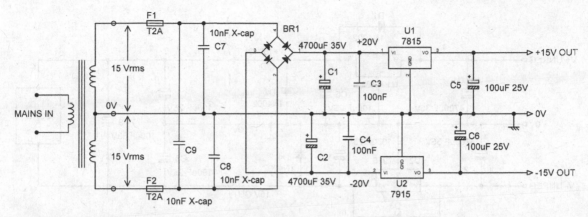

Figure 16.1 A straightforward ±15V power supply using IC regulators

withstand continuous AC stress. The capacitors C3, C4 are to ensure HF stability of the regulators, which like a low AC impedance at their input pins, but these are only required if the reservoir capacitors are not adjacent to the regulators, i.e. more than 10 cm away. C5, C6 are not required for regulator stability with the 78/79 series—they are there simply to reduce the supply output impedance at high audio frequencies.

There are really only two electrical design decisions to be made; the AC voltage of the transformer secondary and the size of the reservoir capacitors. As to the first, you must make sure that the unregulated supply is high enough to prevent the rails dropping out (i.e. letting hum through) when a low mains voltage is encountered but not so high that either the maximum input voltage of the regulator is exceeded or it suffers excessive heat dissipation. How low a mains voltage it is prudent to cater for depends somewhat on where you think your equipment is going to be used, as some parts of the world are more subject to brown-outs than others. You must consider both the minimum voltage drop across the regulators (typically 2V) and the ripple amplitude on the reservoirs, as it is in the ripple troughs that the regulator will first "drop out" and let through unpleasantness at 100 Hz.

In general, the RMS value of the transformer secondary will be roughly equal to the DC output voltage.

The size of reservoir capacitor required depends on the amount of current that will be drawn from the supply. The peak-to-peak ripple amplitude is normally in the region of 1 to 2 volts; more ripple than this reduces efficiency because the unregulated voltage has to be increased to allow for unduly low ripple troughs, and less ripple is usually unnecessary and gives excessive

reservoir capacitor size and cost. The amount of ripple can be estimated with adequate accuracy by using Equation 16.1

$$Vpkpk = \frac{I \cdot \Delta t \cdot 1000}{C} \qquad 16.1$$

where:

Vpk-pk is the peak-to-peak ripple voltage on the reservoir capacitor.

I is the maximum current drawn from that supply rail in amps.

Δt is the length of the capacitor discharge time, taken as 7 milliseconds.

C is the size of the reservoir capacitor in microfarads. The "1000" factor simply gets the decimal point in the right place.

Note that the discharge time is strictly a rough estimate and assumes that the reservoir is being charged via the bridge for 3 msec and then discharged by the load for 7 msec. Rough estimate it may be, but I have always found it works very well.

The regulators must be given adequate heatsinking. The maximum voltage drop across each regulator (assuming 10% high mains) is multiplied by the maximum output current to get the regulator dissipation in watts, and a heatsink selected with a suitable thermal resistance to ambient (in °C per watt) to ensure that the regulator package temperature does not exceed, say, 90 °C. Remember to include the temperature drop across the thermal washer between regulator and heatsink.

Under some circumstances it is wise to add protective diodes to the regulator circuitry, as shown in Figure 16.2.

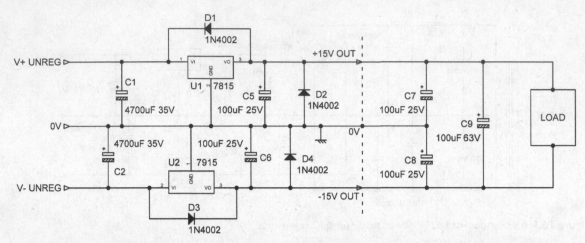

Figure 16.2 Adding protection diodes to a ±15V power supply. The load has decoupling capacitors to both ground (C7, C8) and between the rails (C9); the latter can cause start-up problems. DO NOT FIT C9.

The diodes D1, D3 across the regulators are reverse-biased in normal operation, but if the power supply is driving a load with a large amount of capacitance, it is possible for the output to remain higher in voltage than the regulator input as the reservoir voltage decays. D1, D3 prevent this effect from putting a reverse voltage across the regulators. Such diodes are not usually required with normal opamp circuitry, as the amount of rail decoupling, shown as C7, C8 in Figure 16.2, is usually modest.

The shunt protection diodes D2, D4 are once again reverse-biased in normal operation. D2 prevents the +15V supply rail from being dragged below 0V if the −15V rail starts up slightly faster, and likewise D4 protects the −15V regulator from having its output pulled above 0V. This can be an important issue if rail-to-rail decoupling such as C9 is in use; such decoupling can be useful because it establishes a low AC impedance across the supply rails without coupling supply rail noise into the ground, as C7, C8 are prone to do. However, it also makes a low-impedance connection between the two regulators. D2, D4 will prevent damage in this case but leave the power supply vulnerable to start-up problems; if its output is being pulled down by the −15V regulator, the +15V regulator may refuse to start. This is actually a very dangerous situation, because it is quite easy to come up with a circuit where start-up will only fail one time in twenty or more, the incidence being apparently completely random but presumably controlled by the exact point in the AC mains cycle where the supply is switched on and other variables such as temperature, the residual charge left on the reservoir capacitors, and the phase of the moon. Do *not* fit C9.

If even one start-up failure event is overlooked or dismissed as unimportant, then there is likely to be serious grief further down the line. *Every power supply start-up failure must be taken seriously.*

Designing a ±17V Supply

There are 15 V IC regulators (7815, 7915) and there are 18 V IC regulators (7818, 7918), but there are no 17 V IC regulators. This problem can be effectively solved by using 15 V regulators and adding 2 volts to their output by manipulating the voltage at the REF pin. The simplest way to do this is with a pair of resistors that divide down the regulated output voltage and apply it to the REF pin, as shown in Figure 16.3a. (The transformer and AC input components have been omitted in this and the following diagrams, except where they differ from those shown earlier). Since the regulator maintains 15V between the OUT and REF pin, with suitable resistor values the actual output with respect to 0V is 17V.

The snag with this arrangement is that the quiescent current that flows out of the REF pin to ground is not well controlled; it can vary between 5 and 8 mA, depending on both the input voltage and the device temperature. This means that R1 and R2 have to be fairly low in value so that this variable current does not cause excessive variation of the output voltage, and therefore power is wasted.

If a transistor is added to the circuit as in Figure 16.3b, then the impedance seen by the REF pin is much lower. This means that the values of R1 and R2 can be increased by an order of magnitude, reducing the waste

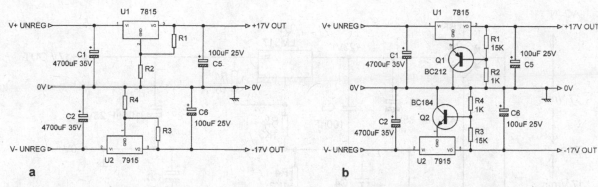

Figure 16.3 Making a ±17V power supply with 15V IC regulators. a) Using resistors is inefficient and/or inaccurate; b) adding transistors to the voltage-determining resistor network makes the output voltage more predictable and reduces the power consumed in the resistors.

of regulator output current and reducing the heat liberated. This sort of manoeuvre is also very useful if you find that you have a hundred thousand 15 V regulators in store, but what you actually need for the next project is an 18 V regulator, of which you have none.

What about the output ripple with this approach? I have just measured a power supply using the exact circuit of Figure 16.3b, with 2200 uF reservoirs, and I found −79 dBu (87 uV rms) on the +17 V output rail and −74 dBu (155 uV rms) on the 17 V rail, which is satisfyingly low for inexpensive regulators and should be adequate for almost all purposes; note that these figures include regulator noise as well as ripple. The load current was 110 mA. If you *are* plagued by ripple troubles, the usual reason is a rail decoupling capacitor that is belying its name by coupling rail ripple into a sensitive part of the ground system, and the cure is to correct the grounding rather than design an expensive ultra-low ripple PSU. Note that doubling the reservoir capacitance to 4400 uF only improved the figures to −80 dBu and −76 dBu respectively; just increasing reservoir size is not a cost-effective way to reduce the output ripple.

Using Variable-Voltage Regulators

It is of course also possible to make a ±17V supply by using variable output voltage IC regulators such as the LM317/337. These maintain a small voltage (usually 1.2V) between the OUTPUT and ADJ (shown in figures as GND) pins and are used with a resistor divider to set the output voltage. The quiescent current flowing out of the ADJ pin is a couple of orders of magnitude lower than for the 78/79 series, at around 55 uA, and so a simple resistor divider gives adequate accuracy of the output voltage, and transistors are no longer needed to

absorb the quiescent current. A disadvantage is that this more sophisticated kind of regulator is somewhat more expensive than the 78/79 series; at the time of writing they cost something like 50% more. The 78/79 series with transistor voltage-setting remains the most cost-effective way to make a non-standard-voltage power supply at the time of writing.

It is clear from Figure 16.4 that the 1.2 V reference voltage between ADJ and out is amplified by many times in the process of making a 17 V or 18 V supply; this not only increases output ripple but also output noise, as the noise from the internal reference is being amplified. The noise and ripple can be considerably reduced by putting a capacitor C7 between the ADJ pin and ground. This makes a dramatic difference; in a test PSU with a 650 mA load, the output noise and ripple was reduced from −63 dBu (worse than 78xx series) to −86 dBu (better than 78xx series), and so such a capacitor is usually fitted as standard. If it is fitted, it is then essential to add a protective diode D1 to discharge C7, C8 safely if the output is short-circuited, as shown in Figure 16.5.

The ripple performance of the aforementioned test PSU, with a 6800 uF reservoir capacitor and a 650 mA load, is summarised for both types of regulator in Table 16.3. Note that the exact ripple figures are subject to some variation between regulator specimens.

Improving Ripple Performance

Table 16.3 shows that the best noise and ripple performance that can be expected from a simple LM317 regulator circuit is about −86 dBu (39 uV rms), and this still contains a substantial ripple component. The reservoir

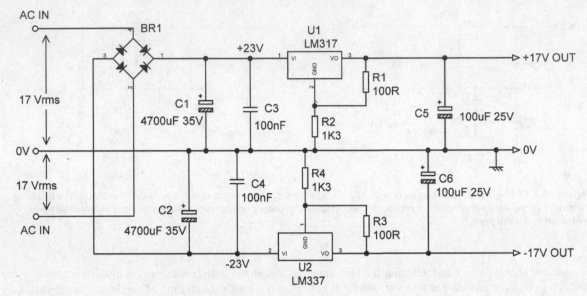

Figure 16.4 Making a ±17V power supply with variable-voltage IC regulators

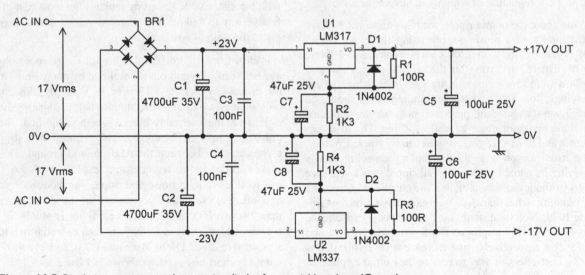

Figure 16.5 Ripple improvement and protective diodes for a variable-voltage IC regulator

capacitors are already quite large at 4700 uF, so what is to be done if lower ripple levels are needed? The options are:

1) Look for a higher-performance IC regulator. They will cost more, and there are likely to be issues with single sourcing.

2) Design your own high-performance regulator using discrete transistors or opamps. This is not a straightforward business if all the protection that IC

regulators have is to be included. There can also be distressing issues with HF stability.

3) Add an RC input filter between the reservoir capacitor and the regulator. This is simple and pretty much bullet-proof and preserves all the protection features of the IC regulator, though the extra components are a bit bulky and not that cheap. There is some loss of efficiency due to the voltage drop across the series

resistor; this has to be kept low and the capacitance large.

Table 16.3 Comparing the noise and ripple output of various regulator options

	7815 + transistor dBu	LM317 dBu	LM317 uV
No C on LM317 ADJ pin	−73 dBu (all ripple)	−63 dBu (ripple & noise)	549 uV
47 uF on LM317 ADJ pin	−73 dBu (all ripple)	−86 dBu (ripple & noise)	39 uV
Input filter 2.2Ω & 2200 uF	−78 dBu (ripple & noise)	−89 dBu (mostly noise)	27 uV
Input filter 2.2Ω & 4400 uF	−79 dBu (mostly noise)	−90 dBu (all noise)	24 uV

The lower two rows of Table 16.3 show what happens. In the first case the filter values were 2.2 Ω and 2200 uF. This has a −3 dB frequency of 33 Hz and attenuates the 100 Hz ripple component by 10 dB. This has a fairly dramatic effect on the output ripple, but the dB figures do not change that much as the input filter does not affect the noise generated inside the regulator. Increasing the filter capacitance to 4400 uF sinks the ripple below the noise level for both types of regulator.

Dual Supplies From a Single Winding

It is extremely convenient to use third-party "wall-wart" power supplies for small pieces of equipment, as they come with all the safety and EMC approvals already done for you, though admittedly they do not look appropriate with high-end equipment. The problem is that the vast majority of these supplies give a single AC voltage on a two-pole connector, so a little thought is required to derive two supply rails. Figure 16.6 shows how it is done in a ±18V power supply; note that these voltages are suitable only for a system that uses 5532s throughout. Two voltage-doublers of opposite polarity are used to generate the two unregulated voltages. When the incoming voltage goes negative, D3 conducts and the positive end of C1 takes up approximately 0V. When the incoming voltage swings positive, D1 conducts instead and the charge on C1 is transferred to C3. Thus the whole peak-to-peak voltage of the AC supply appears across reservoir capacitor C3. In the same way, the peak-to-peak voltage, but with the opposite polarity, appears across reservoir C4.

Since voltage-doublers use half-wave rectification, they are not suitable for high current supplies, but this arrangement should be more than adequate for most phono preamplifiers. When choosing the value of the reservoir capacitor values, bear in mind that the discharge time in Equation 16.1 must be changed from 7 msec to 17 msec. The input capacitors C1, C2 should be the same size as the reservoirs.

Power supplies for discrete circuitry

One of the main reasons for using discrete audio electronics is the possibility of handling larger signals than can be coped with by opamps running off ±17 V rails. The use of ±24 V rails allows a 3 dB increase in headroom, which is probably about the minimum that justifies the extra complications of discrete circuitry. A ±24 V supply can be easily implemented with 7824/7924 IC regulators. On the other hand you have to consider that stages downstream may have opamps running off ±17 V

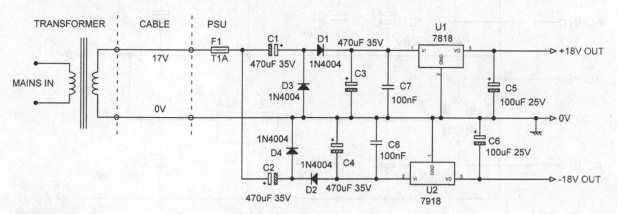

Figure 16.6 A ±18V power supply powered by a single transformer winding

rails, and you don't want to supply signal voltages that will blow them up. Hi-fi equipment rarely if ever has over-voltage protection.

A slightly different approach was used in my first published preamplifier design.[1] This preamp in fact used two LM7824 +24V regulators connected as shown in Figure 16.7 because, at the time, the LM7924 −24V regulator had not yet reached the market. The use of a second positive regulator to produce the negative output rail looks a little strange at first sight, but I can promise you it works. It can be very useful in the sort of situation described earlier; you have a hundred thousand +15V regulators in store but no −15V regulators . . . I'm sure you see the point.

Note that this configuration requires two completely separate transformer windings; it cannot be used with a centre-tapped secondary.

Mutual Shutdown Circuitry

It is an awkward quirk of 5532 opamps that if one supply rail is lost and collapses to 0V, while the other rail remains at the normal voltage, they can under some circumstances get into an anomalous mode of operation that draws large supply currents and ultimately destroys the opamp by over-heating. To prevent damage from this cause, which could be devastating to a large mixing console, the opamp supplies are very often fitted with

a mutual shutdown system. Mutual shutdown ensures that if one supply rail collapses because of overcurrent, over-temperature, or any other cause, the other rail will be promptly switched off. The extra circuitry required to implement this is shown in Figure 16.8, which is an example of a high current supply using 7.5 amp regulators.

The extra circuitry to implement mutual shutdown in Figure 16.8 is very simple; R5, D3, R6 and Q1 and Q2. Because R5 is equal to R6, D3 normally sits at around 0 V in normal operation. If the +17V rail collapses, Q2 is turned on by R6, and the REF pin of U2 is pulled down to the bottom rail, reducing the output to the reference voltage (1.25 V). This is not completely off, but it is low enough to prevent any damage to opamps.

If the −17 V rail collapses, Q1 is turned on by R5, pulling down the REF pin of U1 in the same way. Q1 and Q2 do not operate exactly symmetrically, but it is close enough for our purposes.

Note that this circuit can only be used with variable output voltage regulators because it relies on their low reference voltages.

Microcontroller and Relay Supplies

It is very often most economical to power relays from an unregulated supply. This is perfectly practical, as

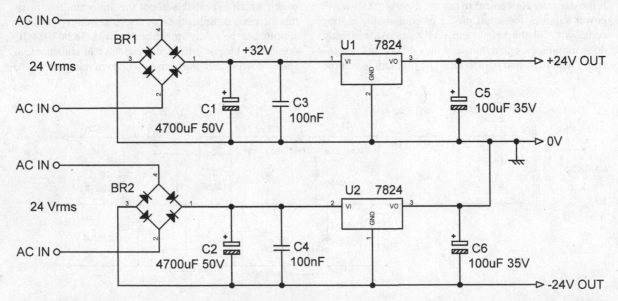

Figure 16.7 A ±24V power supply using only positive regulators

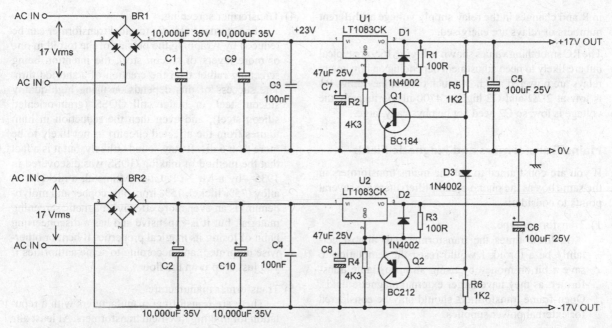

Figure 16.8 A high current ±17V power supply with mutual shutdown circuitry

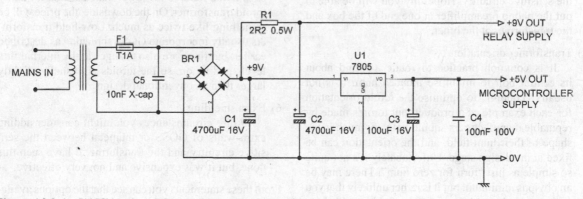

Figure 16.9 A +5V PSU with an RC smoothed +9V relay supply

relays have a wide operating voltage range. If 9 V relays are used, then the same unregulated supply can feed a +5 V regulator to power a microcontroller, as shown in Figure 16.9.

Hum induced by electrostatic coupling from an unregulated relay supply rail can be sufficient to compromise the noise floor; the likelihood of this depends on the physical layout, but inevitably the signal paths and the relay supply come into proximity at the relay itself. It is therefore necessary to give this rail some degree of smoothing, without going to the expense of another

regulator and heatsink. (There must be no possibility of coupling between signal ground and relay power ground; these must only join right back at the power supply.) This method of powering relays is more efficient than a regulated rail because it does not require a voltage drop across a regulator that must be sufficient to prevent drop-out and consequent rail ripple at low mains voltages.

Simple RC smoothing works perfectly well for this purpose. Relays draw relatively high currents, so a low R and a high-value C are used to minimise voltage losses in R and changes in the relay supply voltage as different numbers of relays are energised.

The RC smoothing values shown in Figure 16.9 are typical but are likely to need adjustment depending on how many relays are powered and how much current they draw. R1 is low at 2.2Ω and C2 high at 4700 uF; fortunately the voltage is low, so C2 need not be physically large.

Mains Transformers and Magnetic Fields

If you are constrained to put the mains transformers in the same box as the phono preamplifier, there are several points to consider:

1) Transformer type.

In these times the transformer will almost certainly be a toroid. I would resist the temptation to save a bit on money by using an open-frame transformer, as they have larger external magnetic fields. Open-frame transformers should only be considered for external power supplies.

2) Transformer position.

Put the transformer as far away as possible from the sensitive circuitry. Hopefully you will be able to put the phono preamplifier at one end of the box and the transformer at the other.

3) Transformer orientation.

It is common practice to rotate a toroid about its central bolt to minimise induced hum. It is not usually economic to optimise the toroid orientation for each example of a product, but toroids made by reputable manufacturers should not vary much in the shape of their hum field, and the orientation can be fixed at the design stage. Unfortunately it's not quite so simple as just "turn for zero hum". There may be an obvious minimum, but it is rather unlikely that you will get zero hum; the transformer field is unlikely to be getting into the electronics at just one point.

If the susceptible electronics is spread out over space, perhaps with left and right channels on opposite sides of the enclosure (not a good idea for preamplifiers), then with dreadful certainty it will be found that the hum minimum for one channel is something like the maximum for the other. Even if the two channels are immediately adjacent, you are unlikely to be able to get a good minimum on both at once.

Some toroids have single-strand secondary lead-outs, which are too stiff to allow rotation; for experimental toroid-turning these can be cut short and connected to suitably large flexible wire such as 32/02, with carefully sleeved and insulated joints.

4) Transformer screening.

The external field of a toroidal transformer can be reduced by wrapping the outside of the toroid in one or more layers of silicon steel, the intention being screening rather than the creation of a shorted turn. The success of this depends on using high-quality silicon steel, or better still GOSS (grain-oriented silicon steel), and even then the reduction in hum figures from the affected circuitry is not likely to be more than 6 dB. It may sound unlikely, but it is a fact that the method of making GOSS was discovered in 1935—by a Mr N. P. Goss. Mumetal, a nickel-iron alloy (75% nickel, 15% iron, plus copper and molybdenum) is an even more effective magnetic screening material, but it is expensive and has a disconcerting habit of losing its magical properties if bent or otherwise deformed, and according to some authorities if you just drop it on the floor.

5) Transformer manufacture.

There are transformer manufacturers with a reputation for making low-field transformers. At least one toroid manufacturer specialises in low-field designs for audio applications, and their products can be have an external field 10 dB lower than a standard-quality toroid transformer. On the downside, the price will be something like twice as much. Low-field transformers usually incorporate GOSS screening as described earlier, but there are also changes to the internal flux levels and so on, so the toroids are usually slightly larger than a conventional design.

6) Extra shielding.

In dire circumstances you might consider adding extra walls of GOSS or mumetal between the sensitive circuitry and the transformer. I have seen this done, but it was expensive and not very effective.

From these statements you can see that the options available for reducing transformer hum are strictly limited. The transformer field might be called the transmitting end of the problem, while the receiving end is the electronics of the phono preamplifier, and it is obviously helpful to minimise its sensitivity to magnetic fields. The basic principle is simple—make sure that the loop area between signal conductors and the signal ground is as small as possible. This is relatively easy for input connections, where it is of course crucial, but much harder to do on a PCB full of components, and with NFB paths and so on. I have spent time cutting tracks and replacing them with bits of solid-core that could be rerouted, and my conclusion is that it is a thankless task and unlikely to give much improvement. Every new layout requires a PCB iteration to test it properly, and that takes time and money.

External Power Supplies

Even if you follow all the precautions in the previous section, it is very difficult to keep all traces of transformer-induced hum out of the signal circuitry. It is highly irritating to find that despite the cunning use of low-noise circuitry, the noise floor is defined by the deficiencies of a component—for the ideal transformer would obviously have no external field—rather than the laws of physics as articulated by Johnson. The authoritative solution is to put the mains transformer in a separate box which can be placed a metre or more away from the preamplifier unit, which is powered through an umbilical lead. A cable that is captive at the power supply end saves the cost of a mating pair of connectors, which may be considerable. The supply voltages involved will probably be below the 50 volt limit set by the Low Voltage Directive, so there is no need to take elaborate precautions to ensure that the connector contacts cannot be touched.

Advantages:

- The transformer field hum problem is completely solved.
- It will appeal to some potential customers as a "serious" approach to high-end audio.
- There are no high-voltage AC mains inside the preamp box. If you are going to encourage the customer to open the box to change input loading capacitance, etc., this is essential for safety reasons. In general you don't want customers opening boxes; in the world of safety regulations, if you need a tool such as a screwdriver to open the box, that should give you a hint you shouldn't be doing it.

Disadvantages:

- The cost of an extra enclosure plus an extra cable and connectors, power indicator lights etc. The connectors will probably have to be multi-pole. The transformer box must have fuses or other means of protection in case of short-circuits in the cable.
- In my experience a significant proportion of users will, exhortations to the contrary notwithstanding, promptly place the amplifier box directly on top of the transformer box, immediately defeating the whole object. This is particularly likely if the two boxes have the same footprint and so look as if they *ought* to be stacked together. However, all may not be lost in this situation, as the transformer is still physically further away from the sensitive electronics (though if the transformer has a large field emerging from its

ends things may actually be worse), and there are now two extra layers of steel interposed (assuming the boxes are made of steel, that is). Unfortunately plain sheet steel is not good at stopping magnetic fields and may even increase the coupling.

The cheapest and simplest external supply is a plug-type AC adapter, i.e. a wall-wart. Higher-power versions are usually fitted in the cable and are called "line lumps". The adapter produces low voltage AC either with a conventional transformer or a small switch-mode supply. The latter should be avoided if you want power free from high-frequency interference. The output is isolated from the mains, and to underline the point, the ground pin (I'm thinking here of UK 13 amp plugs) is usually plastic, and its only function is to lift the safety shutter over the live and neutral contacts so the plug can be inserted. These adapters come with safety and EMC approvals done for you, with the important proviso that you choose a reputable supplier. There are dodgy versions available that may be covered with fake safety approvals and CE marks but are actually a serious safety hazard; please take this issue seriously. Fraudulent marking of substandard adapters with the name of reputable manufacturers is known,[2] and a trustworthy supplier is essential.

Most AC adapters produce a single voltage on a two-pin connector; generating + and − supply rails from this is no problem so long as you only need a small amount of power; see the earlier section "Dual Supplies From a Single Winding". Centre-tap outputs on three-pin connectors are available but noticeably more expensive due to much lower production levels.

The use of an AC adapter naturally means that mains-frequency power is brought inside the preamp box, and rectification, smoothing, and regulation take place in there too. While this is perhaps not ideal, the problems are tiny compared with the magnetic field of a transformer, and there is no reason why the audio performance should be in any way compromised given suitable care in design. There is the disadvantage that the current in the umbilical cable will consist of short but relatively large charging pulses at 100 or 120 Hz, so the cable needs to kept away from audio leads. The pulses give a greater voltage drop in the cable resistance than a steady current but also give rise to much greater I^2R heating. At these low power levels the latter is unlikely to cause problems in the cable itself but can be fatal to the contacts of connectors. Speaking from bitter experience, I can warn you that connectors that appear to have a more than adequate safety margin can fail under these conditions, so be cautious with ratings and do plenty of testing.

If you are building your own power supply from scratch in its separate box, there are more options:

1) Put not just the mains transformer but also the rectifiers and reservoir capacitors in the power supply box. The current in the umbilical cable is now rectified and smoothed DC, which is much less likely to interfere with anything, and it is much easier to specify connectors to cope with it. The regulators are placed in the preamp box, so their outputs can be directly connected to the audio electronics, and the impedance of the DC supply cable is not an issue. Depending on the design of the connectors, it may be necessary to provide short circuit protection for the smoothed DC, either with plain fuses or, better, with positive-tempco polyfuses that self-reset.

2) Put the mains transformer, rectifiers, reservoir capacitors and regulators in the power supply box. The current in the umbilical cable is now regulated DC, which cannot interfere with anything, and there are no connector issues. The downside is that the supply cable impedance is now in series with the supply to the audio circuitry. Some sizable decoupling capacitors may be needed at the preamp end. Short circuit protection will be provided by the regulators.

My final thought here is that I would rather use a cheap Chinese wall-wart with a metre-long lead than have the most super-expensive toroid in the world inside the preamp box.

References

1. Self, D. "An Advanced Preamplifier Design" *Wireless World*, Nov 1976.

2. https//en.wikipedia.org/wiki/AC_adapter Accessed Sept 2016.

Chapter 17

Moving-Magnet Inputs

Practical Designs

The closely observed designs given here will be handy for immediate use, and they also demonstrate many of the various techniques discussed in this book, with extra practical details added. There is much information on how to interface the stages properly and avoid bad loading effects. Input and output networks, and all practical details, are given so that each circuit is complete in itself.

Project 1: A Simple Single-Stage +40 dB Phono Amplifier

While this book contains much about achieving the highest quality, sometimes economy is the watchword. This design is intended to give that economy. It is a basic one-stage phono amplifier with a gain of +40 dB (1 kHz) intended for driving preamplifiers directly; the nominal output is 500 mV rms for 5 mV rms in; the maximum input is 100 mV rms, giving a +26 dB overload margin, which is not stupendous but adequate for all but the hottest cartridges and vinyl.

In the interests of economy, this circuit, unlike most in this book, has all resistors restricted to single E24 components, i.e. 1xE24 format. Single capacitors are also used, and Configuration-C employed to minimise total RIAA capacitance. The relatively high gain of +40 dB means that the HF correction pole can be omitted with only small errors introduced at the extreme HF; hence the absence of R3. This also leads to a low value for R0, so its noise contribution will be very small indeed; see Chapter 9 on noise. The only subsonic filtering is provided by a restricted value for C0. Its original value was 470 uF which gave −1.2 dB at 8 Hz, −13 dB at 1 Hz. When the resistor values were converted to the nearest E24, this introduced a response rise in the LF which could be partly countered by further reducing the value

of C0 to 330 uF. This means of course that the response is to a small extent at the mercy of the C0 tolerances, and this is one of the many compromises inevitable in an economy design. You will recall that I cautioned against using electrolytics to set time-constants not only because of tolerance issues but because they distort if significant signal voltage appears across them; since the nominal signal level is 5 mV rms, this is not a problem here.

The exact component values that were the basis of this design can be found in Table 7.31; they were calculated for +40 dB (1 kHz) by adjusting the value of R0 so that so that C1 comes out as exactly 33 nF.

DC drain resistor R4 is very much noncritical, so it has been made 68 kΩ to match Rin and R1; this sort of thing makes component procurement easier. R5 is the output isolating resistor, which ensures stability into cable capacitance and is also non-critical so long as it is greater than 47 Ω, so it is set to 68 Ω to match R0. R4 comes before R5 to avoid a tiny but irritating loss in level of 0.009 dB.

Figure 17.2 shows the RIAA response. The error is within ±0.2 dB (just), which I think is pretty good for so simple a circuit. The central trace is for C0 = 330 uF exactly, and those either side show the effect of a ±20% tolerance. C0 makes the response −2.6 dB at 8 Hz and −17 dB at 1 Hz; not exactly proper subsonic filtering but better than nothing. If desired the LF roll-off can be eliminated by making C0 1000 uF 6V3. The largest RIAA errors are introduced by converting C2 from 11.408 nF to the E12 value of 12 nF and R2 from 9.5723 kΩ to the E24 value 9.1 kΩ. The dotted line at −0.41 dB shows that the overall gain has been slightly reduced to +39.4 dB (1 kHz) compared with the original design value of +40 dB.

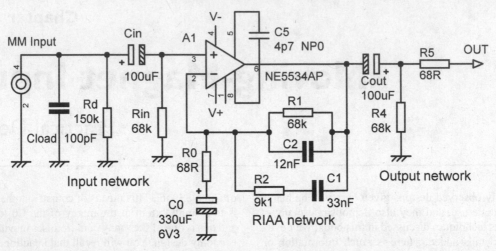

Figure 17.1 Project 1: Economy MM input with +40 dB (1 kHz) of gain, with 1xE24 resistors

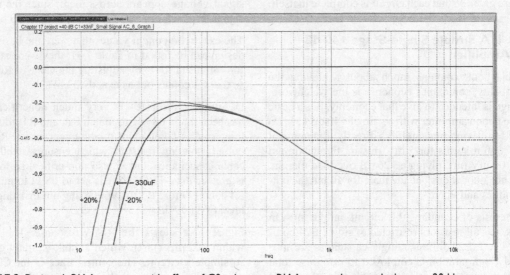

Figure 17.2 Project 1: RIAA accuracy with effect of C0 tolerances. RIAA spec only extends down to 20 Hz

Project 2: A Simple Single-Stage +40 dB Phono Amplifier Upgraded

In Project 1 we saw that the worst effects on the RIAA accuracy were due to forcing C2 to an E12 value and R2 to an E24 value. If we allow just a little more complexity, but without making every RIAA component multiple, we can make a remarkable improvement. The exact value of C2 is 11.408 nF, and we can get very close to that with 3x3n3 and 1x4n7 in parallel, giving 11.3 nF. These are small polystyrene capacitors and can be obtained inexpensively at 1% tolerance.

The exact value of R2 is 9.5723 kΩ; 16 kΩ and 24 kΩ in parallel are only +0.29% high, with an effective tolerance of 0.72%.

The RIAA accuracy is much improved, as shown in Figure 17.4. The gain at 1 kHz is now only 0.12 dB below the +40 dB aimed for, and relative to average gain the errors are within ±0.05 dB. However, bear in mind the effect of C0 tolerance. The response with exact component values is shown for comparison.

Most of the circuits in this book use multiple components for anything that is critical. Here however is an

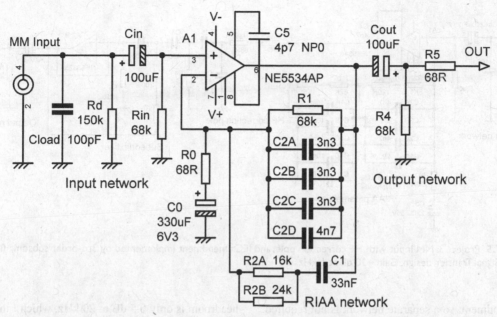

Figure 17.3 Project 2: Economy MM input with +40 dB (1 kHz) of gain, with multiple components for R2 and C2

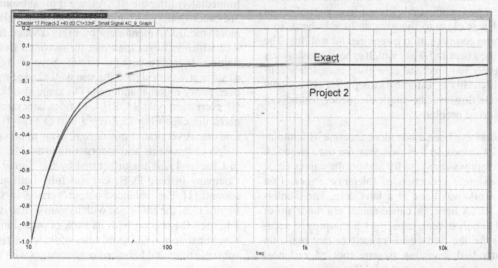

Figure 17.4 Project 2: RIAA accuracy compared with response for exact component values. RIAA spec only extends down to 20 Hz.

example of good results obtained with a halfway house between the 1xE24 and 2xE24 approaches.

Project 3: An MM Amplifier With Subsonic Filter

The next design in Figure 17.5 has a lower gain of +30 dB (1kHz), giving a really first-class overload margin of

36 dB. It is based on the MM section of the Signal Transfer MM/MC phono amplifier.[1] I have used this circuit for many years, and it has given complete satisfaction to many customers, though in the light of the latest knowledge its Configuration-A could be changed to Configuration-C to economise on precision capacitors; this is done in Project 4. It includes a typical subsonic filter which is designed with a slow initial roll-off that implements the

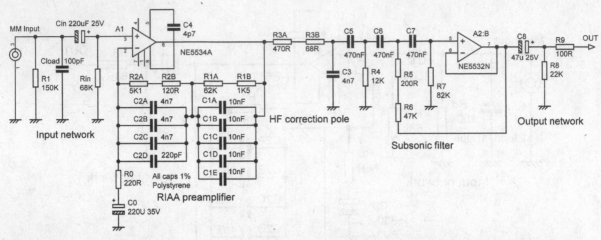

Figure 17.5 Project 3: MM input with HF correction pole, and IEC amendment implemented by 3rd-order subsonic filter. Based on Signal Transfer design. Gain +30 dB at 1 kHz.

IEC amendment, so a separate network is not required. A 5534A is used at the input stage to get the best possible noise performance. A 5534A without external compensation has a minimum stable closed-loop gain of about 3 times; that is close to the gain at 20 kHz here, so a touch of extra compensation is required for stability. The capacitor used here is 4.7 pF, which experience shows is both definitely required and also gives completely reliable stability. This is tested by sweeping a large signal from 20 kHz downwards; single-frequency testing can miss this sort of problem.

The resistors have been made more accurate by combining two E24 values. In this case they are used in series, and no attempt was made to try and get the values equal for the maximum reduction of tolerance errors. That statistical work was done at a later date. The Configuration-A RIAA network capacitances are made up of multiple 1% polystyrene capacitors for improved accuracy. Thus for the five 10 nF capacitors that make up C1, the standard deviation (square root of variance) increases by the square root of 5, while total capacitance has increased 5 times, and we have inexpensively built an otherwise costly 0.44% close-tolerance 50 nF capacitor. You will note that 5 × 10 nF capacitors are required, whereas a Configuration-C RIAA network can do the same job with 4 × 10 nF. The RIAA accuracy is within ± 0.1 dB 20 Hz–20 kHz.

C2 is essentially composed of three 4n7 components, and its tolerance is improved by √3, to 0.58%. Its final value is tweaked by the addition of C15. An HF correction pole R8, R9, C13 is fitted; the resultant loss of HF

headroom is only 0.5 dB at 20 kHz, which I think I can live with.

Immediately after the RIAA stage is the subsonic filter, a 3rd-order Butterworth highpass filter which also implements the IEC Amendment by using a value for R11 + R12 which is lower than that for maximal Butterworth flatness. In this respect the circuit shows its age a little, as since it was designed the use of IEC appears to have been pretty much abandoned. The stage also buffers the HF correction pole R8, R9, C13 from later circuitry and gives the capability to drive a 600 Ω load, if you can find one. A version of this design, using appropriate precision components and incorporating the MC amplifier in Chapter 11, is manufactured by the Signal Transfer Company in bare PCB, kit, and fully built and tested formats.[1] The balanced-output version is shown in Figure 17.6; the MC stage with its gain switch is at upper left, the MM section with its bank of precision polystyrene capacitors is in the middle, while lower right is the subsonic filter and balanced output stages with two XLR connectors.

Project 4: An MM Amplifier With Subsonic Filter: Upgraded

The next project is an update of Project 3, with the following changes:

1) The RIAA network has been converted to Configuration-C, and so the total capacitance required has been reduced from 64 nF to 43 nF without increasing the impedance of the RIAA network, by using the

Figure 17.6 The Signal Transfer Company MC/MM phono amplifier with balanced outputs

capacitance more efficiently. The gain is unchanged at +30 dB (1 kHz).

2) Critical resistors have been converted to optimal 2xE24 parallel pairs.

3) C3 reduced and R3 increased to reduce the loading of the HF correction pole on opamp A1 output. This has a consequence in that the loading effect of the subsonic filter on R3 is now greater, causing a 0.4 dB loss of gain above 40 Hz. This gentle slope in the response is completely obliterated by the start of the subsonic filter roll-off and is of no account, but the loss of 0.4 dB of headroom is not ideal. It can be avoided by placing a unity-gain buffer stage between C3 and C5; the LM4562 is recommended. This is a

good example of the need to be careful when connecting stages together.

4) IEC amendment removed; the subsonic filter now has a standard 3rd-order Butterworth response −3 dB at 20 Hz; see Chapter 12.

5) Subsonic filter capacitors C5, C6, C7 reduced to 220 nF to reduce size and cost. Since there is still a very low-impedance path through the filter at audio frequencies, noise is not degraded.

6) A2 converted from 5532/2 to LM4562/2 to reduce distortion.

7) Output capacitor C8 made nonpolar to withstand DC offsets if doubtful equipment is being driven.

The updated circuit is shown in Figure 17.7. The basic RIAA accuracy is now well within ± 0.05 dB from 20 Hz–20 kHz. With the subsonic filter included it is within ± 0.05 dB from 43 Hz–20 kHz; the filter is −3 dB at 20 Hz.

Project 5: A High-Spec MM Amplifier

The MM/MC input system shown in Figure 17.8 is based on my recent Elektor preamp design,[2] which was a no-holds-barred attempt at getting the best possible performance. The input is switchable between a phono socket for the connection of MM cartridges and an MC head amplifier with a flat gain of +30 dB (see Chapter 11). The switched-gain stage allows every MC and MM cartridge on the market to be catered for without compromise on noise or headroom. A1 will always clip before the MC stage.

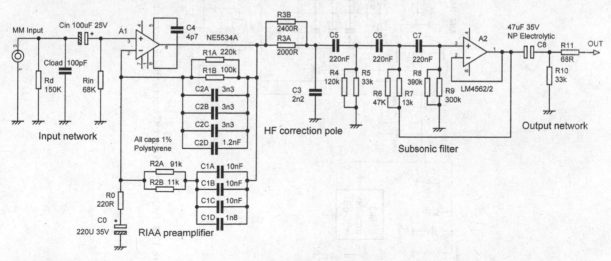

Figure 17.7 Project 4: Upgraded MM input using Configuration-C, with HF correction pole and 3rd-order subsonic filter. IEC amendment omitted.

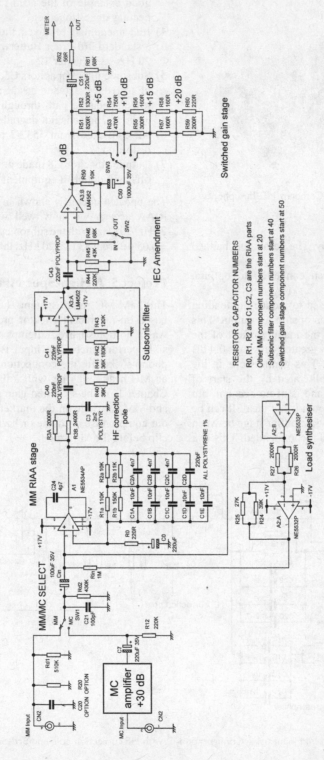

Figure 17.8 Project 5: MM and MC input with HF correction pole, switchable IEC amendment 3rd-order subsonic filter. Gain minimum +30 dB at 1 kHz.

The MM RIAA stage is in Configuration-A with a gain of +30 dB (1 kHz) using 5 × 10 nF polystyrene capacitors to obtain the required value and improve RIAA accuracy. Multiple RIAA resistors R1, R2, and R3 are used to improve accuracy in the same way, using the 2xE24 format. The value of C0 is large at 220 uF, as the IEC amendment is not implemented in this stage, and tweaking the IEC amendment to compensate for its less-than-infinite value (see Chapter 7) was not considered necessary. The +30 dB (1 kHz) gain of the MM stage requires that an HF correction pole is required for accurate RIAA characteristic; this is implemented by R3A, R3B, and C3, which is polystyrene to avoid capacitor distortion.

A 5534A is used here for IC3, as it is quieter than half a 5532 and considerably quieter than an LM4562, with its higher current noise.

A load-synthesis circuit IC4 is used to make an electronic version of the required 47 kΩ loading resistor from the 1M resistor R16. The Johnson noise of the resistor is however *not* emulated, and so noise due to the rising impedance of the MM cartridge inductance is much reduced. R16 is made to appear as 47 kΩ by driving its bottom end in anti-phase to the signal at the top. IC4B prevents loading on the MM input, while IC4A is the inverting stage. Paired resistors R19, R20 and R17, R18 are used to improve the gain accuracy of the inverting stage and therefore the accuracy of the synthesised impedance. There is much more on the load-synthesis technique in Chapter 9.

The subsonic filter is a two-stage 3rd-order Butterworth highpass filter that is −3 dB at 20 Hz, with 2xE24 paired resistors again used to improve accuracy. It is a two-stage filter in that it consists of an under-damped 2nd-order highpass stage (C40, C41, R40–R43 etc.) combined with a 1st-order stage (C43, R44, R45) to give the maximally flat Butterworth response. My previous preamp designs have used a single-stage 3rd-order Butterworth, but I have found the two-stage configuration is preferred when seeking the best possible distortion performance. [3] An LM4562 is used in the filter as it significantly reduces distortion. The low impedance of the 220 nF capacitors means its higher current noise makes a negligible contribution to the overall noise performance.

The IEC Amendment can be switched in by placing an extra resistance R46 across the subsonic filter resistances R44, R45. This is something of an approximation but saves an opamp stage and is accurate to ± 0.1 dB down to 29 Hz. Below this the subsonic filter roll-off begins, and the IEC Amendment accuracy is irrelevant.

The switched-gain stage comes last in the signal path. It allows every MC and MM cartridge currently on the market, including the very low output MC models made by Audio Note, to receive the amount of gain required for optimal noise and headroom. The switched-gain stage is fully described in Chapter 5; gain is varied in 5 dB steps by a switch SW3 which selects the desired tap on the negative-feedback divider R51–R60. Each divider step is made in 2xE24 format to get the exact dB and improve effective tolerances. R35 provides continuity of DC feedback when the switch is altered.

Setting the correct gain can be done by reference to cartridge sensitivities and so on, but it is far more convenient to have some sort of level indication for guidance, hence the meter output shown in Figure 17.2. The Log-Law Level LED (LLLL) is my attempt to get as much useful information as possible from a single LED; it is, to the best of my knowledge, a new idea, and is fully described in Chapter 15 on metering.

Project 6: A High-Spec MM Amplifier Upgraded

Project 5 is a recent design and is not susceptible to much improvement (at present). The only obvious step is to convert the Configuration-A RIAA network to Configuration-C to reduce parts cost and PCB area occupied. The network required is exactly the same as in Project 4.

References

1. Signal Transfer Company SignalTransfer MM/MC phono amplifier www.signaltransfer.freeuk.com/RIAA bal.htm Accessed Nov 2016.

2. Self, D. "Elektor Preamplifier 2012" *Elektor*, Apr, May, June 2012.

3. Billam, P. "Harmonic Distortion in a Class of Linear Active Filter Networks" *Journal of the Audio Engineering Society*, Volume 26, No. 6, June 1978, p. 426.

Component Series E3–E96

Values in the E3, E6, E12, E24, and E96 component series, with ratio pairs for E24 and E96

E3	E6	E12	E24	E96	E96	E96	E96
100	100	100	100	100	178	316	562
			110	102	182	324	576
		120	120	105	187	332	590
			130	107	191	340	604
	150	150	150	110	196	348	619
			160	113	200	357	634
		180	180	115	205	365	649
			200	118	210	374	665
220	220	220	220	121	215	383	681
			240	124	221	392	698
		270	270	127	226	402	713
			300	130	232	412	732
	330	330	330	133	237	422	750
			360	137	243	432	768
		390	390	140	249	442	787
			430	143	255	453	806
470	470	470	470	147	261	464	825
			510	150	267	475	845
		560	560	154	274	487	866
			620	158	280	499	887
	680	680	680	162	287	511	909
			750	165	294	523	931
		820	820	169	301	536	953
			910	174	309	549	976
1000	1000	1000	1000				1000

E24: 6 pairs in 1:2 ratio		E24: 4 pairs in 1:3 ratio		E96: 16 pairs in 1:2 ratio	
100	200	100	300	100	200
110	220	110	330	105	210
120	240	120	360	113	226
150	300	130	390	137	274
180	360			140	280
750	1500	**E24: 2 pairs in 1:4 ratio**		147	294
		300	1200	158	316
		750	3000	162	324
				174	348
		E24: 6 pairs in 1:5 ratio		187	374
		150	750	196	392
		200	1000	221	442
		220	1100	232	464
		240	1200	590	1180
		300	1500	732	1464
		360	1800	750	1500

No pairs in 1:3 ratio

No pairs in 1:4 ratio

E96: 16 pairs in 1:5 ratio

Phono Amplifier Articles
in *Linear Audio*

Jan Didden's *Linear Audio* is one of the prime journals for the publication of articles on new audio design and hardware. Here is a list of those dealing with phono preamplifiers, plus a few notes; you can see there is much thinking going on in this area. Not all of these are referenced in the main body of the book.

"Correcting Transducer Response With an Inverse Resonance Filter"
Steven van Raalte
Volume 3
Looks highly ingenious; have not had the opportunity to explore it myself.

"A Tube-Based Phono Preamplifier"
Marcel van de Gevel
Volume 4 p105
Using pentodes and triodes.

"VinylTrak—A Full Featured MM/MC Phono Preamp"
Bob Cordell
Volume 4 p131
Discrete and opamp circuitry. RIAA equalisation depends on cartridge inductance.

"RIAA Revisited, or How to Better Judge Documented Figures"
Hans Polak
Volume 6 p47
This article does not consider current noise.

"Proteus—A Current Input Moving Coil Preamp"
Erno Borbely & Sigurd Ruschkowski
Volume 6 p109
It is claimed that running an MC cartridge into a virtual-earth reduces its magnetic distortion.
Input devices are JFETs.

"The High-Octane Phono Preamp"
Hannes Allmaier
Volume 6 p157
Elaboration of the three-transistor discrete stage developed by H. P. Walker and myself.

"Optimising RIAA Realisation"
Douglas Self
Volume 7 p43
Using expensive precision capacitance efficiently. See Chapter 7.

"The Equal Opportunity—A Balanced Moving Magnet Phono Stage. (Part 1)"
Stuart Yaniger
Volume 7 p117
JFETS, triodes, and MOSFETS, and wholly passive equalisation.

"The Equal Opportunity—A Balanced Moving Magnet Phono Stage. (Part 2)"
Stuart Yaniger
Volume 8 p71
As above.

"Gramophone Preamplifier Noise Calculations—The 3852 Hz Rule Revisited."
Marcel van de Gevel
Volume 8 p129
The rule says design for spot MM noise at 3852 Hz to optimise RIAA eq'd noise. Good detailed stuff.

"The RIAA Phono-Amp Engine II"
Burkhard Vogel
Volume 9 p113
Triodes and opamps. NB This is the second version of the engine, not the second half of an article.

"The Ins and Outs of Turntable Dynamics—and How They Mess Up Your Vinyl Playback"
Hannes Allmaier
Volume 10 p9
If this doesn't put you off vinyl nothing will.

"Record Replay RIAA Correction in the Digital Domain"
Scott Wurcer
Volume 10 p37
A detailed examination.

"RIAA Equalisation for Displacement—Sensitive Phono Cartridges"
Gary Galo
Volume 11 p7
IE, strain-gauge cartridges. An excellent examination of the equalisation problems.

"The Devinyliser"
Douglas Self
Volume 11 p77
Rumble filtering by LF cancellation. See the last third of Chapter 12.

"An Attempt to Design a Non-Plus Ultra Phono Amp"
Hans Polak
Volume 12 p71
Balanced-input MC design. The input device is the AD797. Passive RIAA.

"Rumble Filtering—Like You Really Mean It"
Douglas Self
Volume 12 p129
Elliptical highpass filters. See the middle third of Chapter 12.

Index